AF361933

Sustainable Structural Design for High-Performance Buildings and Infrastructures

Sustainable Structural Design for High-Performance Buildings and Infrastructures

Editors

Chiara Bedon
Mislav Stepinac
Marco Fasan
Ajitanshu Vedrtnam
Maged A. Youssef

MDPI • Basel • Beijing • Wuhan • Barcelona • Belgrade • Manchester • Tokyo • Cluj • Tianjin

Editors

Chiara Bedon
Department of Engineering
and Architecture
University of Trieste
Trieste
Italy

Mislav Stepinac
Faculty of Civil Engineering
University of Zagreb
Zagreb
Croatia

Marco Fasan
Department of Engineering
and Architecture
University of Trieste
Trieste
Italy

Ajitanshu Vedrtnam
Department of Mechanical
Engineering
Invertis University
Bareilly
India

Maged A. Youssef
Department of Civil and
Environmental Engineering
The University of Western
Ontario
London, Ontario
Canada

Editorial Office
MDPI
St. Alban-Anlage 66
4052 Basel, Switzerland

This is a reprint of articles from the Special Issue published online in the open access journal *Sustainability* (ISSN 2071-1050) (available at: www.mdpi.com/journal/sustainability/special_issues/High-Performance).

For citation purposes, cite each article independently as indicated on the article page online and as indicated below:

LastName, A.A.; LastName, B.B.; LastName, C.C. Article Title. *Journal Name* **Year**, *Volume Number*, Page Range.

ISBN 978-3-0365-4328-4 (Hbk)
ISBN 978-3-0365-4327-7 (PDF)

Cover image courtesy of Brandon Lee

© 2022 by the authors. Articles in this book are Open Access and distributed under the Creative Commons Attribution (CC BY) license, which allows users to download, copy and build upon published articles, as long as the author and publisher are properly credited, which ensures maximum dissemination and a wider impact of our publications.

The book as a whole is distributed by MDPI under the terms and conditions of the Creative Commons license CC BY-NC-ND.

Contents

sustainability

MDPI

Article

Shear Strength Estimation of Reinforced Concrete Deep Beams Using a Novel Hybrid Metaheuristic Optimized SVR Models

Mosbeh R. Kaloop [1,2,3], Bishwajit Roy [4], Kuldeep Chaurasia [4], Sean-Mi Kim [1], Hee-Myung Jang [1], Jong-Wan Hu [1,2,*] and Basem S. Abdelwahed [5]

[1] Department of Civil and Environmental Engineering, Incheon National University, Incheon 22012, Korea; mosbeh@mans.edu.eg (M.R.K.); mook@hdec.co.kr (S.-M.K.); jhm924@kistep.re.kr (H.-M.J.)
[2] Incheon Disaster Prevention Research Center, Incheon National University, Incheon 22012, Korea
[3] Public Works and Civil Engineering Department, Mansoura University, Mansoura 35516, Egypt
[4] School of Computer Science Engineering and Technology, Bennett University, Greater Noida 201310, India; bishwajit.cse15@nitp.ac.in (B.R.); kchaurasia.nitb@gmail.com (K.C.)
[5] Structural Engineering Department, Mansoura University, Mansoura 35516, Egypt; bs_hadi@mans.edu.eg
* Correspondence: jongp24@inu.ac.kr

Abstract: This study looks to propose a hybrid soft computing approach that can be used to accurately estimate the shear strength of reinforced concrete (RC) deep beams. Support vector regression (SVR) is integrated with three novel metaheuristic optimization algorithms: African Vultures optimization algorithm (AVOA), particle swarm optimization (PSO), and Harris Hawks optimization (HHO). The proposed models, SVR-AVOA, -PSO, and -HHO, are designed and compared to reference existing models. Multi variables are used and evaluated to model and evaluate the deep beam's shear strength, and the sensitivity of the selected variables in modeling the shear strength is assessed. The results indicate that the SVR-AVOA outperforms other proposed and existing models for the shear strength prediction. The mean absolute error of SVR-AVOA, SVR-PSO, and SVR-HHO are 43.17 kN, 44.09 kN, and 106.95 kN, respectively. The SVR-AVOA can be used as a soft computing technique to estimate the shear strength of the RC deep beam with a maximum error of ±3.39%. Furthermore, the sensitivity analysis shows that the deep beam's key parameters (shear span to depth ratio, web reinforcement's yield strength, concrete compressive strength, stirrups spacing, and the main longitudinal bars reinforcement ratio) are efficiently impacted in the shear strength detection of RC deep beam.

Keywords: reinforced concrete; deep beam; shear strength; support vector regression; metaheuristic optimization

Citation: Kaloop, M.R.; Roy, B.; Chaurasia, K.; Kim, S.-M.; Jang, H.-M.; Hu, J.-W.; Abdelwahed, B.S. Shear Strength Estimation of Reinforced Concrete Deep Beams Using a Novel Hybrid Metaheuristic Optimized SVR Models. *Sustainability* **2022**, *14*, 5238. https://doi.org/10.3390/su14095238

Academic Editors: Chiara Bedon, Maged A. Youssef, Mislav Stepinac, Marco Fasan and Ajitanshu Vedrtnam

Received: 22 March 2022
Accepted: 20 April 2022
Published: 26 April 2022

Publisher's Note: MDPI stays neutral with regard to jurisdictional claims in published maps and institutional affiliations.

Copyright: © 2022 by the authors. Licensee MDPI, Basel, Switzerland. This article is an open access article distributed under the terms and conditions of the Creative Commons Attribution (CC BY) license (https://creativecommons.org/licenses/by/4.0/).

1. Introduction

In many high-rise reinforced concrete (RC) buildings, as the use of areas is changed from one story to another, some columns in the upper stories are not permitted to reach the foundation. To solve this conflict, transfer girders with a considerable thickness named deep beams are required [1,2]. Furthermore, deep beams are used in many other critical structures and play a significant role in delivering heavy loads to the bearing elements [2,3]. Deep beams have high flexural stiffness, and the shear diagonal failure is the predominant mechanism as the loads are mainly transferred from their action points to the supports locations through a direct diagonal strut [1–4]. Concrete compressive strength, the provided top/bottom reinforcements, and web reinforcements in terms of amount and spacing all form the shear resistance of these deep beams [1,2,4,5]. In literature, plenty of analytical and numerical studies have been focused on the ultimate shear strength assessment of such beams with large depths compared to their spans [2,3,6–9]. Unavoidable discrepancies were found with the implementation of both analytical/numerical methods compared to the experimental results [4]. This study aims to propose a novel soft computing approach that can be used to predict an accurate shear strength of RC deep beams based on a wide range of test results collected from different a experimental studies.

Many researchers have used different regression methods to estimate the shear strength of RC deep beams [8,10]. Recently, soft computing techniques have been proposed and improved the prediction techniques of shear strength of beams [3,7,11–15]. A conventional artificial neural network (ANN) was applied to estimate the shear strength of the RC deep beam, and the accuracy of the designed model was high [7]. The ultimate shear strength of the RC deep beam was also estimated using ANN and compared to different building codes, and the proposed model provided an accurate prediction of shear capacity [6]. ANN, adaptive network-based fuzzy inference system (ANFIS), and group method of data handling (GMDH) approaches were used in predicting the shear strength of RC beam-column joints, and the performance of these models was high at a different range of shear strength [15]. A probabilistic model was applied to estimate the shear strength of beams, and the determination of shear strength was shown to be accurate [14]. Integrated genetic programming and simulated annealing (GSA) outperformed American concrete institute (ACI) and Canadian standard association (CSA) codes in modeling the shear strength of RC deep beams [2]. The shear strength of beams reinforced by fiber was calculated using hybrid support vector regression (SVR) and firefly optimization algorithm (FFA), and the designed model was shown to be robust in shear strength prediction [13]. ANN was integrated with the adaptive harmony search optimization (AHS) technique for modeling the shear strength of RC walls, and the proposed model accuracy was high [12]. Multivariate adaptive regression splines (MARS) and artificial bee colony (ABC) were also integrated to design a model for predicting the shear strength of RC deep beams, and the performance of MARS-ABC was higher than different building codes in shear strength estimation [16]. Generally, parameters of machine learning (ML) models are tuned using metaheuristic algorithms to improve the prediction efficiency of ML models [17–22].

Meanwhile, novel optimization algorithms have recently been developed, such as the African Vultures optimization algorithm (AVOA), particle swarm optimization (PSO), and Harris Hawks optimization (HHO). Although these techniques are used in different engineering applications [17–19], the AVOA optimization technique is not applied yet in shear strength prediction based on our literature. PSO was integrated with an adaptive neuro-fuzzy inference system (ANFIS) to predict the shear strength of high strength concrete for a slender beam, and the ANFIS-PSO attained the best modeling accuracy over ANFIS -ant colony optimizer (ANFIS-ACO), -differential evolution (ANFIS-DE), and -genetic algorithm (ANFIS-GA) [20]. Teaching–learning-based optimization (TLBO), PSO, and HHO were integrated with SVR, and the results of SVR-PSO, SVR-HHO, and SVR-TLBO were robust and can be used to estimate an accurate shear strength prediction of RC shear walls [21].

Based on the above literature, the hybrid SVR models are more robust for predicting the shear strength of RC deep beams [3,16,21,23]. This study aims to evaluate a new hybrid technique (SVR-AVOA) in predicting the shear strength of RC deep beams. To benchmark the proposed SVR-AVOA model, the hybrid known models SVR-PSO and SVR-HHO are proposed and compared; in addition, the recent mathematical studies and building codes are compared to the proposed model to assess its accuracy of it in modeling shear strength of RC beam. SVR-AVOA, SVR-PSO, and SVR-HHO are developed using different scenarios of input variables. For this study, 202 datasets, including 19 variables of experimental studies, were collected from literature to design and evaluate the proposed models. The sensitivity analysis of optimum input variables is proposed and evaluated.

2. Background of Variables Impacts the Shear Strength of RC Deep Beams

Figure 1i presents a real case of deep beam function in load transfer of buildings and variables that impact the shear strength value. Figure 1ii demonstrates the main parameters of the deep beam. Figure 1iii illustrates the failure mode of deep beams. In the figures, V represents the deep beam shear capacity, a is the horizontal distance from the load to the support, and d denotes the deep beam depth. Here, V depends on (1) Concrete quality f_c' (for the diagonal strut), (2) Main steel yield strength f_y (for the main tie), and (3) Web reinforcement (horizontal and vertical). As some variables have a big role in forming the deep beam's shear strength, the main variables considered in this study are the shear span to depth ratio, the main reinforcement, ratio and yield strength, concrete compressive strength, and web reinforcement characteristics.

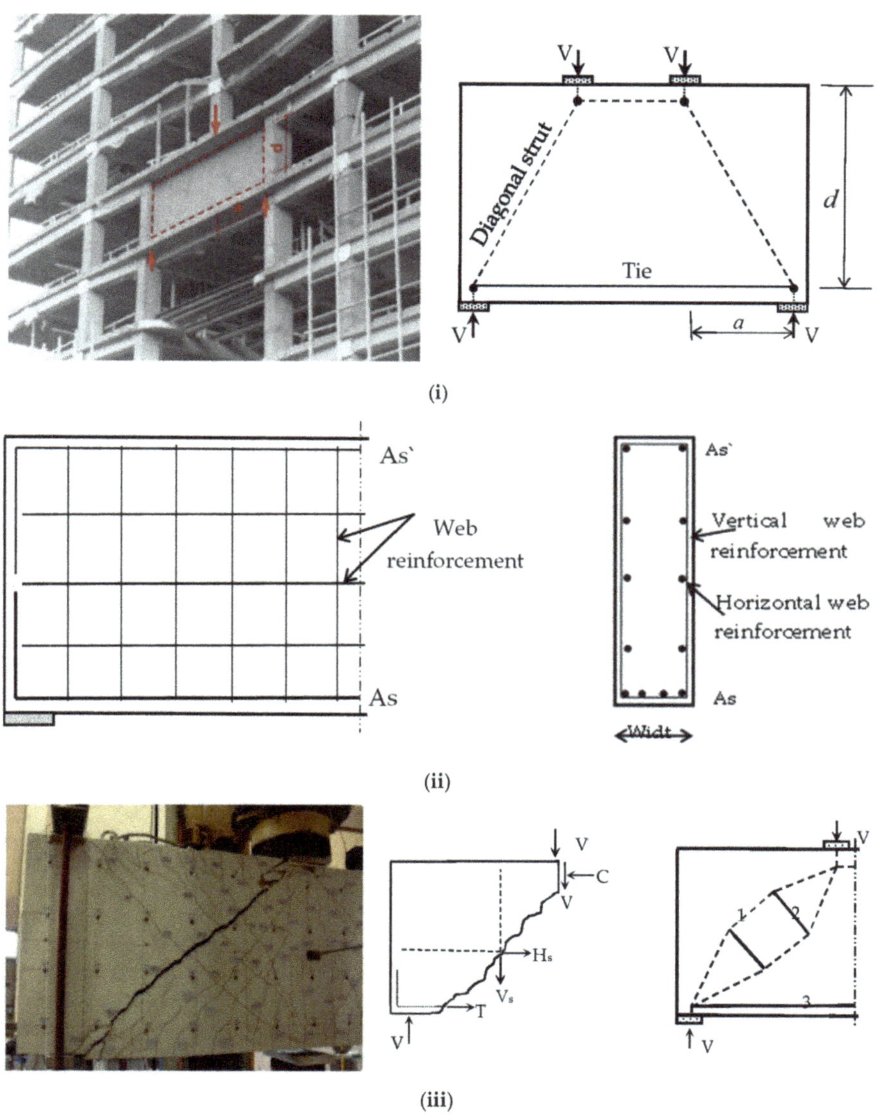

Figure 1. (i) Real case of using deep beams (**left**) and deep beam terminology (**right**); V: Shear strength, a: Shear span, d: Effective depth of beam. (ii) Basic reinforcement details of simple RC deep beam. (iii) Failure pattern in deep beam (**left**), Different mechanism components (**middle**), and Flow of forces in deep beam (**right**); where, C = compression force in concrete, T = tensile force in the main steel, Hs = tensile force in horizontal web reinforcement, Vs = tensile force in vertical web reinforcement; 1: Main diagonal strut, 2: Splitting tension force, 3: Main tensile force.

Many researchers investigated the deep beam's shear strength evaluation and predictions [1–4,24,25]. They concluded that the compressive strength of concrete (f_c'), the shear span to the beam's depth ratio (a/d), bottom longitudinal reinforcement ratio (ρ), and web reinforcement ratio (both vertical ρv and horizontal ρh) are the main key parameters. The previous analytical/experimental studies focused on the role of f_c' in forming V. In the case of beams with smaller a/d ratios, the role of the truss mechanism in transferring loads to the support location diminishes, and the direct diagonal strut is the main transferring load mechanism to the support's location. The effectiveness of such struts has a significant impact on V values [3]. Smith and Vantsiotis [26] and Ahmed [27] observed an increase in V of the deep beam with increasing fc of concrete, but the relationship was not linearly proportional. In addition, they observed no improvement in V if f_c was above a certain limit. The non-proportional increase in shear strength compared to the increase in concrete compressive strength can be attributed to two reasons. First, the limited contribution of the aggregate interlock mechanism in members with high strength concrete compared to the one with normal strength concrete, as the cracks cross the aggregate particles in high strength concrete and do not go around them as in normal strength concrete. Second, the formed tensile strains perpendicular to the main diagonal strut work on reducing the benefits of using high strengths. Oh and Shin [28] noticed a brittle failure of deep beams with concrete of 74 MPa without any warning, which is different from the failure of other beams with 23 MPa. They also observed a decrease in the rate of increase in the ultimate strength of beams with high-strength concrete.

The inclination angle of the main diagonal strut plays a significant role in determining the concrete efficiency in the diagonal strut. This angle is directly dependent on the shear span to depth ratio a/d. As this angle increases, the forces can go directly inside the diagonal strut to the support. The previous studies [3,29] noticed that the increase in shear strength could be detected by decreasing the a/d ratio. Kim and Park [30] found a trivial impact of this ratio on the shear strength of beams with ratios greater than three, and the contrary was noticed for beams with ratios less than three. Oh and Shin [28] concluded that the ratio of a/d is the governing key parameter in determining the shear strength of a deep beam with high-strength concrete. In addition to resisting the induced tensile force of the main horizontal tie, the main reinforcement bars play an important role in controlling the width enlargement of the diagonal main cracks by dowel action mechanism and enable the aggregate interlock to work more effectively. Many researchers investigated the impact of the main reinforcement ratio on the deep beam's shear strength [3,31,32]. They observed a significant increase in the shear strength with increasing the main reinforcement ratio but up to a certain limit. Above a ratio of 1.5%, Ashour et al. [33] noticed a local concrete crushing damage due to compressive stress concentration at the top strut far from the main diagonal strut without enabling the diagonal strut to reach its ultimate resistance.

Web reinforcement has an important role in confining the concrete and delivering the tensile stresses at the main shear diagonal crack to the intact zones around the crack, which consequently increases the deep beam's shear strength [2]. Both vertical and horizontal web reinforcement has a key role in resisting shear stresses and limiting the enlargement of the width of the main diagonal crack [3]. In deep beams with higher a/d ratios, the contribution of vertical reinforcement is more obvious than the horizontal reinforcements, and the contrary is noticed for beams with a/d less than 1.0. As the deep beam's shear strength is dependent on many parameters, the increase of horizontal and vertical reinforcement above a certain limit does not influence the ultimate shear strength as other parameters may govern the situation without reaching the maximum capacity of the provided web reinforcement.

Table 1 presents the existing models used in this study compared to the developed models. The performance of these models was used significantly in shear strength determination for the RC deep beams of structures.

Table 1. Summary of existing models.

Ref.	Equation	Explanation
ACI [10]	$V_u = 0.17\sqrt{f'_c}bd + \dfrac{Av f_y d(\sin\theta + \cos\theta)}{s}$	θ is the angle between the stirrups and the beam longitudinal axis
Russo [8]	$V_u = 0.545\left(kX f'_c \cos\alpha + 0.25\rho_h f_{yh}\cot\alpha + 0.35\frac{a}{d}\rho_v f_{yv}\right)bd$	$k = \sqrt{(n\rho)^2 + 2n\rho} - n\rho$ $\tan\alpha = \frac{a}{0.9d}$ $X = 0.74\left(\frac{f'_c}{105}\right)^3 - 1.28\left(\frac{f'_c}{105}\right)^2 + 0.22\left(\frac{f'_c}{105}\right) + 0.87$
Liu [4]	$V_u = V_{CLZ} + V_{ci} + V_s + V_d$	V_{CLZ} is the shear resisted at the critical loading zone, V_{ci} represents the contribution of aggregate interlock, V_s is the shear resisted by web reinforcement and V_d is the dowel action in the main longitudinal bars.

where: f'_c is the compressive strength of concrete, b is the beam width, d is the beam effective depth, Av is the vertical web reinforcement, f_{yv} and f_{yh} are the yield strength of vertical and horizontal web reinforcement respectively, s is the spacing between the vertical web reinforcement, ρ_h and ρ_v are the ratio of horizontal and vertical web reinforcement respectively, n is the modular ratio, ρ is the ratio of the main longitudinal bars.

3. Material and Data Collection

The Supplementary Material (Table S1) presents the data collected from the literature. For this study, 202 datasets were collected from the literature. In the current study, 19 input variables were used and divided into two categories, the main (8 variables) and other (11 variables) variables, as presented in Table S2. Here, the main variables were considered based on our literature in the previous section, Section 2; the other variables were considered while the impact on shear strength calculation was high. The direct relationship between each variable and the ultimate shear strength (Vu) of the RC deep beam is presented in Figure S1. Exponential, linear, logarithmic, and power functions were used to estimate the best direct relationship equation between Vu and input variables. Table 2 presents the summary of these functions.

Table 2. Direct relationship functions between Vu and input variables.

Variable	Equation (R^2)	Variable	Equation (R^2)	Variable	Equation (R^2)
a/d	$y = 358.43x^{-0.803}$ (0.26)	b	$y = 117.38 \times 10^{0.0052x}$ (0.21)	Ag	$y = 476.32x^{-0.164}$ (0.03)
ρ	$y = 65.95\ln(x) + 345.74$ (0.01)	d	$y = 0.7899x + 35.304$ (0.35)	Std	$y = 165.06 \times 10^{0.0731x}$ (0.05)
f_y	$y = 396\ln(x) - 2024.7$ (0.16)	h	$y = 0.6985x + 32.425$ (0.33)	Bd	$y = 524.3x^{-0.262}$ (0.03)
f'_c	$y = 495.94\ln(x) - 1234.7$ (0.39)	a	$y = 259.88 \times 10^{0.0003x}$ (0.015)		
ρ_v	$y = 325.96 \times 10^{-0.159x}$ (0.01)	Lp	$y = 169.42 \times 10^{0.0054x}$ (0.15)		where:
s	$y = 0.2678x + 344.21$ (0.06)	Sp	$y = 169.42 \times 10^{0.0054x}$ (0.15)		y represents the Vu x represents input variables
f_{yv}	$y = 19.734x^{0.4588}$ (0.07)	V/P	$y = 199.58x + 199.66$ (0.010)		R^2 is the coefficient of determination
ρ_h	$y = -371.18x + 429.11$ (0.10)	# bars	$y = 483.29\ln(x) - 199.43$ (0.37)		

From Table 2, it can be observed that the power function has the best correlation with Vu in the case of using a/d and f_{yv} of the main variables. The relationship of ρ_v with Vu is exponential. The linear correlation can be detected between Vu and (s and ρ_h). ρ, f_y, and f'_c are correlated with Vu based on logarithmic functions. The best R^2 between Vu and the main variables is 0.39 for the f'_c variable. These results indicate that the relationship between the main variables and Vu cannot be estimated directly, and a complex relationship may be detected by using all main variables. Similarly, for the other variables, the relationship between Vu and variables varies. The best R^2 is estimated using a number of main bars, $R^2 = 0.37$. The variation in R^2 indicates the complex relationship between Vu and all variables. Table 2 and Figure S1 show the increase of the resistance with beam effective width and height, number of main bars, and concrete strength up to a certain limit.

The statistical evaluation, range (RA), average (M), standard deviation (SD), kurtosis (KU), and skewness (SK) of the used datasets is presented in Table 3. From the table, the range of datasets varies and will affect the models' performances, so the normalized datasets are used to overcome the range change of variables. The data is normalized between 0 and 1 in this study. In addition, in the proposed models, it is recommended to use the given ranges of input variables. The average and standard deviation values show that the distortion of datasets is high. The kurtosis and skewness values indicate that the distribution of datasets is not normal. Figure 2 presents the histogram and distribution of main variables and Vu. The figure shows positive skewness for whole variables is observed; the distribution is also supported by the presented values in Table 3.

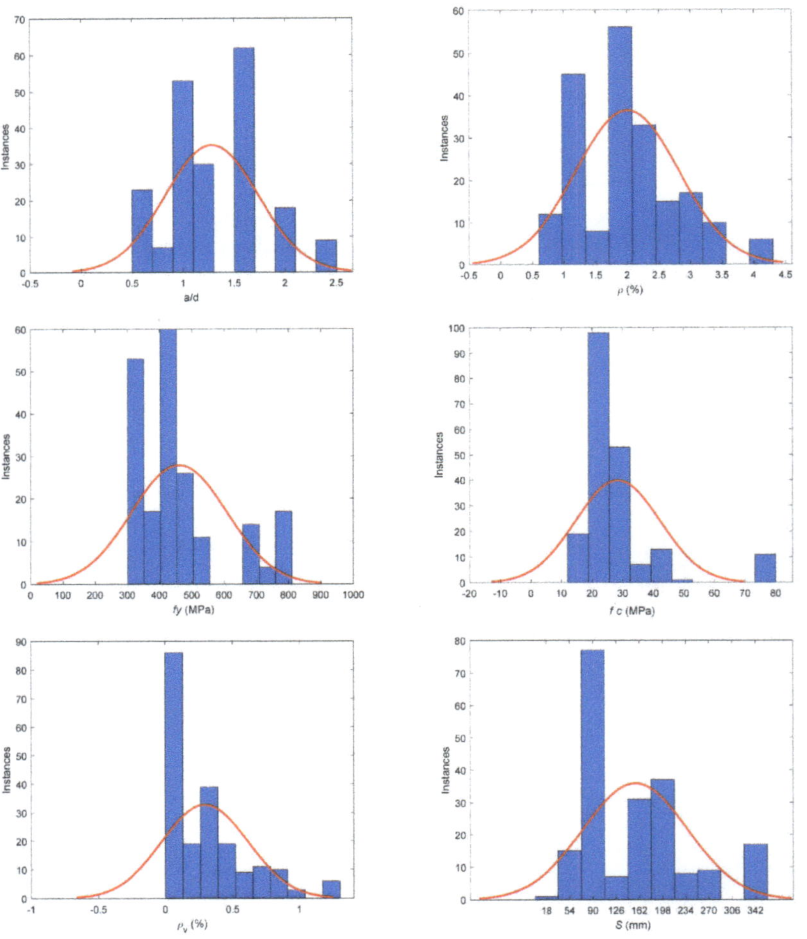

Figure 2. *Cont.*

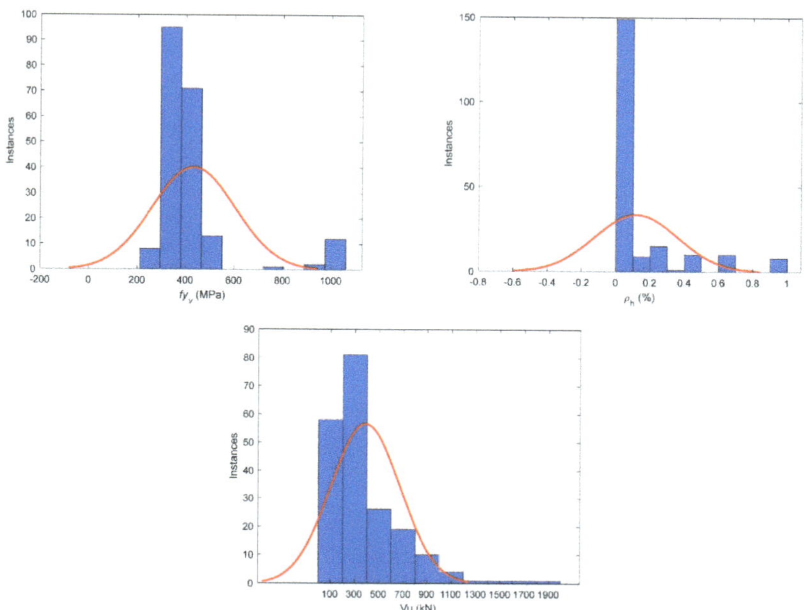

Figure 2. Distribution histogram of input and output variables.

Table 3. Statistical analysis of input and output variables.

Variable	RA	M	SD	KU	SK	Variable	RA	M	SD	KU	SK
a/d	1.93	1.28	0.46	−0.03	0.38	b (mm)	200.00	188.18	66.50	−0.94	0.28
ρ (%)	3.50	2.00	0.82	0.16	0.65	d (mm)	1374.00	443.74	212.19	11.52	3.14
f_y (MPa)	502.00	459.71	147.09	0.50	1.26	h (mm)	1550.00	505.91	235.73	12.91	3.32
f_c' (MPa)	66.10	28.33	13.75	7.04	2.64	a (mm)	1600.00	543.97	242.31	2.87	1.09
ρ_v (%)	1.25	0.29	0.32	0.69	1.10	Lp (mm)	210.00	113.11	45.63	3.05	2.04
s (mm)	330.00	155.33	80.63	0.54	1.07	Sp (mm)	210.00	113.11	45.63	3.05	2.04
f_{yv} (MPa)	791.00	430.68	171.05	6.12	2.55	V/P	0.50	0.93	0.16	2.96	−2.18
ρ_h (%)	0.91	0.12	0.24	3.58	2.15	#bars	10.00	3.61	1.70	10.50	2.95
Vu (kN)	1869.00	385.80	285.02	6.25	2.09	Ag (mm)	22.00	14.20	5.67	0.53	1.29
						Std (mm)	12.70	8.67	2.42	−0.38	−0.11
						Bd (mm)	6.20	7.66	2.08	−0.74	−0.65

4. Methods and Development Models

4.1. Support Vector Regression

Pal and Deswal [23] and Mozumder et al. [34] proposed the SVR formulas and theory in shear strength to predict RC beams. It was found to be a powerful computation technique for predicting the shear strength of deep beams [3,23]. Here, a summary of SVR is presented. SVR is the regression category of support vector machine (SVM), aiming to find a function that represents the relationship between inputs features to forecast the corresponding value when a new input is used. In SVR, "a fixed mapping procedure to map its input to n-dimensional feature space; then nonlinear functions are used to fit the high-dimensional features [35]". Vapnik [35] proposed a loss function to allow the concept of SVM margin to be used for regression solutions. The following equation represents the mathematical formula for the SVRs' approximation function [23,34]:

$$f(x) = w\varphi(x) + b \tag{1}$$

$$C = \frac{1}{2}w + C\frac{1}{n}\sum_{i=1}^{n} L(x,d) \tag{2}$$

where w and φ represent the weight vector and transformation functions, respectively; x and d are the input and output vectors, and b denotes a scalar. In Equation (1), the w and b are used to determine the normal and scalar vector, respectively, for the high-dimensional space, which is determined through $\varphi(x)$. Terms $\frac{1}{2}w$ and $C\frac{1}{n}\sum_{i=1}^{n} L(x,d)$ in Equation (2) are the standard error and the penalty terms, respectively. The ε-insensitive loss function introduced by Vapnik [34] is commonly used to estimate the Equation (1) parameters through the following minimization function [23,34]:

$$minimize\ 0.5\| w \|^2 + C\sum_{i=1}^{n}(\xi_i + \xi_i^*) \quad subjected\ to \begin{cases} d_i - (wx_i) - b \leq \varepsilon + \xi_i \\ (wx_i) + b - d_i \leq \varepsilon + \xi_i^* \\ \xi_i, \xi_i^* \geq 0 \end{cases} \tag{3}$$

where $C > 0$, which controls the trade-off between the model complexity and the amount up to which deviations larger than ε are tolerated. Equation (3) can be transformed to a dual space using Lagrange multipliers solution. This solution can be expressed as follows [23,34]:

$$\begin{aligned} maximize\ L = \quad & -\varepsilon \sum_{i=1}^{n}\left(\alpha_i^* + \alpha_i\right) + \sum_{i=1}^{n} d_i\left(\alpha_i^* - \alpha_i\right) \\ & -0.5 \sum_{i=1}^{n}\sum_{j=1}^{n}\left(\alpha_i^* - \alpha_i\right)\left(\alpha_j^* + \alpha_j\right)\left(x_i - x_j\right) \quad subjected\ to \begin{cases} \sum_{i=1}^{n}\left(\alpha_i - \alpha_i^*\right) = 0 \\ 0 \leq \alpha_i^* \leq C \\ 0 \leq \alpha_i \leq C \end{cases} \end{aligned} \tag{4}$$

where L denotes the Lagrangian and α_i and α_i^* represent the Lagrange multiplier. Once Equation (4) is used to estimate the parameters of Equation (2), Equation (1) can be rewritten as [23,34]:

$$f(x) = \sum_{nsv}(\alpha_i^* - \alpha_i)(x_k.x) + b \tag{5}$$

where nsv represents the number of support vectors (x_r, x_s). Here, the solution of this equation depends on the training pattern of Lagrange multipliers, which are only applied to estimate the w and b. Therefore, the kernel function is commonly used to solve the nonlinear regression problems in SVR. The Kernel functions can transform the nonlinear problems into linear problems, as presented in Yaseen et al. [36], which allows the SVR to solve more complex problems. The nonlinear SVR can be expressed as follows:

$$f(x) = \sum_{nsv}(\alpha_i^* - \alpha_i)K(x_ix) + b \tag{6}$$

where K is the kernel function; $K(x_ix) = (\varphi(x_i)\varphi(x))$. In the current study, the radial basis kernel (RBF) is applied.

It should be mentioned that the SVR is built with statistical theory and based on the minimization of structural risk principle [37]. It is a popular method for a small count of data, high dimensional, and non-linear problems. Therefore, SVR is used in many applications [13,38,39] for prediction tasks. Generally, SVR is a type of convex optimization technique to search a local solution within a problem domain [37]. The tuning of learning parameters of SVR greatly impacts the evaluation quality. Therefore, finding optimal values of SVR parameters from global searched cost is a difficult task [40]. Nature-inspired algorithms are proved to be successful in finding the local best solution from the global one.

This work applied and showed PSO, HHO, and AVOA for optimizing SVR parameters with faster convergence capability to provide better prediction accuracy of the SVR model in the deep beam's shear strength prediction.

4.2. Optimization Methods

To optimize the best parameters (C, ε, and KernelScale (γ)) of SVR, the PSO, HHO, and AVOA algorithms are used separately in the current work.

4.2.1. PSO

Kennedy and Eberhart (1995) proposed a metaheuristic population-based optimization algorithm named particle swarm optimization (PSO). The mechanism of the PSO is inspired by the fish schooling and foraging of the flock of birds while exploring an unknown region. Further, PSO is defined as a swarm of particles moving nearby the problem space by the influence of its global (G_{best}) and local best (P_{best}) position [41]. Therefore, a population search algorithm is used in PSO in a nature-inspired manner to analyze the input data features. The best position of the whole swarm is required to optimize the parameters. These can be performed through nature-inspired behaviors and learning experiences of population particles. PSO was found to be a robust integrated technique with SVR and ANN to model the shear strength of concrete [36,41,42]. PSO algorithm may be summarized as follows:

- First, it initializes the particle of the swarm, then defines the maximum number of iterations, and finally defines the cost function.
- After defining the cost function, it evaluates the swarm in order to identify the global and local best.
- Lastly, it calculates the velocity of each particle and then updates its position using the following equations:

$$v_{ik} = wv_{ik} + coef_1 rand_1 \left(P_{best,ik} - y_{ik} \right) + coef_2 rand_2 \left(G_{best,\ ik} - y_{ik} \right) \tag{7}$$

$$y_{ik} = y_{ik} + v_{ik} \tag{8}$$

where y_i denotes the i-th particle, k = the k-th dimension of the particle, $coef_1$ and $coef_2$ represent the acceleration coefficients, w refers to the inertia weight, $rand_1$ and $rand_2$ represent the random coefficients, which are randomly limited between zero and one. More details for PSO theory can be found in [43,44].

4.2.2. HHO

Heaidari and Mirjalili (2019) proposed a gradient-free, population-based optimization technique named Harris hawks optimization (HHO) [45]. The main inspiration behind HHO is surprised pounce, i.e., the chasing style and cooperative behavior of Harris' hawks in nature. According to the HHO optimization technique, numerous hawks cooperate to surprise a prey by pouncing it from multiple directions. They display a variety of pursuit patterns based on different scenarios and escaping patterns of the prey. The mechanism of HHO is that it mimics the Harris' hawk behavior in that trace, encircle, flush out, and attack the prey. It has been integrated with SVR and ANN to model the concrete characteristics and other engineering applications [21,46–49]. The main phases in the attacking of hawks are exploration (phase 1), transferring (phase 2), and exploitation (phase 3). In phase 1, the hawk depends on its position from the prey based on his waiting, seeking, and discovering; this can be expressed as follows:

$$Y(iter + 1) = \begin{cases} Y_{ranm}(iter) - r_1 | Y_{ranm}(iter) - 2r_2 Y_{ranm}(iter) \ if\ n \geq 0.5 \\ \left(Y_{prey(iter)} - Y_m(iter) \right) - r_3 (LB + r_4 (UL - LL))\ if\ n < 0.5 \end{cases} \tag{9}$$

where Y_{ranm} and Y_{prey} indicate the random position for the selected hawk and prey's position, respectively. UL and LL indicate the upper and lower range; r_i indicates a random number, having a value between 0 and 1; and $Y_m = 1/N \sum_1^N Y_i(iter)$.

In phase 2, the prey energy is modeled as $E = 2E_0\left(1 - \frac{iter}{T}\right)$, where T and $E_0 \in (-1,1)$, and they indicate that energy falls for the prey with their escapes. Thus, the hawk can decide the solution based on the E computation and starting in phase 3 when $|E| \geq 1$, and exploiting the neighborhood when $|E| < 1$. Once starting phase 3, hawks decide to apply a soft or hard besiege. $|E| \geq 0.5$ indicates the prey still has enough energy to escape, but maybe some misleading jumps occur in it to fail, so a soft besiege works. On the other hand, in the case of $|E| < 0.5$, the prey is too fatigued to escape, so hard besiege works. Here, the HHO is used to optimize the SVR parameters.

4.2.3. AVOA

Abdollahzadeha and Gharehchopogh (2021) recently introduced a metaheuristic algorithm named African vultures' optimization algorithm (AVOA) [17]. The inspiration behind the development of the AVOA algorithm is the competing and searching behavior of vultures to acquire a large amount of food. To acquire a large amount of food, these vultures, 'N' (N denotes the population of vulture), were divided into categories based on their fitness to find food and eat. The solution with the highest fitness value is treated as the first-best vulture and the second-best solution as the second-best vulture. The rest of the vultures were trying to approach the best vulture. This is formulated as follows.

Step 1: To determine the best vulture in the group. Fitness of all solutions is determined, and the best solution is selected as the best vulture of the group and other solutions will move towards the best solution using:

$$R(i) = \begin{cases} vulture_{best1} \ if \ p_i = K_1 \\ vulture_{best2} \ if \ p_i = K_2 \end{cases} \tag{10}$$

where the value of K_1 and K_2 lies between 0 and 1 with their sum equal to 1.

Step 2: Starvation rate of vultures. The starvation rate is the rate at which the vultures are satiated or hungry. The satiated rate has a declining trend, and to model, behavior Equation (11) is used,

$$F = (2ranm_1 + 1)P\left(1 - \frac{iter_i}{iter_{max}}\right) + t \tag{11}$$

$$t = h\left(\sin^w\left(0.5\pi \frac{iter_i}{iter_{max}}\right) + \cos\left(0.5\pi \frac{iter_i}{iter_{max}}\right) - 1\right) \tag{12}$$

where F indicates the vultures are satiated. If the value of $|F|$ is greater than 1, vultures search for food in different areas, and the algorithm enters the exploration phase, whereas if the value is less than 1, AVOA enters the exploitation phase, and vultures search for food in the neighborhood. $iter_i$ Indicates the iteration number, $iter_{max}$ indicates the total number of iterations, $ranm_1$ has a random value between 0 and 1. Here, z and h indicate random numbers with values lying between -1 to 1 and -2 to 2, respectively. If the value of z is less than 0, it indicates the vulture is starved, and if it increases to 0, it indicates the vulture is satiated. Here, w indicates the optimization operation disrupts the exploration and operation phases. By increasing the value of w, the probability of entering the exploration phase in the final optimization stages increases, and vice versa for decreasing the value of w.

Step 3: Exploration phase. In this phase, different random areas can be examined using two different strategies. To select the strategies in the $ranm_1$ exploration phase, a random number between 0 and 1 is generated. This procedure is shown in Equation (13)

$$P(i+1) = \begin{cases} R(i) - |XR(i) - P(i)|F \ if \ P_1 \geq ranm_{P1} \\ R(i) - F + ranm_2((UB - LB)ranm_3 + LB) \ if \ P_1 < ranm_{P1} \end{cases} \tag{13}$$

where $P(i+1)$ indicates the vulture position vector in the next iteration, F indicates the rate of vulture being satiated, and $R(i)$ indicates one of the best vultures, which is selected at step 1 in the current iteration. X indicates a coefficient vector that increases the random motion by changing in each iteration and is obtained using the formula $X = 2 \times ranm$, where $ranm$ is a random number between 0 and 1.

Step 4: Exploitation phase. In this phase, the efficiency stage of the algorithm is investigated. If the value of $|F|$ is less than 1, the algorithm enters the exploitation phase. Here, the exploitation phase is categorized based on the $|F|$ value. If $|F|$ is between 1 and 0.5, the rotating flight strategy of the vulture is processed based on parameter P_2. This can be processed as follows:

$$P(i+1) = \begin{cases} |XR(i) - P(i)|(F + ranm_4) - (R(i) - P(i)) \; if \; P_2 \geq ranm_{P2} \\ R(i) - \left[\left(R(i)(\cos(P(i))) \left(\frac{ranm_5 P(i)}{2\pi} \right) \right) + \left(R(i)(\sin(P(i))) \left(\frac{ranm_6 P(i)}{2\pi} \right) \right) \right] \; if \; P_1 < ranm_{P2} \end{cases} \quad (14)$$

If $|F|$ is less than 0.5, the two vultures' movements accumulate several types of vultures over the food sources, and the siege and aggressive strife to find food are implemented using parameter P_3. This can be defined as follows:

$$P(i+1) = \begin{cases} \frac{A_1 + A_2}{2} \; if \; P_3 \geq ranm_{P3} \\ R(i) - (|R(i) - P(i)|)(F)(LF(d)) \; if \; P_3 < ranm_{P3} \end{cases} \quad (15)$$

where,

$$A_1 = vulture_{best1}(i) - F \frac{vulture_{best1}(i)P(i)}{vulture_{best1}(i) - P(i)^2}; and \; A_2 = vulture_{best2}(i) - F \frac{vulture_{best2}(i)P(i)}{vulture_{best2}(i) - P(i)^2} \quad (16)$$

$$LF(x) = 0.01 \frac{\mu\sigma}{|v|^{1/\beta}}, \; \sigma = \left(\frac{\Gamma(1+\beta)\sin(\pi\beta/2)}{\Gamma(1+\beta^2)\beta2((\beta-1)/2)} \right)^{1/\beta} \quad (17)$$

In which, $vulture_{best1}(i)$ and $vulture_{best2}(i)$ are the best vulture of the first and second groups, respectively, in the iteration i; d is the problem dimensions, μ and v denotes a random number between 0 and 1, and $\beta = 1.5$.

In this work, AVOA is used to tune the SVR parameter set to find the efficient performance of the SVR model.

4.3. Models' Development and Accuracy Assessment

This study proposed a new hybrid AVOA metaheuristic algorithm-based SVR model to find the optimal parameters (C, ε, and γ) of SVR and compare its performance with other two metaheuristics nature-inspired algorithms (called PSO and HHO)-based SVR model. Generally, the robustness of the SVR model depends on an appropriate selection of the parameters named as the penalty parameter/"BoxConstraint" (C), insensitive loss function/"epsilon" (ε), and the kernel parameter/"KernelScale" (γ/gamma). The range of these parameters is large, and it is difficult to search for the optimal set of values for these three parameters. Therefore, this optimization task may be solved using optimization algorithms. The authors of this article used three metaheuristic algorithms (AVOA, PSO, and HHO) to find the optimal parameter set of the SVR model. Figure 3 shows the process flow of the proposed technique. Initially, the missing value from the dataset is replaced using the k-nearest neighbor (KNN) imputer method. This study used a Euclidean distance measure to fill the missing value. The whole dataset is bifurcated into a train (80%) and test (20%) set. The three-metaheuristic algorithm is used to train SVR parameters for all/selected featured datasets separately. Root means squared error (RMSE) is selected as the fitness function of all algorithms. Metaheuristic algorithms are sensitive to their different parameter set. The Hit and trail approach is used to select the initial parameter set of metaheuristic algorithms. Table 4 shows the initial value of all parameters of three algorithms for SVR training. Since the number of epochs and population size affect the convergence rate of metaheuristic algorithms, this research work aims to find a faster convergence rate of metaheuristic algorithm with 15 epochs and five population sizes (Table 4).

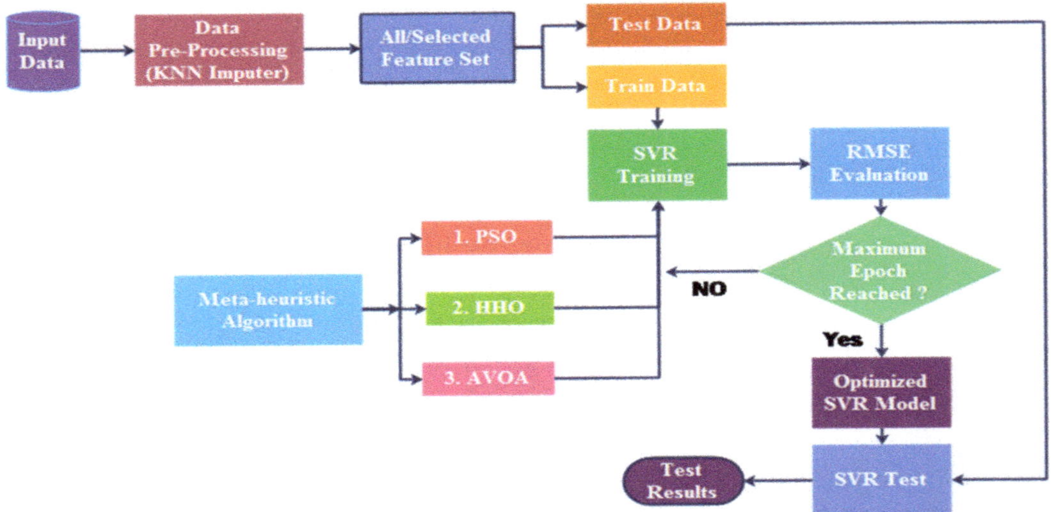

Figure 3. Process flowchart of the proposed method.

Table 4. Initial parameters of metaheuristic algorithms to train the SVR model.

Metaheuristic Algorithm	Parameters	Value
	Population	5
	Iteration	15
	P_1	0.9
	P_2	0.3
	P_3	0.6
AVOA	Alpha	0.8
	Beta	0.2
	Gamma	2.5
	Range of C	$[10^3, 10^{-3}]$
	Range of ε	$[10^3, 10^{-3}]$
	Range of γ	$[10^3, 10^{-3}]$
	Population	5
	Iteration	15
	C_1	1
PSO	C_2	2
	Range of C	$[10^3, 10^{-3}]$
	Range of ε	$[10^3, 10^{-3}]$
	Range of γ	$[10^3, 10^{-3}]$
	Population	5
	Iteration	15
	N	3
HHO	Range of C	$[10^3, 10^{-3}]$
	Range of ε	$[10^3, 10^{-3}]$
	Range of γ	$[10^3, 10^{-3}]$

To evaluate the accuracy of the proposed models, eight statistical indices are used: coefficient of determination (R^2), variance account factor (*VAF*), variance inflation factor (*VIF*), mean absolute error (*MAE*), root mean square error (*RMSE*), performance index (*PI*), mean bias error (*MBE*), and percentage error (*PE*). The R^2 and *VAF* are used to measure the correlation between the measured and predicted values. *VIF* is used to evaluate the collinearity between the measured and predicted values; *VIF* > 10 indicates high collinearity. The models' errors are evaluated using *MAE*, *RMSE*, and *MBE*, and the *PE* is used to estimate the accuracy of the proposed model error in predicting the shear strength of *RC* deep beams. The mathematical expression of these indices can be expressed as follows:

$$R^2 = \frac{\sum_{i=1}^{N}(V_i - V_{mean})^2 - \sum_{i=1}^{N}(V_i - V_{pi})^2}{\sum_{i=1}^{N}(V_i - V_{mean})^2} \tag{18}$$

$$VAF = 100\left(1 - \frac{var(V_i - V_{pi})}{var(V_i)}\right) \tag{19}$$

$$VIF = \frac{1}{1 - R^2} \tag{20}$$

$$RMSE = \sqrt{\frac{\sum_{i=1}^{N}(V_i - V_{pi})^2}{N}} \tag{21}$$

$$MAE = \frac{\sum_{i=1}^{N}|(V_i - V_{pi})|}{N} \tag{22}$$

$$MBE = \frac{1}{N}\sum_{i=1}^{N}(V_i - V_{pi}) \tag{23}$$

$$PI = adj\,R^2 + (0.01\,VAF) - RMSE \tag{24}$$

$$PE = 100\frac{RMSE}{V_{max} - V_{min}} \tag{25}$$

where V_i and V_{pi} represent the measured and predicted shear strength, V_{mean}, V_{max}, and V_{min} are the average, maximum, and minimum, respectively, of measured values, $adj\,R^2$ is the adjustment R^2, and N is the number of the data sample.

4.4. Sensitivity Analysis

Cosine Amplitude Method (CAM) is used to analyze the strength of the relation between input the parameter and power factor [50]. It can also be used to determine the express similarity relation between correlated parameters. To apply CAM, all the data pairs were stated in common X-space. The data pairs used to construct a data array defined K as:

$$K = \{K_1, K_2, K_3, \ldots, K_n\} \tag{26}$$

Every elements i.e., K_i, in the data array K is a vector of lengths j, i.e.,:

$$K_i = \{k_{i1}, k_{i2}, k_{i3}, \ldots, K_{ij}\}_1 \tag{27}$$

Therefore, each of the data pairs is represented as a point in m-dimensional space, where each point requires j-coordinates for its complete description.

5. Results and Discussion

Two scenarios are presented in this section. The first is the evaluation of the proposed models based on all variables and the study of the effect of all variables on Vu estimation. Second, the main variables are considered in modeling and evaluating the sensitivity of input variables on the best selection model. The best solution is compared to existing models in the shear strength determination of RC deep beams.

5.1. All Variables Impact on Vu Estimation

Table 5 presents the statistical indices of the proposed models. The numerical investigation of the statistical indices shows that the performance of the AVOA-SVR model is better to estimate Vu with $R^2 = 0.98$ and RMSE = 32.20 kN in the training stage. The comparison between all models using performance indices of all statistical indices in the training and testing stages shows that the AVOA-SVR model has a high index. However, the performance of HHO-SVR is better in terms of PI, RMSE, and PE in the testing stage. Moreover, Figure 4 illustrates the linear correlation between experimental and predicted Vu values for the proposed models. From Figure 4, it is shown that the performance of the AVOA-SVR model is more accurate than other proposed models in the training and testing stages. The distortion of data points around best fitting is small in modeling Vu with the AVOA-SVR model, and the VIF is higher than in other models, as presented in Table 5. This means that when we used all variables, the AVOA-SVR model can be used to estimate Vu with a model error approach of 6.95%.

Table 5. Statistical evaluation of the proposed models.

Training	R^2	VAF	VIF	PI	RMSE	MAE	MBE	PE
AVOA-SVR	0.984	97.330	64.510	−30.241	32.198	24.377	−0.047	1.723
PSO-SVR	0.813	78.261	5.358	−89.973	91.568	31.605	26.960	4.899
HHO-SVR	0.818	66.278	5.500	−62.003	63.483	105.885	−6.032	3.397
Testing	R^2	VAF	VIF	PI	RMSE	MAE	MBE	PE
AVOA-SVR	0.756	67.921	4.102	−76.076	77.505	101.702	−13.001	6.949
PSO-SVR	0.630	52.981	2.706	−75.687	76.837	106.357	17.850	6.889
HHO-SVR	0.715	45.786	3.514	−46.690	47.856	162.579	−56.320	4.290

The comparison between the AVOA-SVR model and previous studies is presented in Table 6. The statistical evaluation of relative predicted shear strength (Vu measured/predicted (Vm/p)) is presented in Table 6; COV is the coefficient of variation. From this table, it can be observed that the AVOA-SVR model performance is better and slightly better than ACI and Russo algorithms, respectively. Although the distortion around the mean for the proposed model is lower than for the Russo technique, the range of datasets for Russo is better than the proposed models. The Liu technique is better than previous and proposed models when considering the whole variables in modeling shear strength through the AVOA-SVR model. The selected variables are used and evaluated in the next section to estimate more accurate Vu values.

Table 6. Vm/p statistical evaluation for AVOA-SVR and previous studies.

Model	M	Maximum	Minimum	SD	COV
Liu [4]	1.10	1.54	0.65	0.15	0.13
Russo [8]	1.00	1.63	0.48	0.19	0.19
ACI [10]	0.59	2.06	0.09	0.41	0.69
AVOA-SVR	0.95	1.87	0.34	0.16	0.17

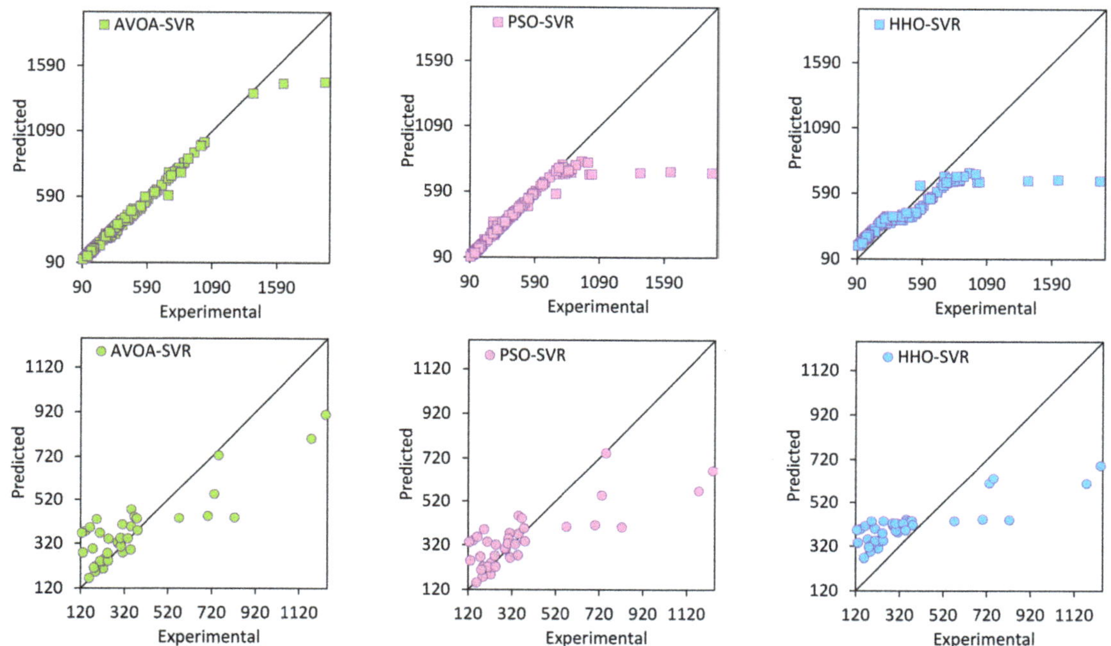

Figure 4. Scatter plot of model's performances in the training (**upper row**) and testing (**lower row**) stages.

5.2. Selected Variables Impact on Vu Estimation

Table 7 and Figure 5 present the performance evaluation of the proposed models. The performance of AVOA-SVR is high in the training stage. A high correlation, $R^2 = 0.97$, and low model error, PE = 2.25%, are observed. In the testing stage, a low distortion around best fitting is observed with the AVOA-SVR. In addition, the statistical correlation factors are high, $R^2 = 0.97$ and VAF = 94.46, as presented in Table 7 and Figure 5. The VIF values of AVOA-SVR in the training and testing stages are higher than other models. This means the accuracy of AVOA-SVR is acceptable with low distortion around the observed values. Although the PI and RMSE of the HHO-SVR models are lower than for the AVOA-SVR model, the distortion of HHO-SVR datasets is high, as presented in Table 7 and Figure 5. Meanwhile, the performance of the proposed models is shown to be low to estimate the high shear strength (as presented in Figures 4 and 5). However, the performance of AVOA-SVR is seen as more robust. This indicates that AVOA-SVR can overcome the variation change in the data used. Figure 6 also shows a faster convergence rate of the AVOA-SVR model compared to the other two hybrid models. Therefore, the AVOA-SVR can be used to estimate the shear strength of RC deep beams with 3.4% model accuracy. The statistical comparison indices in Tables 5 and 7 show that the performance of proposed models with the selected variables is better than that for using all variables in modeling the proposed techniques. This means that the selected variables are more influential in the shear strength of RC deep beams.

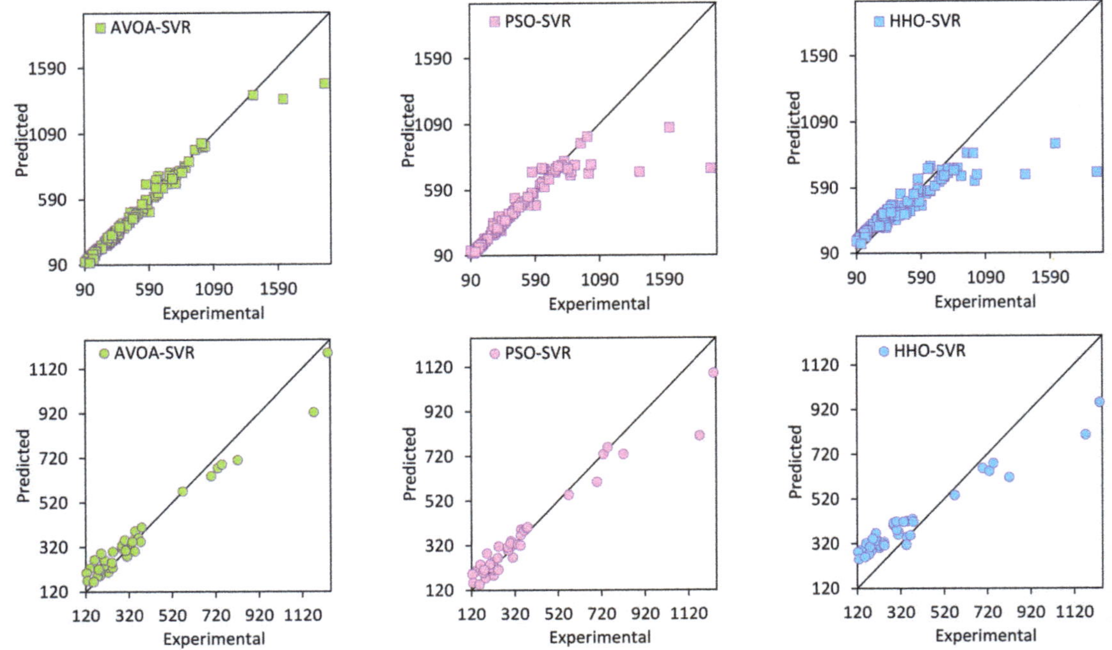

Figure 5. Scatter plot of model's performances in the training (**upper row**) and testing (**lower row**) stages.

Table 7. Statistical evaluation of the proposed models.

Training	R^2	VAF	VIF	PI	RMSE	MAE	MBE	PE
AVOA-SVR	0.974	96.726	39.202	−40.095	42.036	26.728	−0.360	2.249
PSO-SVR	0.834	81.625	6.042	−90.753	92.402	32.755	18.958	4.944
HHO-SVR	0.816	71.805	5.427	−72.442	73.975	92.860	−7.926	3.958
Testing	R^2	VAF	VIF	PI	RMSE	MAE	MBE	PE
AVOA-SVR	0.970	94.460	33.512	−35.876	37.790	43.168	−7.149	3.388
PSO-SVR	0.950	91.774	20.118	−45.091	46.958	44.085	0.475	4.210
HHO-SVR	0.948	79.860	19.147	−33.841	35.586	106.952	−50.633	3.190

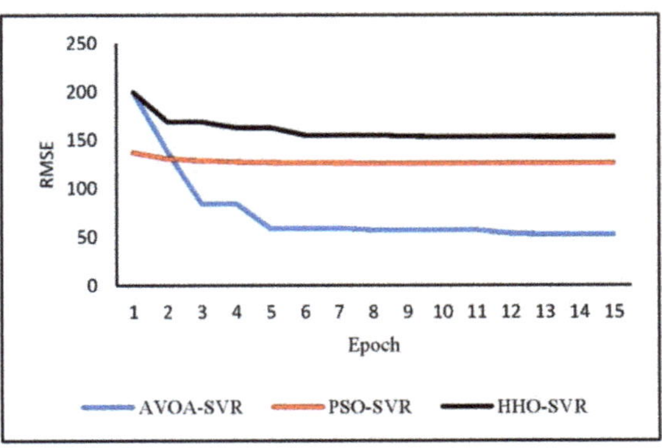

Figure 6. Convergence rate of three models.

5.3. Comparison with Previous Studies and Codes

Table 8 and Figure 7 show the performance of the proposed model compared to the previous studies' formulas. As seen in Table 8, the used selected variable improved the AVOA-SVR performance by 60% in terms of COV. This indicates that the selected variables are significantly affected by the Vu values of the RC deep beams. The comparison between previous formulas and AVOA-SVR shows that the proposed model accuracy is high in estimating the shear strength of RC deep beams. The small range is estimated with AVOA-SVR, and the range is 0.57 kN. The small SD and COV of the statistical indices are observed with AVOA-SVR. This means the accuracy of AVOA-SVR is high compared to other models.

Table 8. Vm/p statistical evaluation for AVOA-SVR and previous studies.

Model	M	Maximum	Minimum	SD	COV
Liu [4]	1.10	1.54	0.65	0.15	0.13
Russo [8]	1.00	1.63	0.48	0.19	0.19
ACI [10]	0.59	2.06	0.09	0.41	0.69
AVOA-SVR	0.98	1.33	0.76	0.07	0.07

The boxplot in Figure 7a shows that the median, red horizontal line of Russo, is close to the true value "1", followed by the AVOA-SVR and Liu models, respectively. The low interquartile range (IQR) value, the height of the box, is observed to be small with the AVOA-SVR model, followed by Liu and Russo models, respectively. The maximum and minimum quartiles, the black horizontal solid lines, are small with the AVOA-SVR model, followed by Liu and Russo models, respectively. The outliers are observed near the median of the AVOA-SVR and far to the median of the ACI model. From the visualization of boxplot results, it can be concluded that the performance of the AVOA-SVR model is more accurate than the previous studies for modeling the shear strength of RC deep beams. In addition, the following model is the Liu model, as this model considers more shear resistance mechanisms and shows a higher normal distribution and lower error than Russo's and ACI's models. The quantile-quantile (Q-Q) plot is presented in Figure 7b for the Liu and AVOA-SVR models for further investigation. The relative shear strength is presented versus the standard normal distribution. From this figure, both models have approximately followed the normal distribution; this indicates that both models can be used to estimate the shear strength of the RC deep beam. The AVOA-SVR model has more correlation with the standard normal distribution, and the VAOA-SVR model is more accurate than the Liu technique in modeling the shear strength of RC deep beams. The scatter plot presented in Figure 7c,d shows that the worst model for estimating the shear strength is the ACI's model. The variation in the best solution "1" is shown as small for Russo, Liu, and AVOA-SVR models, respectively. The most of relative shear strength of the AVOA-SVR model falls within ±20%. The comparison of the AVOA-SVR model and previous models shows that the developed model can be used accurately to model the shear strength of the RC deep beams. Therefore, the AVOA-SVR model is a potential soft computing technique that can be used in predicting the shear strength of RC deep beams.

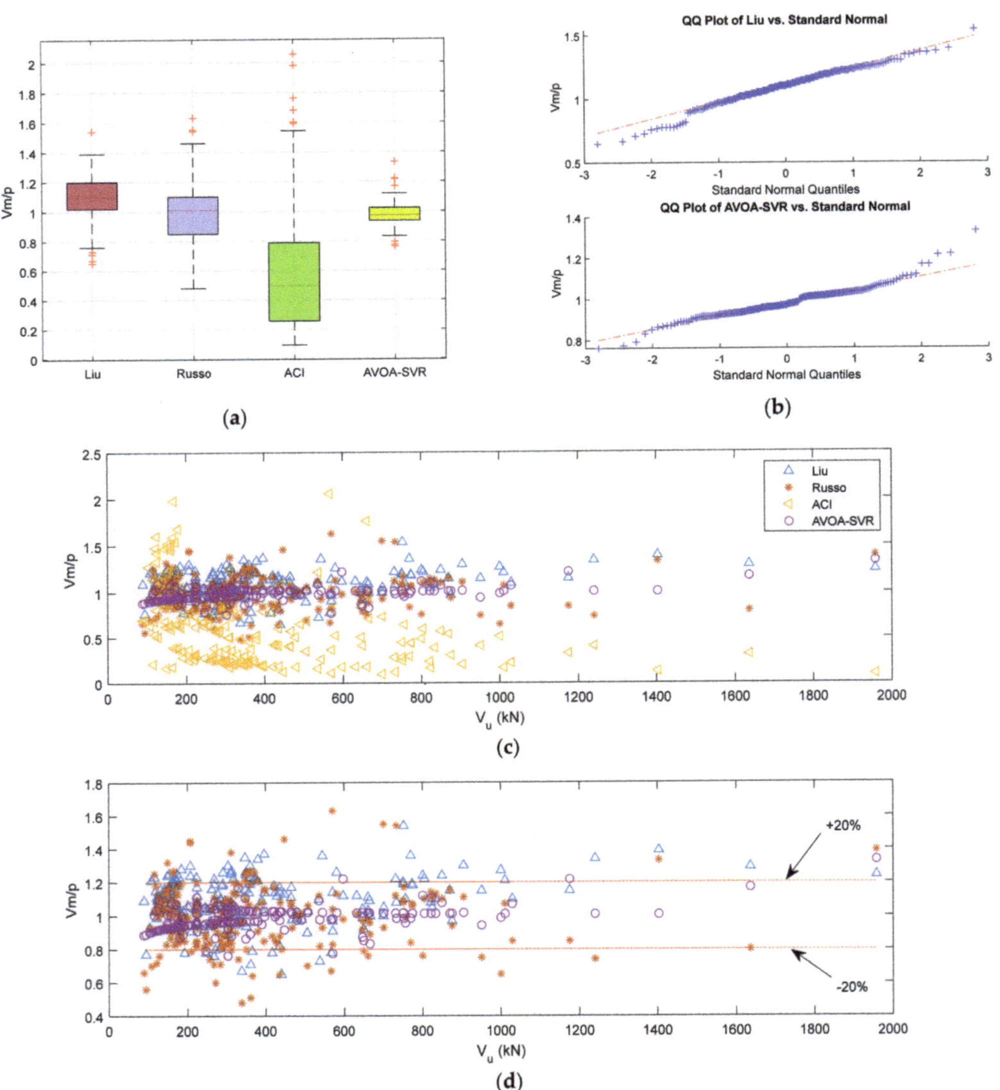

Figure 7. Comparison of model's performances, (**a**) boxplot, (**b**) Q-Q plot, (**c**) scatter plot of relative shear strength with measured shear strength, and (**d**) zoom in for upper plot with +20% limits for the best models.

5.4. Sensitivity Analysis of Input Variables

Figure 8 presents the most influential input variables in modeling the shear strength of the RC deep beam. The impact of the input variables is presented for the three models. From Figure 8, it can be noticed that the significant impact of the ratio of vertical and horizontal web reinforcements is low compared to other variables. The sensitivity of the shear span to depth ratio is high, followed by the yield strength of the main steel, the ratio of the main tensile bars, yield strength of vertical web reinforcement, stirrups spacing, and concrete compressive strength, respectively. The impact of the input variables on output for the other developed models is similar. These results imply that the shear strength of the RC deep beam is highly influenced by the beam geometry, concrete strength, and yield strength of the steel bars. The stirrups spacing also has a large effect on the shear strength of RC deep beams.

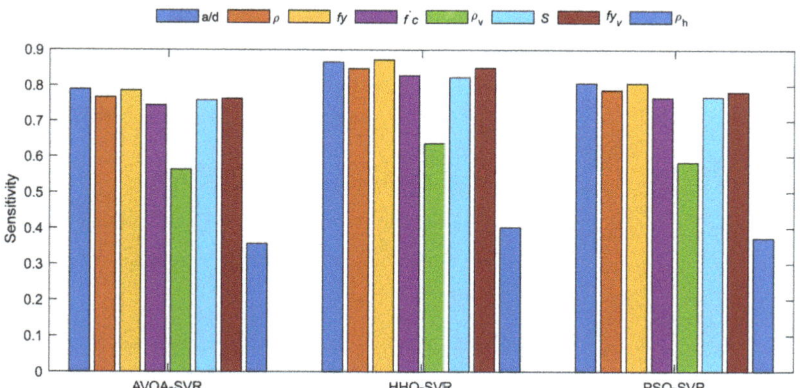

Figure 8. Sensitivity of input variable on model's prediction.

6. Conclusions

This study investigated the use of new metaheuristic optimization algorithms integrated with SVR to model the shear strength of RC deep beams and evaluate the sensitivity of input parameters. SVR-AVOA, -PSO, and -HHO were designed and compared to existing models in the current study. In this study, 202 datasets, including 19 variables of experimental studies, were collected from literature to design and evaluate the proposed models. The common eight parameters (shear span to depth ratio, the ratio of the main tensile bars, yield strength of main bars, concrete compressive strength, the ratio of vertical web reinforcement (stirrups), stirrups spacing, yield strength of vertical web reinforcement, and the ratio of horizontal web reinforcement) are also used to evaluate the performance of the proposed models' in predicting the shear strength of RC deep beams. The performance of SVR-AVOA is high in the cases of the used 19 and 8 parameters for modeling the shear strength. The accuracy of SVR-AVOA is improved by 60%, in COV terms, using the common input variables. Thus, other parameters were found less significant in modeling the shear strength of RC deep beams. The comparison of the SVR-AVOA and the previous studies shows that the accuracy of the proposed model is higher than Liu [4], Russo [8], and ACI [10] by 46%, 63%, and 90%, respectively, in terms of COV. This indicates that SVR-AVOA is the more robust model and can be accurately used in modeling the shear strength of RC deep beams. The sensitivity of the input variables in modeling the shear strength of RC beams with the SVR-AVOA was assessed. This investigation shows the impact of the shear span on the beam's depth ratio, yield strengths of vertical and horizontal web reinforcement, concrete compressive strength, stirrups spacing, and the ratio of the main longitudinal bars on the deep beams' shear strength.

Furthermore, the sensitivity of AVOA algorithm parameters can be tested to balance between exploitation and exploration side for enhancing the SVR performance. In the future, to check the efficiency of the proposed model should be tested on other datasets and other civil engineering application areas. The AVOA algorithm can be combined with other machine learning models like an extreme learning machine, random forest, etc., for prediction tasks.

Supplementary Materials: The following supporting information can be downloaded at: https://www.mdpi.com/article/10.3390/su14095238/s1, Figure S1: Direct relationship between inputs and output; Table S1: Data used in Modeling; Table S2: Modeling variables.

Author Contributions: Conceptualization, M.R.K. and B.S.A.; methodology, M.R.K. and B.R.; software, B.R.; validation, M.R.K., B.R. and B.S.A.; formal analysis, M.R.K. and B.S.A.; investigation, M.R.K., B.R. and B.S.A.; resources, B.S.A.; data curation, M.R.K. and B.S.A.; writing—original draft preparation, M.R.K., B.R. and B.S.A.; writing—review and editing, M.R.K., B.R., K.C. and B.S.A.; visualization, M.R.K., B.R., K.C., S.-M.K., H.-M.J. and B.S.A.; supervision, M.R.K. and J.-W.H.; project administration, M.R.K., J.-W.H. and B.S.A.; funding acquisition, J.-W.H. All authors have read and agreed to the published version of the manuscript.

Funding: This work was supported by Incheon National University Research Concentration Professors Grant in 2021.

Informed Consent Statement: Not applicable.

Data Availability Statement: The data used are available in the manuscript.

Conflicts of Interest: The authors declare no conflict of interest.

References

1. Lee, Y.; Kim, S.; Kim, S. FEM Analysis of RC Deep Beam Depending on Shear-Span Ratio. *Archit. Res.* **2017**, *19*, 117–124.
2. Gandomi, A.; Alavi, A.; Shadmehri, D.M.; Sahab, M. An empirical model for shear capacity of RC deep beams using genetic-simulated annealing. *Arch. Civ. Mech. Eng.* **2013**, *13*, 354–369. [CrossRef]
3. Chou, J.-S.; Ngo, N.-T.; Pham, A.-D. Shear Strength Prediction in Reinforced Concrete Deep Beams Using Nature-Inspired Metaheuristic Support Vector Regression. *J. Comput. Civ. Eng.* **2016**, *30*, 04015002. [CrossRef]
4. Liu, J. Kinematics-Based Modelling of Deep Transfer Girders in Reinforced Concrete Frame Structures. Ph.D. Thesis, Liege University, Liege, Belgium, 2019.
5. Hwang, W.-Y.L.S.; Lee, H. Shear Strength Prediction for Deep Beams. *ACI Struct. J.* **2000**, *97*, 367–376. [CrossRef]
6. Yavuz, G. Shear strength estimation of RC deep beams using the ANN and strut-and-tie approaches. *Struct. Eng. Mech.* **2016**, *57*, 657–680. [CrossRef]
7. Nguyen, T.-A.; Ly, H.-B.; Mai, H.-V.T.; Tran, V.Q. On the Training Algorithms for Artificial Neural Network in Predicting the Shear Strength of Deep Beams. *Complexity* **2021**, *2021*, 5548988. [CrossRef]
8. Russo, G.; Venir, R.; Pauletta, M. Reinforced Concrete Deep Beams-Shear Strength Model and Design Formula. *ACI Struct. J.* **2005**, *102*, 429–437.
9. Dang, T.D.; Tran, D.T.; Nguyen-Minh, L.; Nassif, A.Y. Shear resistant capacity of steel fibres reinforced concrete deep beams: An experimental investigation and a new prediction model. *Structures* **2021**, *33*, 2284–2300. [CrossRef]
10. ACI. *Building Code Requirement for Structural Concrete and Commentary*; ACI: Detroit, MI, USA, 2011; p. ACI-318.
11. Ben Chaabene, W.; Nehdi, M.L. Novel soft computing hybrid model for predicting shear strength and failure mode of SFRC beams with superior accuracy. *Compos. Part C Open Access* **2020**, *3*, 100070. [CrossRef]
12. Keshtegar, B.; Nehdi, M.L.; Kolahchi, R.; Trung, N.-T.; Bagheri, M. Novel hybrid machine leaning model for predicting shear strength of reinforced concrete shear walls. *Eng. Comput.* **2021**, 0123456789. [CrossRef]
13. Al-Musawi, A.; Alwanas, A.A.H.; Salih, S.; Ali, Z.; Tran, M.T.; Yaseen, Z.M. Shear strength of SFRCB without stirrups simulation: Implementation of hybrid artificial intelligence model. *Eng. Comput.* **2018**, *36*, 1–11. [CrossRef]
14. Ning, C.; Li, B. Analytical probabilistic model for shear strength prediction of reinforced concrete beams without shear reinforcement. *Adv. Struct. Eng.* **2017**, *21*, 171–184. [CrossRef]
15. Naderpour, H.; Nagai, K. Shear strength estimation of reinforced concrete beam–column sub-assemblages using multiple soft computing techniques. *Struct. Des. Tall Spéc. Build.* **2020**, *29*, 1–15. [CrossRef]
16. Cheng, M.-Y.; Cao, M.-T. Evolutionary multivariate adaptive regression splines for estimating shear strength in reinforced-concrete deep beams. *Eng. Appl. Artif. Intell.* **2014**, *28*, 86–96. [CrossRef]
17. Abdollahzadeh, B.; Gharehchopogh, F.S.; Mirjalili, S. African vultures optimization algorithm: A new nature-inspired metaheuristic algorithm for global optimization problems. *Comput. Ind. Eng.* **2021**, *158*, 107408. [CrossRef]
18. Dragoi, E.N.; Dafinescu, V. Review of Metaheuristics Inspired from the Animal Kingdom. *Mathematics* **2021**, *9*, 2335. [CrossRef]
19. Bagal, H.A.; Soltanabad, Y.N.; Dadjuo, M.; Wakil, K.; Zare, M.; Mohammed, A.S. SOFC model parameter identification by means of Modified African Vulture Optimization algorithm. *Energy Rep.* **2021**, *7*, 7251–7260. [CrossRef]
20. Sharafati, A.; Haghbin, M.; Aldlemy, M.S.; Mussa, M.H.; Al Zand, A.W.; Ali, M.; Bhagat, S.K.; Al-Ansari, N.; Yaseen, Z.M. Development of Advanced Computer Aid Model for Shear Strength of Concrete Slender Beam Prediction. *Appl. Sci.* **2020**, *10*, 3811. [CrossRef]
21. Parsa, P.; Naderpour, H. Shear strength estimation of reinforced concrete walls using support vector regression improved by Teaching–learning-based optimization, Particle Swarm optimization, and Harris Hawks Optimization algorithms. *J. Build. Eng.* **2021**, *44*, 102593. [CrossRef]
22. Tosee, S.V.R.; Faridmehr, I.; Bedon, C.; Sadowski, Ł.; Aalimahmoody, N.; Nikoo, M.; Nowobilski, T. Metaheuristic Prediction of the Compressive Strength of Environmentally Friendly Concrete Modified with Eggshell Powder Using the Hybrid ANN-SFL Optimization Algorithm. *Materials* **2021**, *14*, 6172. [CrossRef]

23. Pal, M.; Deswal, S. Support vector regression based shear strength modelling of deep beams. *Comput. Struct.* **2011**, *89*, 1430–1439. [CrossRef]
24. Zhang, D.; Shahin, M.; Yang, Y.; Liu, H.; Chang, L. Effect of microbially induced calcite precipitation treatment on the bonding properties of steel fiber in ultra-high performance concrete. *J. Build. Eng.* **2022**, *50*, 104132. [CrossRef]
25. Chen, B.; Zhou, J.; Zhang, D.; Su, J.; Nuti, C.; Sennah, K. Experimental study on shear performances of ultra-high performance concrete deep beams. *Structures* **2022**, *39*, 310–322. [CrossRef]
26. Smith, K.N.; Vantsiotis, A.S. Shear Strength of Deep Beams. *J. Am. Concr. Inst.* **1982**, *79*, 201–213. [CrossRef]
27. Ahmed, A.K.E.-S. Concrete Contribution to the Shear Resistance of FRP-Reinforced Concrete Beams. Ph.D. Thesis, University of Sherbrooke, Sherbrooke, QC, Canada, 2006.
28. Oh, J.-K.; Shin, S.-W. Shear Strength of Reinforced High-Strength Concrete Deep Beams. *ACI Struct. J.* **2001**, *98*, 164–173. [CrossRef]
29. El-Zoughiby, M.E. Z-Shaped Load Path: A Unifying Approach to Developing Strut-and-Tie Models. *ACI Struct. J.* **2021**, *118*, 35–48. [CrossRef]
30. Jin-Keun, K.; Yon-Dong, P. Shear strength of reinforced high strength concrete beam without web reinforcement. *Mag. Concr. Res.* **1994**, *46*, 7–16. [CrossRef]
31. Londhe, R. Shear strength analysis and prediction of reinforced concrete transfer beams in high-rise buildings. *Struct. Eng. Mech.* **2011**, *37*, 39–59. [CrossRef]
32. Mau, S.T.; Hsu, T. Formula for the Shear Strength of Deep Beams. *ACI Struct. J.* **1989**, *86*, 516–523. [CrossRef]
33. Ashour, A.; Alvarez, L.; Toropov, V. Empirical modelling of shear strength of RC deep beams by genetic programming. *Comput. Struct.* **2003**, *81*, 331–338. [CrossRef]
34. Mozumder, R.A.; Roy, B.; Laskar, A.I. Support Vector Regression Approach to Predict the Strength of FRP Confined Concrete. *Arab. J. Sci. Eng.* **2016**, *42*, 1129–1146. [CrossRef]
35. Vapnik, V. *The Nature of Statistical Learning Theory*; Springer: New York, NY, USA, 1995.
36. Yaseen, Z.M.; Tran, M.T.; Kim, S.; Bakhshpoori, T.; Deo, R. Shear strength prediction of steel fiber reinforced concrete beam using hybrid intelligence models: A new approach. *Eng. Struct.* **2018**, *177*, 244–255. [CrossRef]
37. Yap, C.W.; Li, Z.; Chen, Y. Quantitative structure–pharmacokinetic relationships for drug clearance by using statistical learning methods. *J. Mol. Graph. Model.* **2006**, *24*, 383–395. [CrossRef]
38. Du, K.; Liu, M.; Zhou, J.; Khandelwal, M. Investigating the Slurry Fluidity and Strength Characteristics of Cemented Backfill and Strength Prediction Models by Developing Hybrid GA-SVR and PSO-SVR. *Min. Met. Explor.* **2022**, *39*, 433–452. [CrossRef]
39. Liu, Q.; Li, S.; Yin, J.; Li, T.; Han, M. Simulation of mechanical behavior of carbonate gravel with hybrid PSO–SVR algorithm. *Mar. Georesour. Geotechnol.* **2022**, 1–14. [CrossRef]
40. Ortúzar, J. Future transportation: Sustainability, complexity and individualization of choices. *Commun. Transp. Res.* **2021**, *1*, 100010. [CrossRef]
41. Mohammed, A.; Kurda, R.; Armaghani, D.J.; Hasanipanah, M. Prediction of Compressive Strength of Concrete Modified with Fly Ash: Applications of Neuro-Swarm and Neuro-Imperialism Models. *Comput. Concr.* **2021**, *27*, 489–512.
42. Shahbazian, A.; Rabiefar, H.; Aminnejad, B. Shear Strength Determination in RC Beams Using ANN Trained with Tabu Search Training Algorithm. *Adv. Civ. Eng.* **2021**, *2021*, 1639214. [CrossRef]
43. Kennedy, J.; Eberhart, R. Particle Swarm Optimization. In Proceedings of the IEEE International Conference on Neural Networks, Perth, Australia, 27 November–1 December 1995; pp. 1942–1948. [CrossRef]
44. Yang, H.-C.; Zhang, S.-B.; Deng, K.-Z.; DU, P.-J. Research into a Feature Selection Method for Hyperspectral Imagery Using PSO and SVM. *J. China Univ. Min. Technol.* **2007**, *17*, 473–478. [CrossRef]
45. Heidari, A.A.; Mirjalili, S.; Faris, H.; Aljarah, I.; Mafarja, M.; Chen, H. Harris hawks optimization: Algorithm and applications. *Futur. Gener. Comput. Syst.* **2019**, *97*, 849–872. [CrossRef]
46. Golafshani, E.M.; Arashpour, M.; Behnood, A. Predicting the compressive strength of green concretes using Harris hawks optimization-based data-driven methods. *Constr. Build. Mater.* **2021**, *318*, 125944. [CrossRef]
47. Wei, W.; Li, X.; Liu, J.; Zhou, Y.; Li, L.; Zhou, J. Performance Evaluation of Hybrid WOA-SVR and HHO-SVR Models with Various Kernels to Predict Factor of Safety for Circular Failure Slope. *Appl. Sci.* **2021**, *11*, 1922. [CrossRef]
48. Zhang, H.; Nguyen, H.; Bui, X.-N.; Pradhan, B.; Asteris, P.G.; Costache, R.; Aryal, J. A generalized artificial intelligence model for estimating the friction angle of clays in evaluating slope stability using a deep neural network and Harris Hawks optimization algorithm. *Eng. Comput.* **2021**, 1–14. [CrossRef]
49. Sammen, S.; Ghorbani, M.; Malik, A.; Tikhamarine, Y.; AmirRahmani, M.; Al-Ansari, N.; Chau, K.-W. Enhanced Artificial Neural Network with Harris Hawks Optimization for Predicting Scour Depth Downstream of Ski-Jump Spillway. *Appl. Sci.* **2020**, *10*, 5160. [CrossRef]
50. Ji, X.; Liang, S.Y. Model-based sensitivity analysis of machining-induced residual stress under minimum quantity lubrication. *Proc. Inst. Mech. Eng. Part B J. Eng. Manuf.* **2015**, *231*, 1528–1541. [CrossRef]

sustainability

MDPI

Article

Deformation Capacity of RC Beam-Column Joints Strengthened with Ferrocement

M. Zardan Araby [1,2], Samsul Rizal [3], Abdullah [2], Mochammad Afifuddin [2] and Muttaqin Hasan [2,*]

[1] Doctor of Engineering Study Program, Universitas Syiah Kuala, Darussalam, Banda Aceh 23111, Indonesia; zardan_araby@unsyiah.ac.id

[2] Department of Civil Engineering, Universitas Syiah Kuala, Darussalam, Banda Aceh 23111, Indonesia; abdullahmahmud@unsyiah.ac.id (A.); m.afifuddin@unsyiah.ac.id (M.A.)

[3] Department of Mechanical Engineering, Universitas Syiah Kuala, Darussalam, Banda Aceh 23111, Indonesia; samsul.rizal@unsyiah.ac.id

* Correspondence: muttaqin@unsyiah.ac.id

Abstract: Beam-column joints constructed in the pre-seismic building code do not provide transverse reinforcement and good reinforcement detailing within the region. These cause the occurrence of brittle shear failure, which is one of the factors affecting the number of reinforced concrete (RC) moment resistance building structures collapsing during an earthquake. Therefore, in this study a brittle beam-column joint with a non-seismic building code was designed and strengthened by a ferrocement. Four layers of wire mesh with a diameter of 1 mm and a mesh size of 25.4 mm were installed on both sides of the beam-column joint and cement mortar was cast on it. As a comparison, a ductile beam-column joint was also designed following the current building code, which considers seismic effects. The test results by applying reversed cyclic loading at the beam tip showed that strengthening using ferrocement prevents crack propagation, increasing the deformation capacity, ductility, stiffness, and energy dissipation of beam column joint which are higher than those of the beam-column joint which is designed following the current building code. However, the strengthening does not improve the load carrying capacity significantly.

Keywords: beam-column joint; ferrocement; crack; ductility; displacement

Citation: Araby, M.Z.; Rizal, S.; Abdullah; Afifuddin, M.; Hasan, M. Deformation Capacity of RC Beam-Column Joints Strengthened with Ferrocement. *Sustainability* **2022**, *14*, 4398. https://doi.org/10.3390/su14084398

Academic Editors: Maged A. Youssef, Chiara Bedon, Mislav Stepinac, Marco Fasan and Ajitanshu Vedrtnam

Received: 10 March 2022
Accepted: 4 April 2022
Published: 7 April 2022

Publisher's Note: MDPI stays neutral with regard to jurisdictional claims in published maps and institutional affiliations.

Copyright: © 2022 by the authors. Licensee MDPI, Basel, Switzerland. This article is an open access article distributed under the terms and conditions of the Creative Commons Attribution (CC BY) license (https://creativecommons.org/licenses/by/4.0/).

1. Introduction

Reinforced concrete (RC) buildings constructed in the 1970s and 1980s or earlier lacked transverse reinforcement installations in the beam-column joint region. Furthermore, they did not adhere to the requirements of a detailed reinforcement to withstand seismic loads because the building code at that time did not accommodate these effects on the beam-column joint structures. In Indonesia, the building code used to design certain structures during that period was NI-2 [1]. The absence of special provisions regarding reinforcement detailing and transversal reinforcement in beam-column joint region due to the ductile design philosophy has not been adopted yet, led to the failure of the beam-column joints. Globally, it is one of the causes of collapsed buildings during earthquakes [2–14]. During that era, many buildings were also designed without the strong column–weak beam design philosophy which results in the appearance of column hinges in their collapse pattern. However, the column hinges were also found although the structures are designed based on the strong column–weak beam criterion [15].

Generally, the beam-column joint experiences a brittle shear failure. Therefore, efforts are needed to strengthen the structures that were built in the 70s and 80s for their sustainability [16–19]. Several structural strengthening methods have been proposed to withstand seismic loads. These include the use of Fiber Reinforced Polymer (FRP) [20–40], steel jacketing [20,41–44], pre-fabricated composite blocks [45], diagonal steel bars [46], steel haunch strengthening and confining joint reinforcement [44,47], injection of cracks with epoxy and concrete jacket [20,48,49], textile-reinforced engineered cementitious composites [50,51], installing reinforced concrete wing walls [52], and modified reinforcement technique [53]. However, those strengthening systems are expensive and need to be professionally installed.

Ferrocement is a type of thin reinforced concrete usually made of hydraulic cement mortar reinforced with a metallic mesh or similar materials [54] and has been used as a material for the strengthening of reinforced concrete structures. It possesses greater tensile strength and resistance to cracking than conventional reinforced concrete. Despite having similar durability, ferrocement is extremely elastic. In this study, ferrocement was used to strengthen the beam-column joint, because this method is cheap compared to the above-described methods, and its materials are always available, simple, and easy to install. Several studies have been carried out on its use as a beam strengthening material, and it was discovered to increase flexural capacity, stiffness, ductility, and energy dissipation [55–61]. Similarly, as a column strengthening material, it also significantly increases ductility and energy dissipation as well as load-bearing capacity and stiffness [62–65]. However, very limited studies have been carried out on beam-column joint strengthening using ferrocement [66–68].

This study was carried out with the aim to understand the deformation capacity of a beam-column joint designed with a non-seismic building code strengthened with ferrocement under reversed cyclic loading. The reversed cyclic loading was chosen to simulate the earthquake action in building structures since the strengthening method proposed in this study will be applied in building structures in a seismic prone area. The strengthened beam-column joint performance is then compared to the unreinforced one and used to ascertain the increase in its deformation capacity. Furthermore, it is also compared with the performance of beam-column joint designed with the current Indonesian building code [69], which considers the seismic loads effect to discern the strengthening efficiency to withstand such impacts. Therefore, it is expected that this study recommends an economical and practical structural strengthening method that is applicable when strengthening existing buildings. It is important to note that the behaviors of strengthened beam-column joint presented in this paper are only based on the experimental results. The numerical analysis with reliable models of such structure is considered for further study.

2. Experimental Program

2.1. Detail of Specimens

Three beam-column joint specimens were prepared as shown in Table 1. The beams and columns are designed to have similar cross-sectional width and longitudinal reinforcements. The beam cross-section is 300 mm × 400 mm with 8D14 mm longitudinal reinforcements. Furthermore, the transverse reinforcements is in the form of D10 mm stirrups, which are installed every 100 mm distance from the column face to as far as 200 mm in its front. In the other part of the beam, D10 mm stirrups are installed within a distance of 50 mm. The column cross-section is 300 mm × 300 mm with longitudinal reinforcements of 8D14 mm. Conversely, the transverse reinforcements in the form of D10 mm stirrups are installed every 100 mm distance from the beam face to as far as 200 mm in its front. In the other part of the column, D10 mm stirrups are installed within a distance of 50 mm. The installation of a tighter stirrup in those areas is to ensure that cracks do not occur in this area when a load is applied. This is due to the fact the beam-column joint is the only part observed in this study. The weak column–strong beam was selected to represent buildings constructed in the 70s and 80s. The mix proportion and the concrete compressive strength as well as reinforcing bar yield strength and Young's modulus used for the beam-column joints are shown in Tables 2 and 3. The concrete compressive strength presented in Table 2 is the average compressive strength tested on 20 cylinder specimens with diameter of 150 mm and height of 300 mm at the age of 28 days, while the yield strength and elastic modulus of steel bars presented in Table 3 were obtained from the test results on one specimen for each bar diameter.

Table 1. Detail of specimens.

Specimen	Description
BCJ1	Beam column joint designed based on NI-2 [1]
BCJ2	Beam column joint designed based on SNI 2847:2020 [69]
BCJ3	Beam column joint designed based on NI-2 [1] and strengthened with ferrocement

Table 2. Mix proportion of concrete.

Materials	Quantity
Cement (kg)	317
Water (kg)	205
Fine sand (kg)	365
Coarse sand (kg)	729
Split stone d_{max} 20 mm (kg)	729
w/c	0.65
Slump (mm)	12
Compressive strength (MPa)	24.8

Table 3. Yield strength and Young's modulus of reinforcing bars.

Bar Diameter (mm)	Yield Strength, f_y (MPa)	Young's Modulus (GPa)
14	310	200
10	375	200

BCJ1 is a beam-column joint specimen designed according to NI-2 [1] without using transverse reinforcement in the joint region. The beam longitudinal reinforcing bars are not continuously bent towards the upper and lower columns. Figure 1 shows details of specimen BCJ1. Specimen BCJ2 has a similar size and longitudinal reinforcement as BCJI. The difference is that specimen BCJ2 added stirrups in the joint region using bars with a diameter of 10 mm placed at a spacing of 100 mm according to the current Indonesian building code [69]. The beam longitudinal reinforcing bars are also bent towards the upper and lower column with a length of 500 mm therefore their functions as anchors. Comprehensive details of specimen BCJ2 are shown in Figure 2. The specimen BCJ3 was designed in a similar manner as BCJ1. However, it was strengthened using ferrocement on both sides of the beam-column joint, as shown in Figure 3. Subsequently, the strengthening was provided in the following way. The first step involves disassembling the concrete cover of the specimen in the joint area, up to 400 mm in front of the beam and column, with a thickness until the reinforcing bars of the specimen is visible with the thickness of 40 mm. Furthermore, a T-shaped wire mesh with 1 mm diameter and 25.4 mm mesh size, similar to the beam-column joint area was cut. Four layers of wire mesh were installed and mounted on both sides of the specimen with the orientation angle of 0° to beam and column longitudinal axis as shown in Figure 4. Afterward, the wire mesh was anchored to the specimen using 4 dynabolts. Finally, a cement mortar was re-cast (Figure 5). Cement and sand with a volume ratio of 1:4 and water to cement ratio of 0.5 was used for the mortar mixture. A polycarboxylate-based superplasticizer having a specific gravity of 1.06 with a content of 1% of cement weight was also added to the mixture thereby enabling it flows easily when filling the cavities of the wire mesh during casting. Furthermore, the strengthening of the beam-column joint with ferrocement was carried out on both sides of the specimen by strengthening it on one side first, and after the mortar has hardened, the same procedure was performed on the other side.

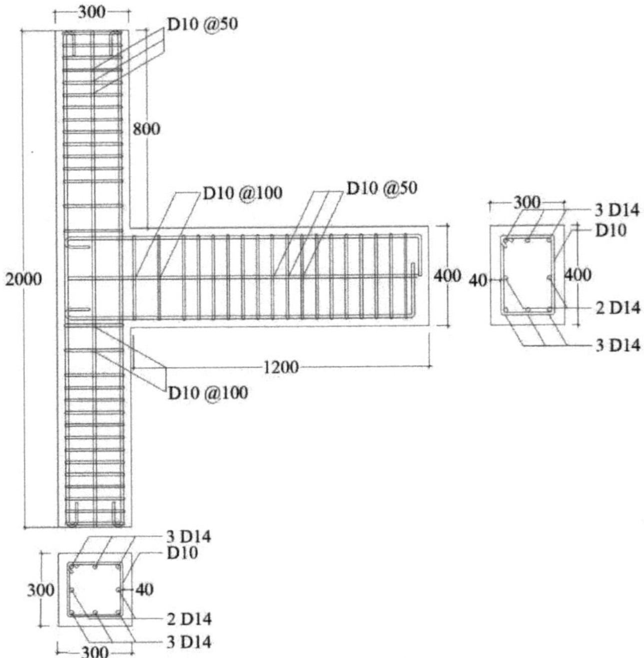

Figure 1. Detail specimen BCJ1.

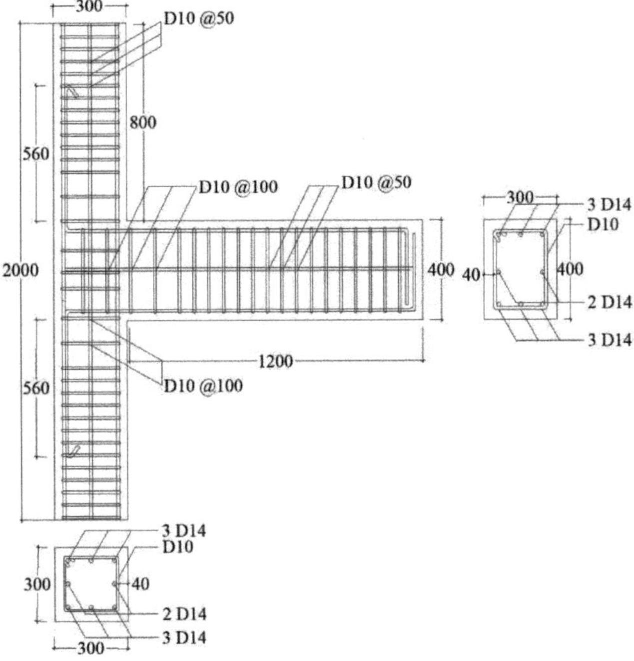

Figure 2. Detail specimen BCJ2.

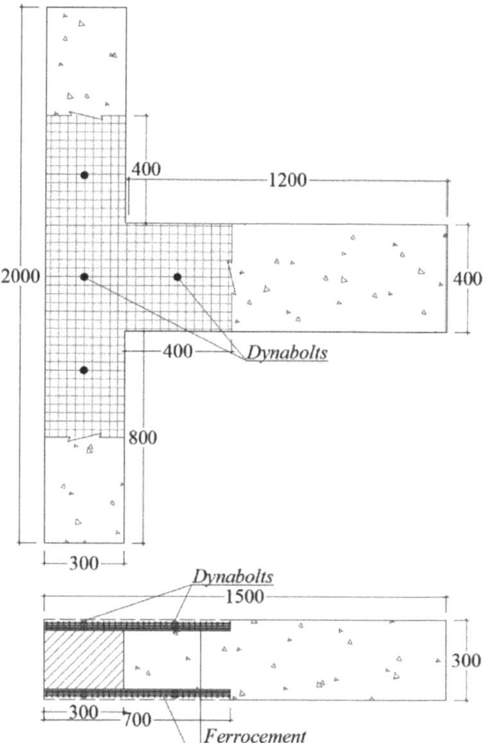

Figure 3. Detail specimen BCJ3.

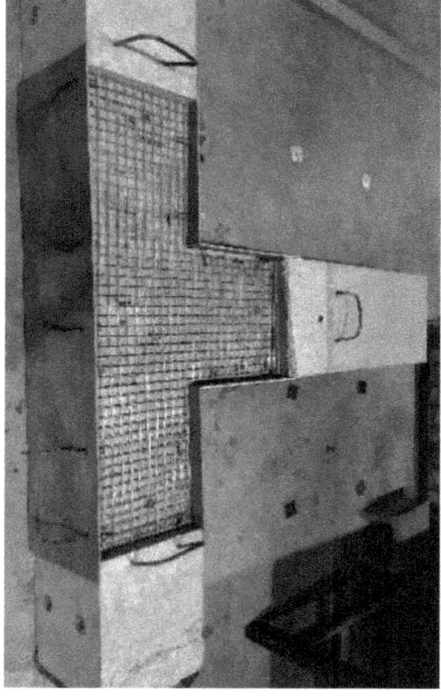

Figure 4. Wire mesh already installed on one side of the beam-column joint.

Figure 5. Re-casting the beam-column joint after installing the wire mesh.

2.2. Loading Procedure

The specimen was first set on the loading frame, as shown in Figure 6. The column was placed horizontally above the loading frame by anchoring it at a distance of 200 mm from both ends using bolts. Furthermore, at both ends of the column, L shape steel is installed, welded to its longitudinal reinforcement, and anchored to the loading frame using bolts. The load is applied through an actuator driven by a hydraulic jack placed on the beam tip. A steel plate is installed on the beam surface to fasten the actuator and specimen, thereby providing a reversed cyclic loading under deformation control. Unloading and reloading were performed at the displacement of 0.75 mm, 1.5 mm, 3 mm, 6 mm, 12 mm, and 24 mm while two loading cycles were applied for each displacement as shown in Figure 7. After 12 cycles, the loading was continued with monotonic until the specimen failed.

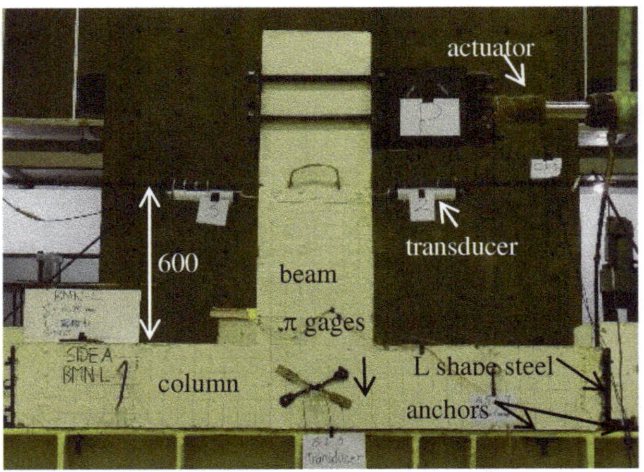

Figure 6. Loading set up.

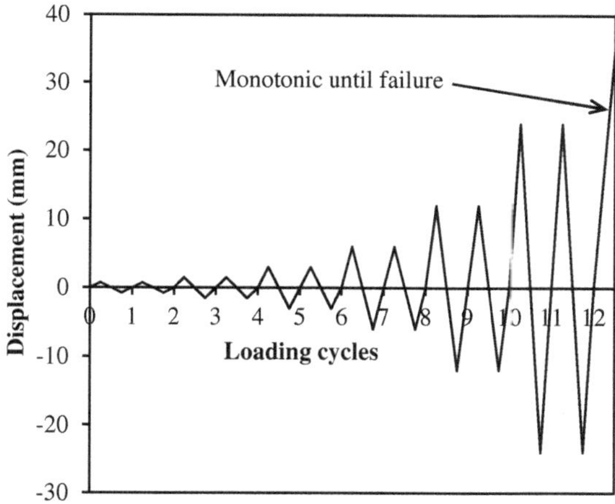

Figure 7. Displacement controlled loading history.

Beam displacement was measured by installing 2 transducers on both sides at a distance of 600 mm from the column face, as shown in Figure 6. In the joint area, two π gages were mounted to measure the width of the crack that occurs. Furthermore, strain gages were installed at the longitudinal reinforcement as well as on the stirrups of the beam and column to measure the strains. Moreover, the applied load was measured with a load cell. All information acquired during loading was recorded with a data logger and entered into a computer. In addition, the crack pattern during the loading process was also observed and drawn until the failure of the specimen.

3. Results and Discussion

3.1. Crack Propagation and Failure Mode

Figure 8 clearly shows an image of the cracks that occurs at the BCJ1 specimen when the displacement was relatively 3 mm, 6 mm, 12 mm, and 24 mm. The crack did not occur until the displacement was approximately 1.5 mm. The first crack occurred at the beam-column joint corner in the longitudinal direction of the column towards the center of the beam in the direction of the push load. Furthermore, when the displacement was relatively 2.3 mm, the length of the first crack was increased to 300 mm. Another crack appeared at the same corner, however it was directed towards the center of the beam-column joint (inclined crack). When the displacement was approximately 3 mm, the crack that appeared first has turned towards the center of the column with a crack length of 100 mm and the inclined crack has reached the center of the beam-column joint as shown in Figure 8a. Figure 8b illustrates that when the displacement has reached 6 mm, 3 more cracks appeared, and with one starting at the corner of the beam-column joint in the longitudinal direction of the beam, it reached the other side of the column. Furthermore, another one started from the opposite side of the column in a slightly inclined direction, and the final crack appeared in the beam on the opposite side with the load applied and propagated to its center. Meanwhile, the first and second appeared cracks only increased in width. Consequently, when the displacement is relatively 12 mm, the second crack, which was an inclined crack, was propagated to the other side of the column until relatively 150 mm in front of the beam, as shown in Figure 8c. This inclined crack, as well as the horizontal one at the column face, increased in width with increasing in displacement. In addition, when the displacement was approximately 24 mm, another inclined crack appeared at another beam-column joint corner and propagated to the reverse side of the column until it was relatively 150 mm measured from the face of the beam. This crack formed an X shape together with the previously occurred inclined crack as shown in Figure 8d. Furthermore, lack of stirrups in the joint region led to a rapid increase in the width of the X crack and horizontal cracks that first appeared, thereby causing a decrease in load when the displacement reached 25.27 mm and structural failure in the subsequent cycle at 32.58 mm. The failure mode of the specimen BCJ1 is shown in Figure 9. This is a typical shear failure in the beam-column joints.

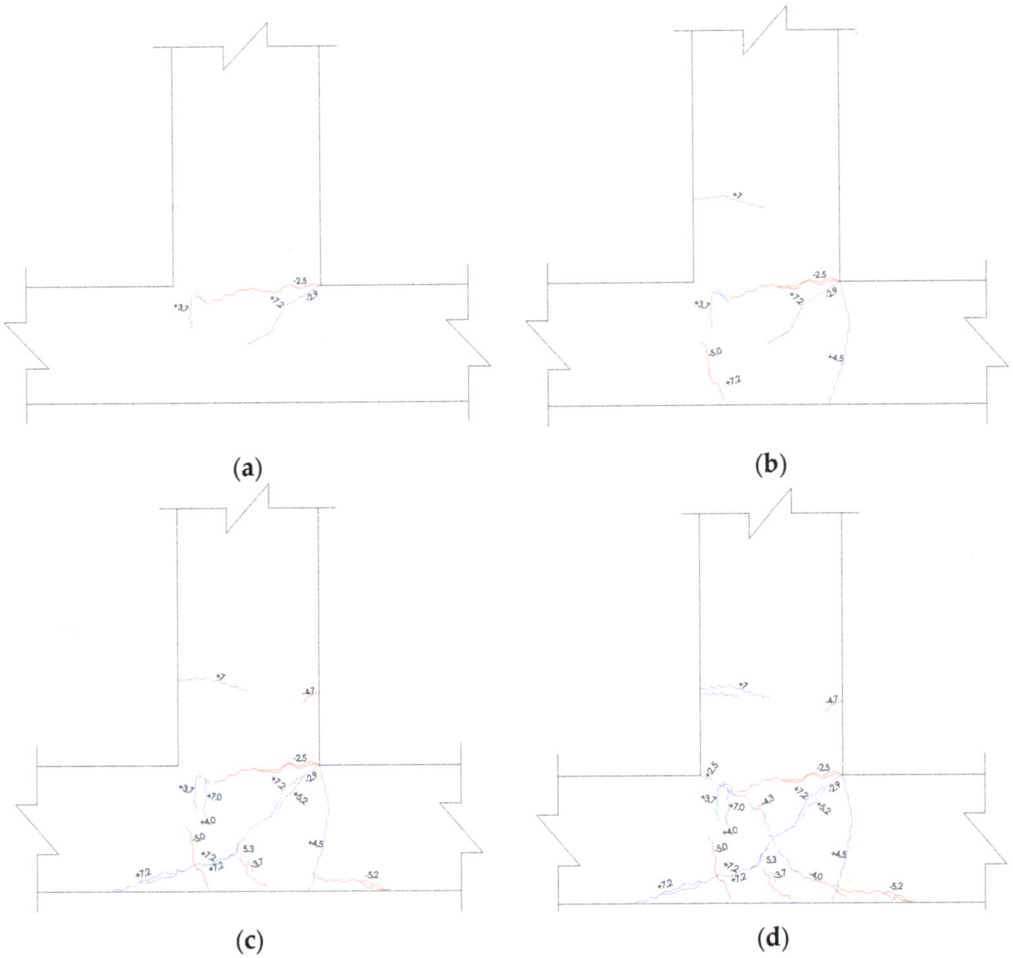

(a)

(b)

(c)

(d)

Figure 8. Crack propagation of specimen BCJ1 at displacement of: (**a**) 3 mm; (**b**) 6 mm; (**c**) 12 mm; and (**d**) 24 mm.

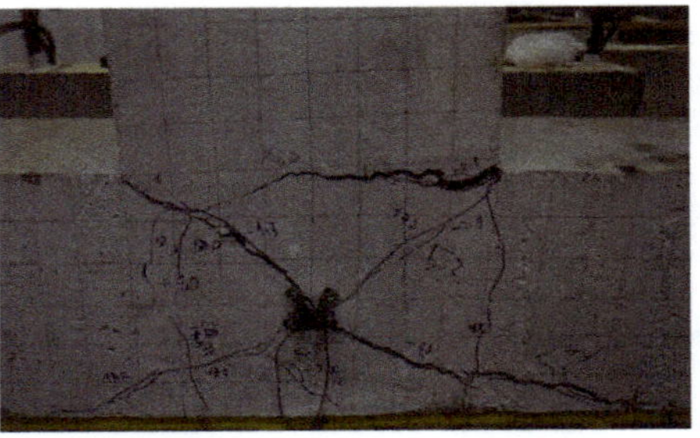

Figure 9. The failure mode of specimen BCJ1.

Crack propagation of specimen BCJ2 at a displacement of 3 mm, 6 mm, 12 mm, and 24 mm is shown in Figure 10. A crack occurred when a displacement exceeded 1.5 mm at the beam-column joint corner on the opposite side of the applied loading side. The crack propagated parallel to the longitudinal direction of the column. However, when the displacement was approximately 3 mm, the crack length was relatively 190 mm, as shown in Figure 10a. Another crack with a length of 30 mm simultaneously appeared on the side of the beam opposite the load. The application of a compressive load in the same cycle led to the formation of new cracks at an opposite corner which was also horizontal towards the previous one, thereby causing the two cracks to meet. At the time of applying the tensile load for the next cycle, a new crack with a length of 220 mm occurred, starting from the first one, which was located 50 mm from the face of the beam towards column's longitudinal axis. Meanwhile, when the compressive load was applied in the same cycle, a similar incident where another crack occurred in the longitudinal axis direction of the beam leading to the inner joint, which started at the previous one with a length of 190 mm, as shown in Figure 10b. Besides, when the displacement was greater than 6 mm, inclined cracks started to occur at the joint. Furthermore, there were also cracks on the other side of the column and jointed with previous cracks in the column axis at the joint region when the displacement is relatively 12 mm, as shown in Figure 10c. In addition, when the crack in the beam propagated, its length was relatively 180 mm apart from the appearance of the other two cracks with similar lengths. Subsequently, when the displacement was approximately 24 mm, the cracks at the joint formed the X shape, as shown in Figure 10d. However, those X shape cracks did not increase in width when the displacement was greater than 24 mm, due to the stirrups provided in the joint region. This was different from specimen BCJ1, where the crack got wider and caused structural failure. In the BCJ2 case, the structure was still capable of deforming up to 52.39 mm and had only failed recently. Structural failure was also not caused by X-shaped cracks that occurred in the joints. However, it was caused by the widening of the cracks at the beam-column joint corner. The failure mode of specimen BCJ2 is shown in Figure 11.

Crack propagation of the specimen BCJ3 at displacement of relatively 3 mm, 6 mm, 12 mm, and 24 mm are shown in Figure 12. The first crack started immediately after the displacement exceeded 1.5 mm. Cracks also started from the beam-column joint corner on the loading side. In contrast to the BCJ1 and BCJ2 specimens, which the first crack was initially directed horizontally, in the BCJ3 specimen, the first crack had an inclined shape directed towards the center of the beam-column joint. Changes in the load direction led to the formation of a second crack on the other beam-column joint corner, which also had an inclined shape towards the beam-column joint center. In the next cycle a displacement of 3 mm was reached, and the 2 inclined cracks met at the middle of the column, in a vertical direction, away from the center of the beam-column joints. There was a cracked branch at a depth of approximately 1/3 of the column height in a horizontal direction towards its center as shown in Figure 12a. This crack failed to propagate and increase its width until the displacement was relatively 6 mm due to the presence of wire mesh. This was installed as a strengthening to prevent the propagation and widening of the crack as shown in Figure 12b. However, when the displacement was approximately 6 mm, another fine crack with a length of 5 mm appeared on the side of the column. As the displacement increased, the newly emerged inclined crack propagated towards the center of the beam-column joints. The initial crack also propagated towards the other side of the column. Furthermore, 2 other inclined cracks were formed, starting with the horizontal one that appeared first. Additionally, vertical and horizontal cracks appeared on the side of the column and beam respectively. There were also 2 vertical cracks in the beam ferrocement section. The shape of the crack when the displacement was relatively 12 mm is shown in Figure 12c. The crack pattern remained the same as shown in Figure 12d till the displacement was relatively 24 mm. There was an increase in the number of cracks on the beam and an additional vertical one on the side of the column. In the beam-column joint area, there was no increase in the length and width of the cracks. It was due to the presence of wire mesh installed on the beam-column joint as a strengthening way to prevent crack growth. This condition occurred in the next load cycle until the connection between the ferrocement and the old concrete surface was damaged at a displacement of 51.37 mm. This is almost similar to the maximum displacement of the BCJ2 specimen. The failure mode of the beam-column joints strengthened with ferrocement was delamination of the ferrocement from the old concrete as shown in Figure 13. Supposing the bond between the old concrete and ferrocement is made stronger, it is believed that the displacement achieved by this beam-column joint may increase, because at the time of failure the crack width in the joint area was less than 1 mm, thereby improves its deformation capacity and ductility.

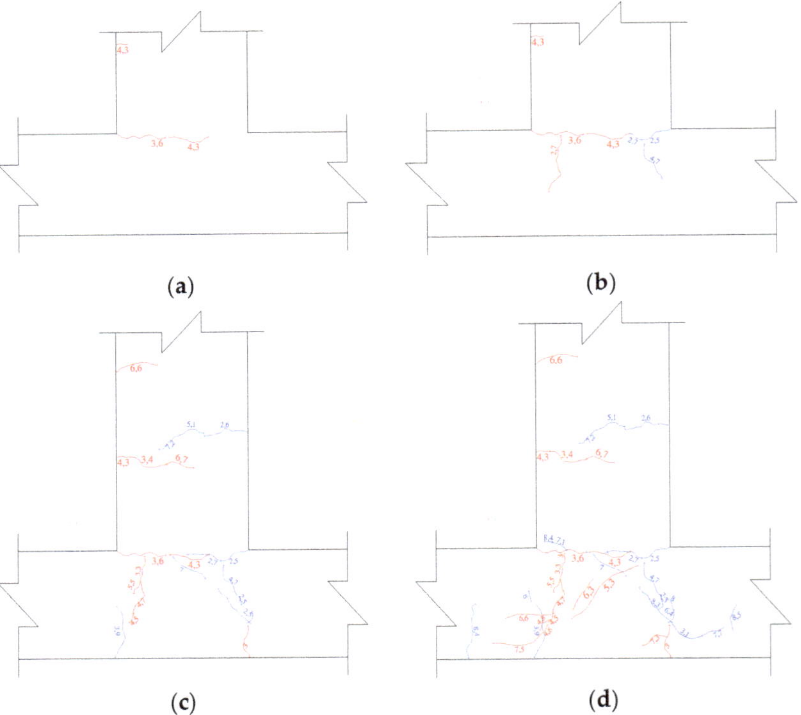

Figure 10. Crack propagation of specimen BCJ2 at displacement of: (**a**) 3 mm; (**b**) 6 mm; (**c**) 12 mm; (**d**) 24 mm.

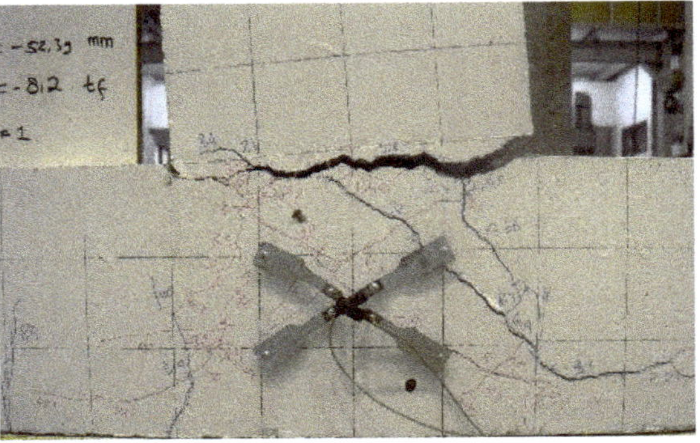

Figure 11. The failure mode of specimen BCJ2.

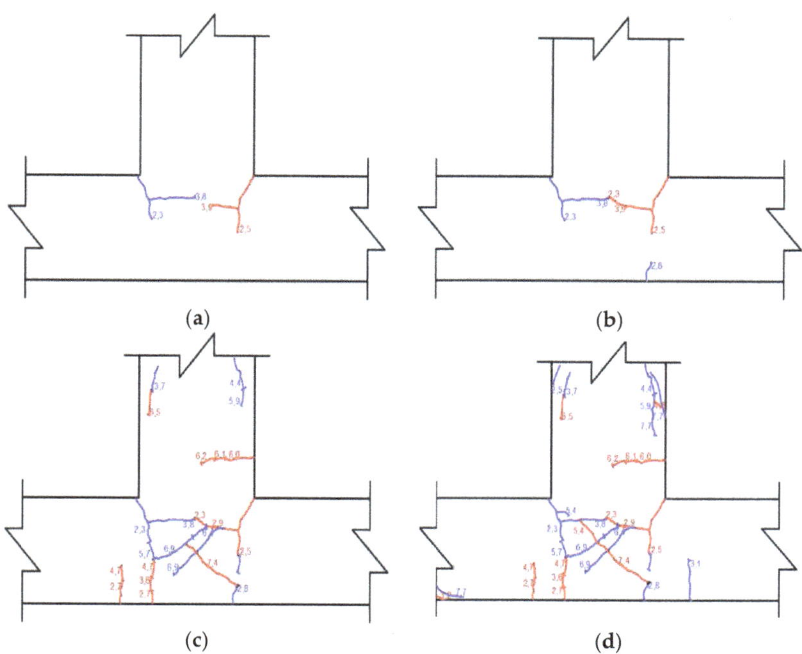

Figure 12. Crack propagation of specimen BCJ3 at displacement of: (**a**) 3 mm; (**b**) 6 mm; (**c**) 12 mm; (**d**) 24 mm.

Figure 13. The failure mode of specimen BCJ3.

3.2. Load and Displacement Relationship

The relationship between the load and displacement measured with transducers mounted on the beam for specimen BCJ1 is shown in Figure 14. The maximum load of this specimen is 73.95 kN at a displacement of 25.74 mm. When displacement was 24 mm, there was an extremely wide crack on the column face with a length has reached along the beam height. This also includes an X-shaped crack in the beam-column joint area, as shown in Figure 8d. Therefore, in the next cycle, the load only reached approximately 51.6 kN, and the specimen failed at the displacement of 32.58 mm. The absence of the anchorage of beam longitudinal reinforcement to column as well as no transverse reinforcement in beam-column joint region of this specimen led to the easy of crack propagation in this specimen, thereby leading to a small deformation capacity and ductility index. The first yield of beam's longitudinal reinforcement of specimen BCJ1 occurred at a displacement of 12.07 mm.

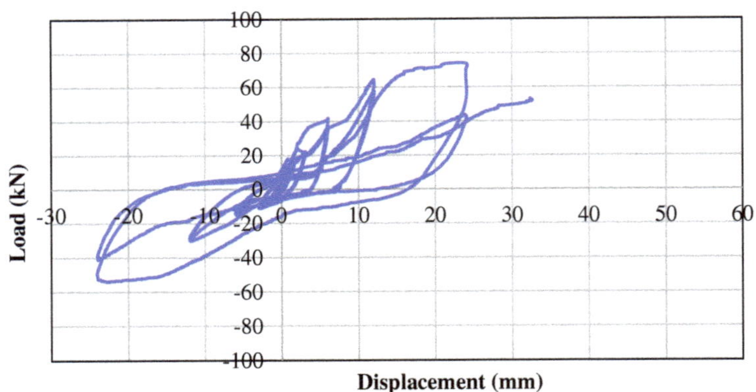

Figure 14. The hysteretic load-displacement curve for specimen BCJ1.

The hysteretic load-deflection curve for specimen BCJ2 is shown in Figure 15. The displacement at the first yield of beam's longitudinal reinforcement of specimen BCJ2 was 13.55 mm, which is similar to that of BCJ1. This is because both specimens have similar cross-sectional size and reinforcement. Due to the fact that specimen BCJ2 has a stirrup in the beam-column joint region, the maximum load is greater than that of the specimen BCJ1, which is 83.48 kN at a displacement of 23.15 mm. It is important to note that the presence of stirrups in the joint region and the anchorage of beam's longitudinal reinforcement to relatively 560 mm into the column, causing delays in the widening and propagation of cracks, therefore the specimen was able to withstand any suddenly load drop as in the case of BCJ1. As a result, this specimen was able to sustain the maximum load in the next cycle. Furthermore, the specimen failure occurred at a displacement of 52.39 mm with a maximum load at failure of 77.89 kN, which was decreased by 6.7%. Therefore, the presence of transverse reinforcement in the beam-column joint region and the anchorage of beam's longitudinal reinforcement in the column cause the deformation capacity to increase.

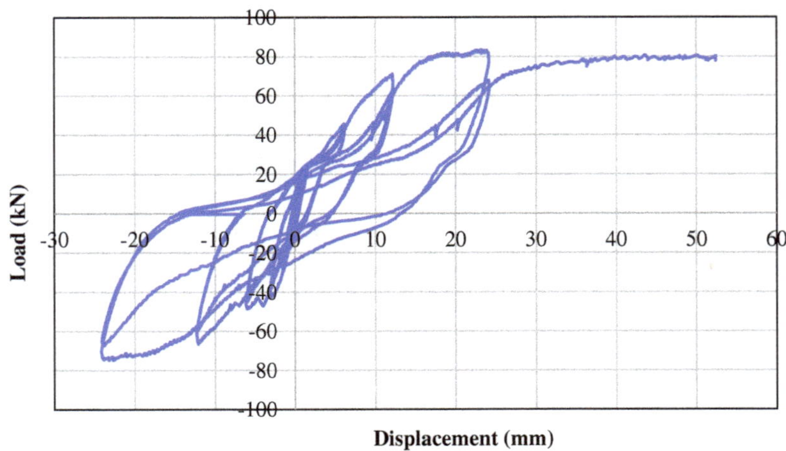

Figure 15. The hysteretic load-displacement curve for specimen BCJ2.

Figure 16 shows the hysteretic load-deflection curve for specimen BCJ3, which is a strengthening of the BCJ1 model using ferrocement. Similar to the cases of specimens BCJ1 and BCJ2, BCJ 3 experienced its first yield of longitudinal reinforcing bars at a deformation of 12.00 mm. Installation of wire mesh in the beam-column joint area had an insignificant effect on the maximum load, although it increased the deformation capacity of specimen BCJ3. The maximum load was similar to BCJ1, namely 75.64 kN at a displacement of 16 mm. In addition, this load remained constant until the displacement was 24 mm. There was no increase or decrease in load after a displacement of 16 mm because the specimen had significant number of cracks without any crack propagation and widening. The propagation and widening of cracks was prevented by wire mesh which was installed as a structural strengthening. As a result of the ferrocement strengthening, in the following cycle, the load did not decrease immediately as was the case of BCJ1, rather BCJ3 was able to undergo deformation even without increasing the load till a failure occurred at displacement of 52.04 mm with a load reduction of only 5%, which was equivalent to 71.86 kN. However, in the opposite direction, the maximum load of BCJ3 was 81.68 kN, which exceeded that of BCJ2 relatively 74.56 kN in the same loading direction. Therefore, the strengthening of beam-column joint designed with non-seismic building code using ferrocement improves the deformation capacity and ductility. The deformation capacity and ductility of strengthened beam-column joint even was higher than those of beam-column joint designed with current code. However, in this case, as previously reported, a failure occurred due to the delamination of the ferrocement from the old reinforced concrete beam-column joint. Furthermore, it is necessary to establish a good bond between the old concrete and the ferrocement, either by increasing the dynabolt anchor numbers or by providing a bonding adhesive between these old concrete and new mortar. Cases with a higher number of anchors and bonding adhesive between the old concrete and new mortar were not reviewed in this study and need to be further investigated.

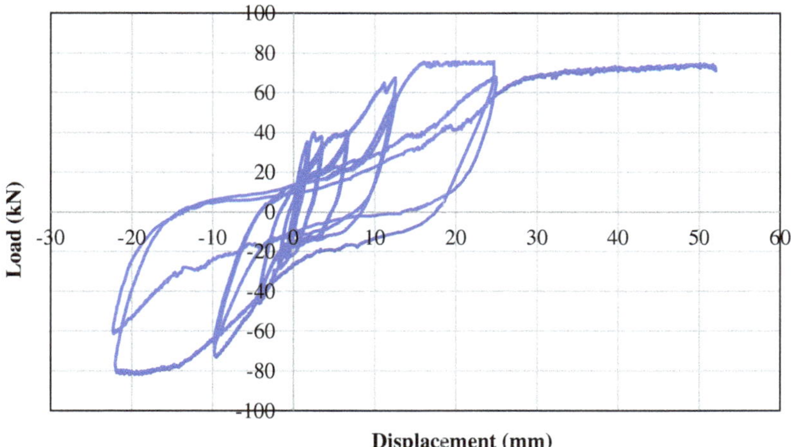

Figure 16. The hysteretic load-displacement curve for specimen BCJ3.

The maximum loads, loads at first crack, loads at first yield of beam's longitudinal reinforcing bars, and loads at failure together with their corresponding displacements of all specimens tested in this study are summarized in Table 4. Table 4 shows that the strengthening of beam-column joint with ferrocement did not improve the load carrying capacity, but enhanced the ability to deform after peak load. Therefore, the deformation capacity of strengthened beam-column joint was increased by 60%, which was almost similar to that was designed with current building code. To improve the load carrying capacity as well as the deformation capacity, it is recommended to use high strength heat-treated steel with fine grains for wire mesh in the future study since such steel has better performance in carrying static and dynamic loads [70,71]. This type of steel may be welded to reinforcing bars using welding heat input [72], so that the delamination of ferrocement that occurred in this study may be prevented.

Table 4. Maximum loads (P_{max}), loads at first crack (P_{cr}), loads at first yield of reinforcing bars (P_y) and loads at failure (P_{fail}) and their corresponding displacements (Δ).

Specimen	Load (kN) and Displacement (mm)								Ratio of Δ_{fail} to $\Delta_{fail,BCJ1}$
	At First Crack		At First Yield		Maximum		At Failure		
	P_{cr}	Δ_{cr}	P_y	Δ_{y1}	P_{max}	Δ_m	P_{fail}	Δ_{fail}	
BCJ1	22.68	1.80	56.85	12.07	73.95	24.01	51.6	32.58	1.00
BCJ2	29.53	1.57	68.38	13.55	83.48	23.15	77.89	52.39	1.61
BCJ3	34.14	1.55	65.14	12.00	75.64	24.00	71.86	52.04	1.60

The enveloped load-displacement curves of the 3 specimens tested in this study were compared as shown in Figure 17. It also shows that strengthening the beam-column joints designed with a non-seismic building code using ferrocement increases the deformation capacity. Therefore it is similar to a beam-column joint designed with a new building code that takes into account the effects of seismic.

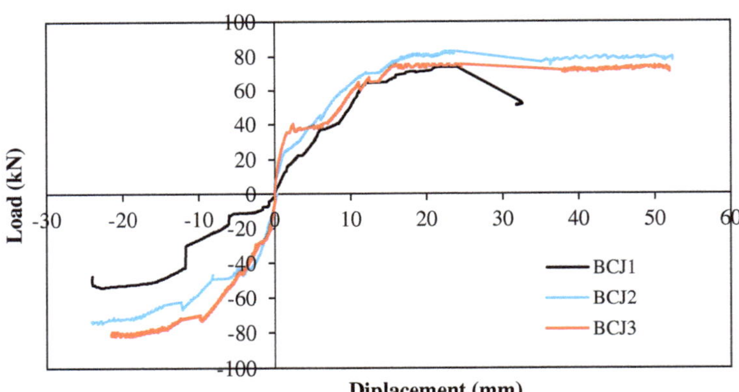

Figure 17. Envelope load-displacement curves.

3.3. Structural Ductility

Structural ductility is defined as the ability of a structure to undergo large inelastic deformation without experiencing significant loss of strength. In this study displacement ductility index was used for assessing the structural ductility. The displacement ductility index (μ) is defined as the ratio of ultimate displacement (Δ_u) to yield displacement (Δ_y) as follows [73]:

$$\mu = \frac{\Delta_u}{\Delta_y} \tag{1}$$

For specimen BCJ1, ultimate displacement is defined as displacement corresponding to 15% drop of maximum load [37,74–77]. Since the maximum load for specimens BCJ2 and BCJ3 at failure dropped only by 6.75 and 5%, respectively, then the ultimate displacement for those specimens is given as failure displacement.

The yield displacement was assessed by three different methods. For the first method, the yield displacement is defined as the displacement at the first yield of reinforcing bars as presented in Table 4. The second method is based on balance of energy. The detail description on how to calculate the yield displacement based on the balance of energy can be found in the references [68,74,75]. The third method is based on the general yielding.

The detail description on how to calculate the yield displacement based on the general yielding can be found in the references [68,74]. Since the yield displacement obtained by those three different methods was almost similar, then the average value was used in calculating the displacement ductility index.

Table 5 presents the yield displacement, ultimate displacement, and displacement ductility index of all specimens tested in this study. The table shows that the structural ductility of the ferrocement-strengthened beam-column joint was improved by 91% which was higher than the ductility of beam-column joint designed with the new seismic building code. This significantly ductility improvement is due to the inhibition of the crack propagation by the resistance of the installed wire mesh. As a result, the failure of the beam-column joint, which was originally brittle, becomes more ductile. These results indicate that the structures constructed before the implementation of the seismic building code may be strengthened by using ferrocement. Therefore it is presumed to have a similar deformation capacity and ductility as building structures designed with the new seismic building code. This aids in preventing sudden collapse due to failure of the beam-column joint during an earthquake.

Table 5. Displacement ductility index.

Specimen	Δ_u (mm)	First Yield, Δ_{y1} (mm)	Balance of Energy, Δ_{y2} (mm)	General Yielding, Δ_{y3} (mm)	$\Delta_{y,avg}$ (mm)	μ	Ratio of μ to μ_{BCJ1}
BCJ1	27.60	12.07	12.42	12.65	12.38	2.23	1.00
BCJ2	52.39	13.55	13.65	13.78	13.66	3.84	1.72
BCJ3	52.04	12.00	12.26	12.43	12.23	4.26	1.91

3.4. Stiffness Degradation

Stiffness is one of the parameter that can show the seismic performance of reinforced concrete members which is the ability to resist deformation [75,76]. In this study, the stiffness was calculated as secant modulus of envelope load-displacement curves in positive direction shown in Figure 17. Figure 18 shows the comparison of the stiffness of all beam-column joint specimens tested in this study. It further shows that the presence of wire-mesh in the beam-column joint strengthened with ferrocement tends to increase its initial stiffness due to the higher elastic modulus of wire-mesh compared to the elastic modulus of concrete. Since there was no crack and the linear load-displacement relationship at the initial stage of loading, the stiffness was constant until the displacement of around 1.5 mm. The presence of cracks that occurred due to loading leads to stiffness degradation along with an increase in beam displacement, as shown in the figure The stiffness degradation of all tested specimens is similar and closely related to the existent crack propagation.

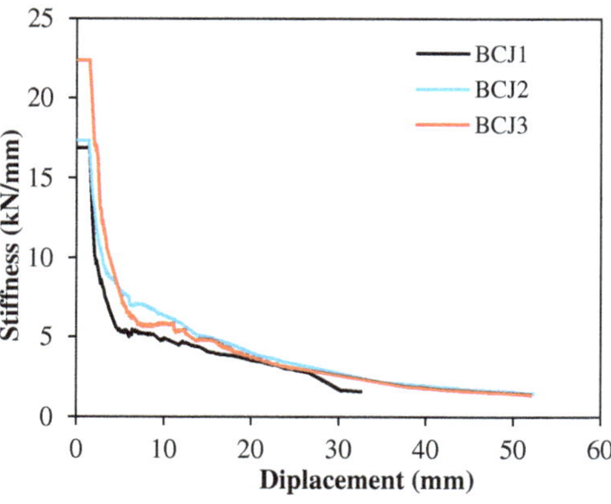

Figure 18. Degradation of stiffness.

3.5. Energy Dissipation

The total energy given to a structure under loading is called the input energy. Some of the input energy given to a structure is absorbed (dissipated) by the structure. Energy dissipation is described as the ability of a structure to absorb energy through the yielding process in the plastic hinge region. It is used in designing an earthquake-resistant building structure that is ductile in the plastic hinge area. This, therefore, leads to plastic deformation that occurs before failure. The greater the energy dissipation of a structure, the greater it is able to withstand earthquake loads.

Energy dissipation at each cycle can be calculated from the enclosed area within load-displacement loop at this cycle. The cumulative energy dissipation is calculated by summating energy dissipated in previous cycles [68]. Figure 19 shows the cumulative energy dissipation of all beam-column joint specimens tested in this study as a function of displacement. Furthermore, when there was slight deformation, the energy dissipation of all the specimens was similar. As the displacement increased, the beam-column joint specimen strengthened with ferrocement provided greater cumulative energy dissipation than the unreinforced. The energy dissipation of ferrocement-reinforced specimens was almost similar to the specimens designed with the new seismic building code. Even at the displacement greater than 20 mm, the strengthened beam-column joint had a greater cumulative energy dissipation, which shows the effectiveness of beam-column joint strengthening using ferrocement in resisting earthquake loads.

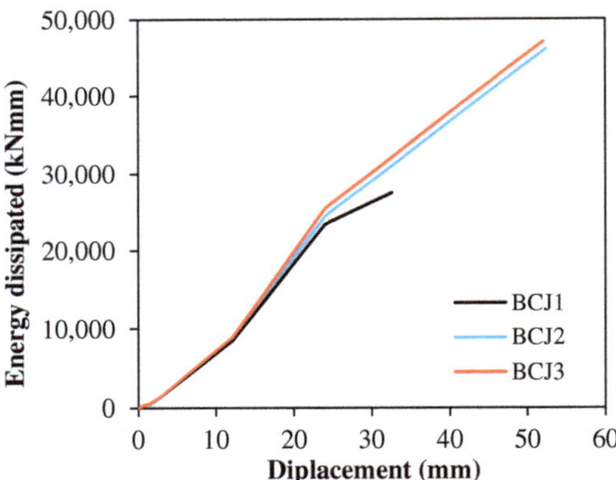

Figure 19. Cumulative energy dissipation.

3.6. Comparison with Previous Studies

Table 6 presents the comparison of improvement of maximum displacement, ductility, initial stiffness, and energy dissipation obtained from this study and the previous studies [66,68] with the difference in strengthening scheme. This table shows that the different strengthening scheme affects the improvement of deformation capacity. The number layer of wire mesh and its orientation angle affect the improvement of maximum displacement and ductility significantly. Meanwhile, the addition of diagonal reinforcement improves energy dissipation significantly.

Table 6. Comparison of improvement of maximum displacement, ductility, initial stiffness, and energy dissipation obtained from this study and the previous studies [66,68].

Strengthening Scheme		Improvement in (%)			
Orientation Angle	Number Layer of Wire Mesh	Maximum Displacement	Ductility	Initial Stiffness	Energy Dissipation
0° (Present study)	4	60	91	29	71
45° [66]	3	18	NA *	12	16
60° [66]	3	20	NA *	68	21
0° with addition of diagonal reinforcement [68]	2	22	17	19	154

NA * = not available.

4. Conclusions

In this research, the deformation capacity of the reinforced concrete beam-column joint designed with a non-seismic building code was investigated. In addition, structural strengthening was analyzed and compared with the deformation capacity of a similar structure designed with the new seismic building code. Strengthening was carried out using ferrocement provided on both sides of the beam-column joint using 4 layers of wire mesh with a diameter of 1 mm and mesh of 25.4 mm. The reversed cyclic loading test results show that the beam-column joint strengthened with ferrocement improved the deformation capacity and ductility. The beam-column joint, which initially experienced brittle shear failure after being strengthened, increased its ductility index from 2.23 to 4.26. This was greater than the ductility index of the beam-column joint designed with the new seismic building code, which was 3.84. The beam-column joint, which initially failed at a displacement of 32.58 mm after being strengthened with ferrocement, failed at a displacement of 52.04 mm and was almost the same as the displacement, which was designed with the new seismic building code. Moreover, the strengthening also significantly improved its stiffness and energy dissipation. Meanwhile the load carrying capacity of the strengthened beam-column joint was 75.64 kN, a slight higher than that of non-strengthened one which was 73.95 kN, but still lower than that was designed with new seismic building code which was 83.48 kN. The failure of the beam-column joint strengthened with ferrocement occurred due to the delamination of ferrocement from the old reinforced concrete beam-column joint.

Author Contributions: Conceptualization, M.Z.A. and A.; methodology, M.Z.A., A., M.A. and M.H.; investigation, M.Z.A.; validation, A., M.A. and M.H.; resources, M.Z.A.; writing—original draft preparation, M.H.; writing—review and editing, M.H.; supervision, S.R. All authors have read and agreed to the published version of the manuscript.

Funding: This research received no external funding.

Institutional Review Board Statement: Not applicable.

Informed Consent Statement: Not applicable.

Data Availability Statement: The data of this research is available upon request to the first author.

Acknowledgments: The authors are grateful to Delfian Masrura, Maulana Lirad, Muhammad Farizki, Mahlil, and M. Nasir for the assistance provided in obtaining the experimental activities needed for data collection.

Conflicts of Interest: The authors declare no conflict of interest.

References

1. *Peraturan Beton Bertulang Indonesia 1971 NI-2*; Departemen Pekerjaan Umum dan Tenaga Listrik: Bandung, Indonesia, 1979.
2. Celep, Z.; Erken, A.; Taskin, B.; Ilki, A. Failure of masonry and concrete buildings during the March 8, 2010 Kovancilar and Palu (Elazig) earthquakes in Turkey. *Eng. Fail. Anal.* **2011**, *18*, 868–889. [CrossRef]
3. Coşgun, C.; Dindar, A.A.; Seçkin, E.; Önen, Y.H. Analysis of building damage caused by earthquakes in Western Turkey. *Gradevinar* **2013**, *65*, 743–752.
4. Damcı, E.; Temur, R.; Bekdaş, G.; Sayin, B. Damages and causes on the structures during the October 23, 2011 Van earthquake in Turkey. *Case Stud. Constr. Mater.* **2015**, *3*, 112–131. [CrossRef]

5. Fikri, R.; Dizhur, D.; Walsh, K.; Ingham, J. Seismic performance of reinforced concrete frame with masonry infill buildings in the 2010/2011 Canterbury, New Zealand earthquakes. *Bull. Earthq. Eng.* **2019**, *17*, 737–757. [CrossRef]

6. Fikri, R.; Dizhur, D.; Ingham, J. Typological study and statistical assessment of parameters influencing earthquake vulnerability of commercial RCFMI buildings in New Zealand. *Bull. Earthq. Eng.* **2019**, *17*, 2011–2036. [CrossRef]

7. Fikri, R.; Derakhshan, H.; Ingham, J. Numerical evaluation of a non-ductile RCFMI building subjected to the Canterbury, New Zealand earthquakes: A case study of the St Elmo Courts building. *Structures* **2020**, *28*, 991–1008. [CrossRef]

8. Sezen, H.; Whittaker, A.S.; Elwood, K.J.; Mosalam, K.M. Performance of reinforced concrete buildings during the August 17, 1999 Kocaeli, Turkey earthquake, and seismic design and construction practise in Turkey. *Eng. Struct.* **2003**, *25*, 103–114. [CrossRef]

9. Hasan, M.; Saidi, T.; Afifuddin, M.; Setiawan, B. The assessment and strengthening proposal of building structure after the Pidie Jaya earthquake in December 2016. *J. King Saud Univ. Eng. Sci.* **2021**; *in press*. [CrossRef]

10. Chang, H.Y.; Lin, K.C. Reconnaissance observations on the buildings damaged by the 2010 Taiwan Kaohsiung earthquake. *Nat. Hazard* **2013**, *69*, 237–249. [CrossRef]

11. Kabeyasawa, T. Damages to RC school buildings and lesson froms from the 2011 East Japan earthquake. *Bull. Earthq. Eng.* **2017**, *15*, 535–553. [CrossRef]

12. Muguruma, H.; Nishiyama, M.; Watanabe, F. Lessons learned from the Kobe earthquake—A Japanese perspective. *PCI J.* **1995**, *40*, 28–42. [CrossRef]

13. Mitchell, D.; DeVall, R.H.; Kobayashi, K.; Tinawi, R.; Tso, W.K. Damage to concrete structures due to the January 17, 1995, Hyogo-ken Nanbu (Kobe) earthquake. *Can. J. Civ. Eng.* **1996**, *23*, 757–770. [CrossRef]

14. Su, N. Structural evaluations of reinforced concrete buildings damaged by Chi-Chi earthquake in Taiwan. *Prac. Period. Struct. Des. Constr.* **2001**, *6*, 119–128. [CrossRef]

15. Zhang, C.; Tao, M.X. Strong-column–weak-beam criterion for reinforced concrete frames subjected to biaxial seismic excitation. *Eng. Struct.* **2021**, *241*, 112481. [CrossRef]

16. De Angelis, A.; Tariello, F.; De Masi, R.F.; Pecce, M.R. Comparison of different solutions for a seismic and energy retrofit of an auditorium. *Sustainability* **2021**, *13*, 8761. [CrossRef]

17. Facconi, L.; Lucchini, S.S.; Minelli, F.; Grassi, B.; Pilotelli, M.; Plizzari, G.A. Innovative method for seismic and energy retrofitting of masonry buildings. *Sustainability* **2021**, *13*, 6350. [CrossRef]

18. Parrella, V.F.; Molari, L. Building retrofitting system based on bamboo-steel hybrid exoskeleton structures: A case study. *Sustainability* **2021**, *13*, 5984. [CrossRef]

19. Cardani, G.; Massetti, G.E.; Riva, D. Multipurpose retrofitting of a tower building in Brescia. *Sustainability* **2021**, *13*, 8877. [CrossRef]

20. Gkournelos, P.D.; Triantafillou, T.C.; Bournas, D.A. Seismic upgrading of existing reinforced concrete buildings: A state-of-the-art review. *Eng. Struct.* **2021**, *240*, 112273. [CrossRef]

21. Attari, N.; Amziane, S.; Chemrouk, M. Efficiency of beam-column joint strengthened by FRP laminates. *Adv. Compos. Mater.* **2010**, *19*, 171–183. [CrossRef]

22. Ghobarah, A.; Said, A. Shear strengthening of beam-column joints. *Eng. Struct.* **2002**, *24*, 881–888. [CrossRef]

23. Beydokhty, E.Z.; Shariatmadar, H. Behavior of damaged exterior RC beam-column joints strengthened by CFRP composites. *Lat. Am. J. Solid Struct.* **2016**, *13*, 880–896. [CrossRef]

24. De Vita, A.; Napoli, A.; Realfonzo, R. Full sace reinforced concrete beam-column joints strengthened will steel reinforced polymer systems. *Front. Mater.* **2017**, *4*, 18. [CrossRef]

25. Ghobarah, A.; Said, A. Seismic rehabilitation of beam-column joints using FRP laminates. *J. Earthq. Eng.* **2001**, *5*, 113–129. [CrossRef]

26. Pohoryles, D.A.; Melo, J.; Rossetto, T.; Varum, H. Seismic retrofit schemes with FRP for deficient RC beam-column joints: State-of-the-art review. *J. Compos. Constr.* **2019**, *23*, 03119001. [CrossRef]

27. Li, B.; Chua, H.Y.G. Seismic performance of strengthened reinforced concrete beam-column joints using FRP composites. *J. Struct. Eng.* **2009**, *135*, 1177–1190. [CrossRef]

28. Mahmoud, M.H.; Afefy, H.M.; Kassem, N.M.; Fawzy, T.M. Strengthening of defected beam-column joints using CFRP. *J. Adv. Res.* **2014**, *5*, 66–77. [CrossRef]

29. Zhao, Y.; Liu, Y.; Li, J. Study of seismic behavior of RC beam-column joints strengthened by sprayed FRP. *Adv. Mater. Sci. Eng.* **2018**, *2018*, 3581458. [CrossRef]

30. Santarsiero, G. FE modelling of the seismic behavior of wide beam-column joints strengthened with CFRP systems. *Buildings* **2018**, *8*, 31. [CrossRef]

31. Katakalos, K.; Manos, G.; Papakonstantinou, C. Seismic retrofit of R/C T-beams with steel fiber polymers under cyclic loading conditions. *Buildings* **2019**, *9*, 101. [CrossRef]

32. Baji, H.; Eslami, A.; Ronagh, A.R. Development of a nonlinear FE modelling approach for FRP-strengthened RC beam-column connections. *Structures* **2015**, *3*, 272–281. [CrossRef]

33. Roy, B.; Laskar, A.I. Cyclic performance of beam-column subassemblies with construction joint in column retrofitted with GFRP. *Structures* **2018**, *14*, 290–300. [CrossRef]

34. Attari, N.; Youcef, Y.S.; Amziane, S. Seismic performance of reinforced concrete beam–column joint strengthening by frp sheets. *Structures* **2019**, *20*, 353–364. [CrossRef]

35. Maheri, M.R.; Torabi, A. Retrofitting external RC beam-column joints of an ordinary MRF through plastic hinge relocation using FRP laminates. *Structures* **2019**, *22*, 65–75. [CrossRef]

36. Tahnat, Y.B.A.; Samaaneh, M.A.; Dwaikat, M.M.S.; Halahla, A.M. Simple equations for predicting the rotational ductility of fiber-reinforcedpolymer strengthened reinforced concrete joints. *Structures* **2020**, *24*, 73–86. [CrossRef]

37. Akhlaghi, A.; Mostofinejad, D. Experimental and analytical assessment of different anchorage systems used for CFRP flexurally retrofitted exterior RC beam-column connections. *Structures* **2020**, *28*, 881–893. [CrossRef]

38. Tafsirojjaman, T.; Fawzia, S.; Thambiratnam, D.P. Structural behaviour of CFRP strengthened beam-column connections under monotonic and cyclic loading. *Structures* **2021**, *33*, 2689–2699. [CrossRef]

39. Davodikia, B.; Saghafi, M.H.; Golafshar, A. Experimental investigation of grooving method in seismic retrofit of beam-column external joints without seismic details using CFRP sheets. *Structures* **2021**, *34*, 4423–4434. [CrossRef]

40. Murad, Y.Z.; Alseid, B.H. Retrofitting interior RC beam-to-column joints subjected to quasi-static loading using NSM CFRP ropes. *Structures* **2021**, *34*, 4158–4168. [CrossRef]

41. Alcocer, S.M.; Jirsa, J.O. Strength of reinforced concrete frame connections rehabilitated by jacketing. *ACI Struct. J.* **1993**, *90*, 249–261.

42. Ghobarah, A.; Aziz, T.S.; Biddah, A. Rehabilitation of reinforced concrete frame connections using corrugated steel jacketing. *ACI Struct. J.* **1997**, *94*, 283–294.

43. Ghobarah, A.; Biddah, A.; Mahgoub, M. Seismic retrofit of reinforced concrete columns using steel jackets. *Eur. Earthq. Eng.* **1997**, *11*, 21–31.

44. Truong, G.T.; Dinh, N.H.; Kim, J.C.; Choi, K.K. Seismic performance of exterior RC beam–column joints retrofitted using various retrofit solutions. *Int. J. Concr. Struct. Mater.* **2017**, *11*, 415–433. [CrossRef]

45. Misir, I.S.; Kahraman, S. Strengthening of non-seismically detailed reinforced concrete beam column joints using SIFCON blocks. *Shadana* **2013**, *38*, 69–88. [CrossRef]

46. Gokdemir, H.; Tankut, T. Seismic strengthening of beam-column joints using diagonal steel bars. *Teknik Dergi* **2017**, *28*, 7977–7992.

47. Adil, D.M.; Subramanian, N.; Sumeet, P.; Dar, A.R.; Raju, J. Performance evaluation of different strengthening masures for exterior beam-column joints under opening moments. *Struct. Eng. Mech.* **2020**, *74*, 243–254. [CrossRef]

48. Yurdakul, Ö.; Avşar, Ö.; Kilinçl, K. Application of different rehabilitation and strengthening methods for insufficient RC beam-column joints. *Adv. Mater. Res.* **2013**, *688*, 222–229. [CrossRef]

49. Engindeniz, M.; Kahn, L.F.; Zureick, A.H. Repair and Strengthening of Reinforced Concrete Beam-Column Joints: State of the Art. *ACI Struct. J.* **2005**, *102*, 187–197.

50. Yan, Y.; Liang, H.; Lu, Y.; Zhao, X. Slender CFST columns strengthened with textile-reinforced engineered cementitious composites under axial compression. *Eng. Struct.* **2021**, *241*, 112483. [CrossRef]

51. Qian, H.; Guo, J.; Yang, X.; Lin, F. Seismic rehabilitation of gravity load-designed interior RC beam-column joints using ECC-infilled steel cylinder shell. *Structures* **2021**, *34*, 1212–1228. [CrossRef]

52. Li, Y.; Sanada, Y. Seismic strengthening of existing RC beam-column joints by wing walls. *Earthq. Eng. Struct. Dyn.* **2017**, *46*, 1987–2008. [CrossRef]

53. Tiwary, A.K.; Singh, S.; Chohan, J.S.; Kumar, R.; Sharma, S.; Chattopadhyaya, S.; Abed, F.; Stepinac, M. Behavior of RC Beam–Column Joints Strengthened with Modified Reinforcement Techniques. *Sustainability* **2022**, *14*, 1918. [CrossRef]

54. *ACI 549.1R-18: Design Guide for Ferrocement*; American Concrete Institute: Farmington Hills, MI, USA, 2018.

55. Paramasivam, P.; Lim, C.T.E.; Ong, K.C.G. Strengthening of RC beams with ferrocement laminates. *Cem. Concr. Compos.* **1998**, *20*, 53–65. [CrossRef]

56. Raghunathapandian, P.; Palani, B.; Elango, D. Flexural strengthening of RC tee beams using ferrocement. *Int. J. Recent Technol. Eng.* **2020**, *8*, 2277–3878.

57. Shakthivel, V.; Pradeep Kumar, C. Experimental investigation of strengthening of RC beam using ferro cement laminates. *Int. J. Eng. Tech.* **2019**, *5*, 186–195.

58. Khan, S.U.; Rafeeqi, S.F.A.; Ayub, T. Strengthening of RC beams in flexural using ferrocement. *IJST Transac. Civ. Eng.* **2013**, *37*, 353–365.

59. Khan, S.U.; Nuruddin, M.F.; Ayub, T.; Shafiq, N. Effects of ferrocement in strengthening the serviceability properties of reinforced concrete structures. *Adv. Mater. Res.* **2013**, *690–693*, 686–690. [CrossRef]

60. Makki, R.F. Response of reinforced concrete beams retrofitted by ferrocement. *Int. J. Sci. Technol. Res.* **2014**, *3*, 27–34.

61. Miah, M.J.; Miah, M.S.; Alam, W.B.; Lo Monte, F.; Li, Y. Strengthening of RC beams by ferrocement made with unconventional concrete. *Mag. Civ. Eng.* **2019**, *89*, 94–105. [CrossRef]

62. Abdullah; Takiguchi, K. An investigation into the behavior and strength of reinforced concrete columns strengthened with ferrocement jackets. *Cem. Concr. Compos.* **2003**, *25*, 233–242. [CrossRef]

63. Mourad, S.M.; Shannag, M.J. Repair and strengthening of reinforced concrete square column using ferrocement jackets. *Cem. Concr. Compos.* **2012**, *34*, 288–294. [CrossRef]

64. Jiang, L.M.; Tang, F.H.; Ou, M.L. Experimental research on the strengthening of RC columns by high performance ferrocement laminates. *Adv. Mater. Res.* **2011**, *243–249*, 1409–1415. [CrossRef]

65. Kaish, A.B.M.A.; Jamil, M.; Raman, S.N.; Zain, M.F.M.; Nahar, L. Ferrocement composites for strengthening of concrete columns: A review. *Constr. Build. Mater.* **2018**, *160*, 326–340. [CrossRef]

66. Shaaban, I.G.; Seoud, O.A. Experimental behavior of full-scale exterior beam-column space joints retrofitted by ferrocement layers under cyclic loading. *Case Stud. Constr. Mater.* **2018**, *8*, 61–78. [CrossRef]
67. Shaaban, I.G.; Said, M. Finite element modeling of exterior beam-column joints strengthened by ferrocement under cyclic loading. *Case Stud. Constr. Mater.* **2018**, *8*, 333–346. [CrossRef]
68. Li, B.; Lam, E.S.S.; Wu, B.; Wang, Y.Y. Experimental investigation on reinforced concrete interior beam–column joints rehabilitated by ferrocement jackets. *Eng. Struct.* **2013**, *56*, 897–909. [CrossRef]
69. *SNI 2847:2019 Persyaratan Beton Struktural Untuk Bangunan Gedung Dan Penjelasan*; Badan Standardisasi Nasional: Jakarta, Indonesia, 2019.
70. Tuz, L. Determination of the causes of low service life of the air fan impeller made of high-strength steel. *Eng. Fail. Anal.* **2021**, *127*, 105502. [CrossRef]
71. Tuz, L. Evaluation of microstructure and selected mechanical properties of laser beam welded S690QL high-strength steel. *Adv. Mater. Sci.* **2018**, *18*, 34–42. [CrossRef]
72. Nowacki, J.; Sajek, A.; Matkowski, P. The influence of welding heat input on the microstructure of joints of S1100QL steel in one-pass welding. *Arch. Civ. Mech. Eng.* **2016**, *16*, 777–783. [CrossRef]
73. Park, R.; Paulay, T. *Reinforced Concrete Structures*; Wiley: Hoboken, NJ, USA, 1975.
74. Dabiri, H.; Kheyroddin, A.; Kaviani, A. A numerical study on the seismic response of RC wide column–beam joints. *Int. J. Civ. Eng.* **2019**, *17*, 377–395. [CrossRef]
75. Kheyroddin, A.; Mohammadkhah, A.; Dabiri, H.; Kaviani, A. Experimental investigation of using mechanical splices on the cyclic performance of RC columns. *Structures* **2020**, *24*, 717–727. [CrossRef]
76. Kheyroddin, A.; Dabiri, H. Cyclic performance of RC beam-column joints with mechanical or forging (GPW) splices; an experimental study. *Structures* **2020**, *28*, 2562–2571. [CrossRef]
77. Dabiri, H.; Kheyroddin, A. An experimental comparison of RC beam-column joints incorporating different splice methods in the beam. *Structures* **2021**, *34*, 1603–1613. [CrossRef]

sustainability

MDPI

Article

Behavior of RC Beam–Column Joints Strengthened with Modified Reinforcement Techniques

Aditya Kumar Tiwary [1], Sandeep Singh [1], Jasgurpreet Singh Chohan [2], Raman Kumar [2], Shubham Sharma [3,*], Somnath Chattopadhyaya [4], Farid Abed [5] and Mislav Stepinac [6,*]

1   Civil Engineering Department, Chandigarh University, Mohali 140413, Punjab, India;
    aditya.civil@cumail.in (A.K.T.); drsandeep1786@gmail.com (S.S.)
2   Mechanical Engineering Department, Chandigarh University, Mohali 140413, Punjab, India;
    jaskhera@gmail.com (J.S.C.); ramankakkar@gmail.com (R.K.)
3   Department of Mechanical Engineering, IK Gujral Punjab Technical University, Main Campus-Kapurthala,
    Kapurthala 144603, Punjab, India
4   Department of Mechanical Engineering, Indian Institute of Technology (ISM),
    Dhanbad 826004, Jharkhand, India; somnathchattopadhyaya@iitism.ac.in
5   Department of Civil Engineering, American University of Sharjah, Sharjah P.O. Box 26666, United Arab Emirates;
    fabed@aus.edu
6   Faculty of Civil Engineering, University of Zagreb, 10000 Zagreb, Croatia
*   Correspondence: shubham543sharma@gmail.com or shubhamsharmacsirclri@gmail.com (S.S.);
    mislav.stepinac@grad.unizg.hr (M.S.)

Abstract: Using a significant number of transverse hoops in the joint's core is one recognized way for achieving the requirements of strength, stiffness, and ductility under dynamic loading in a column joint. The shear capacity of a joint is influenced by the concrete's compressive strength, the anchoring of longitudinal beam reinforcement, the number of stirrups in the joint, and the junction's aspect ratio. Seismic motion on the beam may produce shear capacity and bond breaking in the joint, causing the joint to fracture. Furthermore, due to inadequate joint design and details, the entire structure is jeopardized. In this study, the specimens were divided into two groups for corner and interior beam–column joints based on the joint reinforcement detailing. The controlled specimen has joint detailing as per IS 456:2000, and the strengthened specimen has additional diagonal cross bars (modified reinforcement technique) at the joints detailed as per IS 456:200. The displacement time history curve, load-displacement response curves, load-displacement hysteretic curve, and load cycle vs. shear stress were used to compare the results of the controlled and strengthened specimens. The findings show that adding diagonal cross bars (modified reinforcing techniques) to beam–column joints exposed to cyclic loads enhances their performance. The inclusion of a diagonal cross bar increased the stiffness of the joint by giving an additional mechanism for shear transfer and ductility, as well as greater strength with minimum cracks.

Keywords: beam–column joints; shear capacity; cyclic loading; joint's numerical modeling; interior joint; corner joint; modified reinforcement technique (MRT)

Citation: Tiwary, A.K.; Singh, S.; Chohan, J.S.; Kumar, R.; Sharma, S.; Chattopadhyaya, S.; Abed, F.; Stepinac, M. Behavior of RC Beam–Column Joints Strengthened with Modified Reinforcement Techniques. *Sustainability* 2022, 14, 1918. https://doi.org/10.3390/su14031918

Academic Editor: Francesca Tittarelli

Received: 29 December 2021
Accepted: 4 February 2022
Published: 8 February 2022

Publisher's Note: MDPI stays neutral with regard to jurisdictional claims in published maps and institutional affiliations.

Copyright: © 2022 by the authors. Licensee MDPI, Basel, Switzerland. This article is an open access article distributed under the terms and conditions of the Creative Commons Attribution (CC BY) license (https://creativecommons.org/licenses/by/4.0/).

1. Introduction

The beam–column joint is the crucial zone in a reinforced concrete (RC) frame subjected to large forces during several ground shaking events, and its behavior has a significant influence on the response of the structure. Beam–column joints are the link between horizontal and vertical structural elements, and therefore, the joints are directly involved in the transfer of seismic forces [1]. The strength of the joint's component materials is restricted, and the joint's force-carrying capacity is also restricted. As a result, joints can be severely damaged or even destroyed during an earthquake [2,3]. The primary cause of joint failure is insufficient joint shear strength, which occurs due to insufficient and inadequate reinforcing details at the junction region [4]. Since fixing a fractured joint is challenging, the damage level needs to be minimized at the construction stage using a variety of techniques. An earthquake-resistant frame's column should be stronger than its beam. Joint panel stirrups contribute to the confining pressure and shear strength required to prevent early brittle collapses; the movement between the columns and beams can be transmitted correctly with enough transverse reinforcement [5]. However, structural elements constructed against gravity loads or in line with seismic standards in the Mediterranean region usually lack transverse reinforcements in the junction [6–9].

Consequently, beam–column junctions have been recognized among the principal causes of damage in pre-existing reinforced concrete (RC) structures in various studies conducted in the context of prior significant earthquakes. In most countries vulnerable to seismic events, pre-seismic codes do not meet present standards for reinforced concrete structures [10]. Earthquakes that have occurred recently (e.g., M7.4 Oaxaca (Mexico), M7.0 Aegean Sea (Turkey–Greece), M6.4 Croatia) have mostly caused failures of masonry buildings, but RC buildings were also, in some cases, heavily damaged [11–13]. One reason is that the beam–column joints in moment-resisting RC frame constructions have enough shear strength and ductility [14].

A design guideline for achieving the appropriate strength for beam–column joints is included in the existing standards. These specifications include enough anchoring for both beam and column bars traveling through or terminating in the joint area, and appropriate flexural strength for the beam and column to ensure beam-failure mechanisms [15,16]. Tsonos et al. [17] gave an overview of modern design codes (Eurocode family of codes) for the seismic performance of RC beam–column joints and compared them with the older standards. A simplified model for strengthening RC beam–column internal joints is given by Bossio et al. [1], which could be used by designers of new joints to quantify the performance of new structures, as well as by designers of external strengthening of existing joints to compute the benefits of the retrofit and shift the initial failure mode to a more desirable one.

In order to improve the structural behavior of precast beam–column connections, Hanif and Kanakubo [18] studied the use of fiber-reinforced cementitious composite (FRCC) to grout the joint region. Under reversed cyclic stress, two full-scale internal beam–column junctions were evaluated. In the joint region of the first specimen, aramid fibers cementitious composite was used. At the same time, polypropylene fibers cementitious composite was utilized to grout the junction in the second specimen. The findings were compared to previous research (no fibers, PVA fibers, and steel fibers). In contrast, steel fibers had considerably increased shear capacity and the largest hysteretic region. According to the test findings, PVA fibers outperformed others in terms of fracture width. In contrast, steel fibers had considerably increased shear capacity and the largest hysteretic region.

Li et al. [19] studied the cyclic behavior of joints built using prefabricated beams and columns composed of engineered cementitious composite (ECC). Two large-scale joint specimens made of conventional concrete and three large-scale joint specimens constructed of ECC were fabricated and tested under cyclic stresses till failure. One specimen was manufactured monolithically and served as a control. The effects of bar splicer sleeves and the connection location on the load-carrying capacity, failure mode, and ductility of produced joints were evaluated using three alternative assembly strategies. ECC enhanced the load-carrying capacity and ductility of constructed joints, according to the findings. The inclusion of longitudinal bars and splicer sleeves increased the load-carrying capacity but reduced ductility because the failure mechanism shifted from flexural to shear. The cyclic behavior was indifferent to connection location when ECC was employed.

The volume percentage of steel fibers in concrete and the detailing of reinforced steel in external beam–column joints were studied by Oinam et al. [20]. Regardless of transverse reinforcement in the joints, specimens of longitudinal beam bars positioned diagonally in the beam–column joints revealed interfacial shear fractures, according to the test findings. Those with a straight longitudinal bar in beams, on the other hand, showed a flexural plastic hinge away from the joint location. Steel fiber-reinforced concrete (SFRC) in the joint's region demonstrated outstanding ductility, energy absorption, and consistent hysteresis response, despite increasing the spacing of the hoops in the beam–column joints region.

Ravichandran [21] tested fourteen specimens under cyclic loads, one of which was built according to seismic code IS 13920 [22]. In contrast, the others were built without seismic details according to American Concrete Institute (ACI 318) [23], with HyFRC substituting regular concrete in the joint's location. For each volume fraction, two specimens were cast using the same concrete grade with hybrid fibers (80% steel + 20% polyolefin and 60% steel + 40% polyolefin) (0.5%, 1%, 1.5%, and 2%). The findings revealed that high-strength concrete containing 80% steel and 20% polyolefin improved ductility, energy absorption, and overall strength across all volume fractions. However, when compared to the seismic detail specimen, the hybrid fibers specimen with a volume fraction of 2% (80% steel/20% polyolefin) outperformed the seismic detail specimen in terms of energy absorption capacity and ductility.

Six beam–column knee joint specimens were constructed utilizing five created hybrid synthetic fibers, and one control specimen was evaluated under lateral cyclic stress by Zainal et al. [24]. Ferro-Ultra (F6U3), Ferro-Super (F6S3), Ferro-Econo (F6E3), and Ferro-Nylo (F6N3) were used to cast the hybrid fiber-reinforced concrete (HyFRC) joint area (FFC). According to the findings, the HyFRC joints showed substantial improvements in energy dissipation capacity, stiffness degradation rate, and displacement ductility. Compared to the reference specimens, the F6U3 generated the most augmentation, while the FFC produced the least. All hybrid specimens were numerically simulated using the finite-element method. The average margin error for peak load capacity, peak load displacement, and maximum displacements was 25.89%, 3.45%, and 0.18%, respectively.

Dehghani et al. [25] built a 3D finite-element model to analyze the impact of employing ECC in various patterns of beam–column connections. The model's validity was determined based on Yuan et al.'s findings [26]. The ECC material enhanced the load-carrying capacity and ductility of the beam–column connections but did not affect their initial stiffness. Furthermore, employing ECC outside of the plastic hinge areas was useless, as most tensile and shear cracks are found within the joints. The findings also showed that ECC was ineffective for preventing the diagonal shear crack in the joint area. Alwash et al. [27] and Bossio et al. [28] studied the corrosion of RC joints when exposed to bending moment and axial forces. Their study was oriented on the loss of integrity, a decrease of load-bearing capacity, stiffness, and serviceability due to corrosion, and the on joints' rehabilitation with the patch repair technique (PRT).

The steel plate energy absorption device (SPEAD) system, presented by Giuseppe Santarsiero et al. [29], is a revolutionary strengthening approach that aims to boost the flexural strength of beam and column components in RC frame constructions. The updated SPEAD model produced a 50-percent increase in strength, as well as a significant reduction in bond-slip effects in the joint panel region. This, in turn, resulted in an increase in ductility, which was good.

The concrete compressive strength, joint aspect ratio, and a number of lateral connections inside the joint are the most critical parameters determining the shear capacity of RC beam–column joints. A modified reinforcing technique to increase the shear capacity of cyclically loaded RC beam–column junctions is a viable option. The primary concerns discovered in the literature examined are the anchoring length requirements for beam bars, the provision of transverse reinforcements, and the involvement of stirrups in shear transmission at the joint. Research evaluating the use of extra cross-inclined bars at the joint core found that the inclined bars contribute a novel method of shear transfer, reducing the risk of a diagonal cleavage fracture at the joint. According to major international standards, diagonal cross bars have no effect on the shear strength of a joint. The goal of this study was to enhance core concrete confinement while avoiding reinforcing congestion in joints. The inclusion of diagonal bars adds another mechanism for shear transmission.

1.1. Interior Beam–Column Joint

In most instances, the breakdown of inner beam–column junctions triggered the failure mechanism in buildings. The statement mentioned above is proven by a thorough analysis of numerous collapsed or seriously damaged pre-seismic code-designed RC-framed structures after moderate or major earthquakes [2,3]. As a result, the weakest connection in existent RC movement-resistant frames was identified as the beam–column joints. These joints' inadequate shear strength has been established as the primary cause of joint failure. This lack in strength is commonly caused by insufficient and poorly specified joint reinforcements [30–32].

Furthermore, joint brittleness develops due to deficient reinforcement, especially the joint's transverse reinforcement causing a reduction in the overall ductility of the construction. The modified reinforcing technique aids in transferring shear or provides an extra shear-transfer mechanism at the joint. The force exerted on bars at the column faces causes the bond force to be dispensed by one of the top beam bars.

$$\text{Bond Force} = \frac{\Pi d2}{4}(fy + f's) \tag{1}$$

where $f's$ is the compression steel stress at the far face of the joint, fy is the yield stress of steel, and d is the diameter of steel.

1.2. Corner RC Beam–Column Joint

The joints are usually situated near the roof level in movement-resistant RC structures. Suppose these joints are simply intended for gravity loads and are built according to pre-seismic regulations. In that case, they may sustain significant harm during seismic events because of insufficient shear strength in the corner beam–column junction [10,33,34]. Internal forces created at corner joints may induce joint failure before the beam or column—whichever is weaker—reaches its maximum strength. In earthquake-prone nations such as Japan, Mexico, and China, several approaches for repairing and strengthening corner beam–column joints that earthquakes have damaged have been documented [35].

Only the following requirements may be expected to provide appropriate strength for the corner joint [35]:

i.   Around the corner, the tension steel is persistent, i.e., there is no lapping in the joint;
ii. The tension bars have to be curved into a radius that prevents the bars from bending or breaking. Nominal transverse bars are placed beneath the bent bars;
iii. Only a certain quantity of tension reinforcement is allowed [32].

$$\rho \leq 6\sqrt{f'_c}/f_y \tag{2}$$

The stresses are measured in pounds per square inch (psi).

2. Detailing Recommendations for Joints

The below suggestions are provided in regard to the need for anchorage, confinement, and shear inside the core of joints in earthquake-resistant structures [22]:

i.   Anchorage: Due to loss of bond at the inner face of an exterior joint, the development length of the beam reinforcement should be computed from the beginning of the 90° bend, rather than the face of the column. In wide columns, any portion of the beam bars within the outer third of the column could be considered for the computed development length. For shallow columns, the use of stub beams will be imperative. A large-diameter bearing bar fitted along the 90° bend of the beam bars should be beneficial in distributing bearing stresses. In deep columns, and whenever straight beam bars are preferred, mechanical anchorage could be advantageous. Joint ties should be so arranged that the critical outer-column bars and the bent-down portions of the bars are held against the core of the joint;

ii. Shear Strength: When the computed axial compression on the column is small, the contribution of the concrete shear resistance should be ignored, and shear reinforcement for the entire joint-shearing force should be provided. In exterior joints, only the ties that are situated in the outer two-thirds of the length of the potential diagonal failure crack, which runs from corner to corner of the joint, should be considered to be effective. The joint shear to be carried by the ties is calculated as:

$$A_v = \frac{1.5\, VsS}{dFy},\tag{3}$$

where Vs = joint shear carried by the ties, A_v = the total area of tie legs in a pair that makes up one layer of shear reinforcement, and d = the beam's effective depth. For preventing the excessive diagonal compression of core concrete, an upper bound for joint shear, usually stated as a nominal shearing force, must be imposed. The value between $10\sqrt{f'_c}$ and $11.5\sqrt{f'_c}$ (psi) is recommended for beams;

iii. Confinement: Horizontal tie legs are ineffectual for providing constraint against the concrete core volumetric expansion, while shear reinforcement restricts only the joint's corner regions. As a result, extra confining bars at right angles to the shear reinforcement are required. The distance between these bars should not exceed 150 mm.

The IS 13920:1993 (Ductile Detailing) [22] gives the following provisions:

i. Cl-7.4.1 Special confining reinforcement (l_o) (unless shear strength considerations demand a larger amount of transverse reinforcement) should be provided across a span of every joint face, towards the mid-span, and on each side of any area where flexural yielding may occur owing to earthquake pressures. The length 'l_o' should not be less than (a) the member's greater lateral dimension at the section where yielding occurs, (b) 1/6 of the member's clear span, and (c) 450 mm;

ii. Cl-8.1 Unless the joint is confined, the special confining reinforcement necessary at the column end must also be carried through the joint;

iii. Cl-8.2 A connection with beams framing all vertical sides, with each beam having a width of at least 3/4 of the column width, may be given half of the special restricting reinforcement needed at the column's end. The hoops' spacing shall not be more than 150 mm;

iv. In the joint region, diagonal cracking and concrete crushing can be managed by providing large column dimensions and densely packed closed-loop steel ties surrounding the column bars. The ties help resist the shear stress and hold the concrete in the joint, hence preventing concrete cracking and crushing;

v. The transverse loop should continue around the joint region around the column bars. This is cultivated by setting up the instance of all bar supports (both longitudinal bars and stirrups) on top of the shaft formwork at that level and lower into the case;

vi. The building columns in seismic zones III, IV, and V are to be at least 300-mm wide in each direction of the cross-section when the column support beams are longer than 5 m or when these columns are taller than 4 m between floors;

vii. A piece of the top pillar bar is consolidated in the segment that is projected up to the soffit of the bar, and a piece of it overhangs in segments with short widths and huge-breadth shaft bars;

viii. Beam bars may not reach past the soffit of the pillar if the section width is extensive;

ix. Interior joints need the top and base bars to go through the intersection without being cut, and these bars should be set inside the section bars without any twists;

x. The American Concrete Institute suggests a segment width that is no less than multiple times the distance across the longest longitudinal bar in the adjoining pillar.

3. Experimental Program

Interior and corner beam–column joint specimens were divided into two groups: one with changed reinforcing techniques and another without. For both reinforcement frameworks, the specimens were cast with reinforcement details according to IS 456:2000 [36], including extra diagonal cross-bracing reinforcement at the two faces of the joints for joint confinement, in accordance with the reinforcement category.

Details of Specimens

The beam and column dimensions at the corner and interior beam–column joints were similar. The columns were 300-mm deep and 400-mm broad, while the beams were 400-mm deep and 300-mm wide. The beam's span was 3000 mm, while the column's height was 3500 mm. Figure 1 shows the inner beam–column joint reinforcement detailing without changed reinforcement techniques, whereas Figure 2a,b shows the interior beam–column joint reinforcement detailing with modified reinforcement techniques at the joint location. The concrete mix consisted of ordinary Portland cement (43 grade), sand passing through a 4.75-mm IS sieve, and coarse aggregate ranging in size from 10 to 18 mm. The concrete cube's compressive strength after 28 days was 25 N/mm^2. The main reinforcement was made of steel bars with a yield stress of 415 N/mm^2. To account for the pull-out force, the longitudinal beam bars and cross-bracing bars were given enough development lengths, as required by the code. Inside a steel mold, the specimens were cast horizontally.

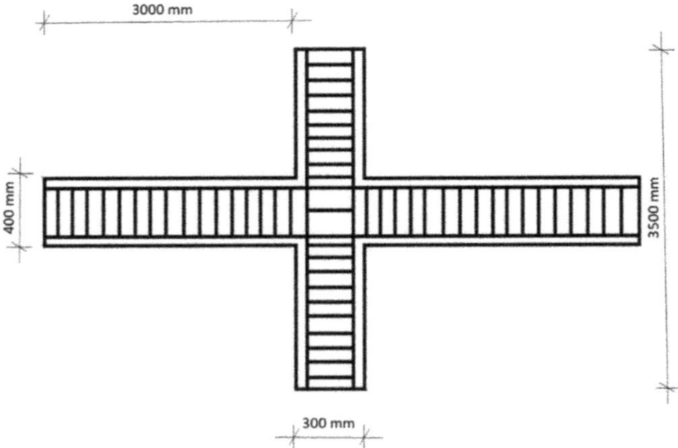

Figure 1. Reinforcement details of controlled specimens.

All of the specimens were put through their paces with a constant axial load and cyclic loading at the beam's end, as shown in Figure 2c. To duplicate the gravity force on the column, a 0–500-kN hydraulic jack was attached vertically to the loading frame and applied a constant column axial load, as shown in Figure 2d. The external hinge support was attached to one end of the column and anchored to the strong reaction floor, while the other end was restrained laterally by roller support. To apply reverse cyclic loading, two 200-kN hydraulic jacks were employed, one connected to the loading frame at the top and the other to the strong reaction floor. At a distance of 50 mm from the free end of the beam section of the assembly, the cyclic load was applied. In a load-controlled test, the specimen was exposed to an increasing cyclic load until failure. The load increment was set at 1.962 kN. The specimens were equipped with a linear variable differential transformer (LVDT) with a least count of 0.1 mm to measure the deflection at the loading point [10]. Figures 1 and 2a,b illustrate the schematic diagrams of the controlled and reinforced specimens, respectively.

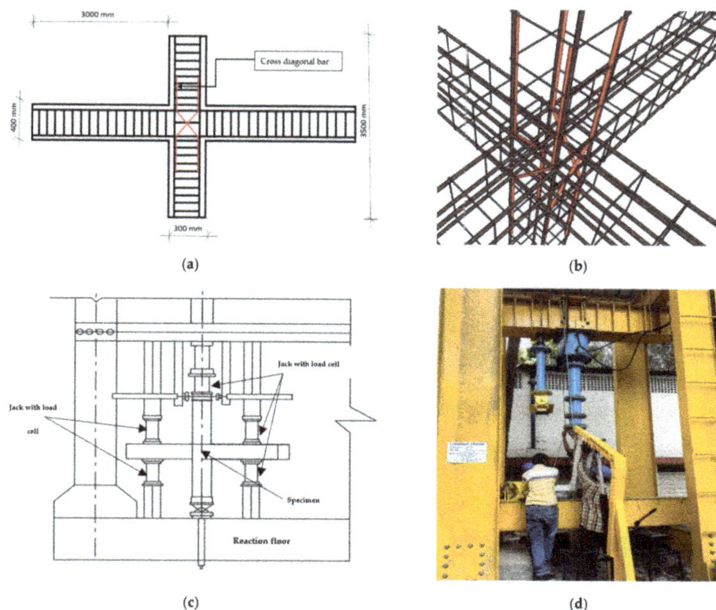

Figure 2. (**a**) Schematic diagram of strengthened specimens; (**b**) reinforcement detailing of strengthened specimens; (**c**) schematic diagram of test set-up; (**d**) specimens loading.

4. Numerical Modeling and Analysis of Beam–Column Joints

In order to appropriately replicate the tested joint sub-assemblies, symmetry boundary conditions are used. Solid 65, Solid 45, and Link8 elements were used to model the beam–column junction. The concrete was modeled with the Solid 65 element, while the hinge support at the base was modeled with the Solid 45 element. There are eight nodes in these elements, each with three degrees of freedom-translations in the nodal x, y, and z directions. The reinforcement was modeled using the Link8 element. There are two nodes in this three-dimensional spar element, each with three degrees of freedom-translations in the nodal x, y, and z directions. For this, the finite-element software ANSYS Workbench V12 was employed [37]. Following that, each material's element specifics are introduced in Tables 1 and 2. The major goal is to stiffen the column to emulate the beam–column junction behavior on the beam under cyclic loads. The finite-element method [30] converts partial differential equations into a series of algebraic linear equations:

$$[K]\{q\} = \{F\}, \tag{4}$$

where K = stiffness matrix, q = the nodal displacement vector, and F = the nodal force vector.

Table 1. Concrete characteristics.

Uniaxial Tensile Strength (MPa)	Poisson's Ratio Value	Ultimate Uniaxial Compressive Strength (Mpa)	Modulus of Elasticity (Mpa)
$0.62\sqrt{f_c}$	0.2	25	$5000\sqrt{f_c}$

Table 2. Steel characteristics.

Poisson's Ratio Value	Transverse Steel Yielding Stress (Mpa)	Longitudinal Steel Yielding Stress (Mpa)	Modulus of Elasticity (Mpa)
0.3	250	415	200,000

Concrete: An 8-noded solid element, or Solid 65 element, is used to simulate the concrete [38]. Every node in the corner and inner beam–column junction solid elements have translations in the nodal planes x, y, and z with a degree of freedom of three. Therefore, plastic deformation, three-dimensional cracking, and crushing are all possible with this element. The concrete's characteristics are shown in Table 1, below.

Steel: Standard Grade Fe 415 Mpa steel is used for the steel reinforcement in both the corner and interior beam–column connections. A Link8 component characterizes the steel reinforcement. This element necessitates the use of two nodes. Every node has a degree of freedom of three that correlates to the translations of the node's x, y, and z coordinates. Table 2 shows the characteristics of the steel.

In the engineering data utilized for the FE analysis of corner and interior joints, the geometric properties were as stated in Table 3.

Table 3. Geometric properties of the corner and interior joints.

Parametric Specifications of Beam	Measurements (mm)	Parametric Specifications of Column	Measurements (mm)
Concrete cover	30	Concrete cover	30
Span	3000	Column's depth	300
Depth	400	Width of column	300
Width	300	Column height	3500
Steel at bottom	4–10	Floor-to-floor height	3250
Steel at top	4–10	Longitudinal steel	4–12
Diameter of Transverse steel	6	Diameter of Transverse steel	6
Spacing of Transverse steel	220	Spacing of Transverse steel	200

ANSYS software's geometry tools model the interior and corner beam–column junction specimens as a 3D model. Figures 3a,b and 4a,b demonstrate the developed geometry as well as usual reinforcing (with steel) details of regulated and strengthened specimens, respectively, where Figure 3a,b represents an interior joint and Figure 4a,b represents a corner joint. In an ideal circumstance, the true binding force between reinforcing steel and concrete needs to be envisioned. However, a perfect bond between the two materials is postulated in this investigation. The Link8 component symbolizes the reinforcing steel bars connected to the nodes of each adjacent solid component of concrete to provide a consistent bonding; hence, both the materials contribute to the same node. A square mesh is used to obtain good results from the Solid65 element. As a result, the meshing is configured to produce square or rectangular mesh segments. This ensures that the dimensions of the components in the concrete support are compatible with the components and nodes in the model's concrete sections. The specimen is modeled with a square concrete element and a mesh size of 50 mm [35].

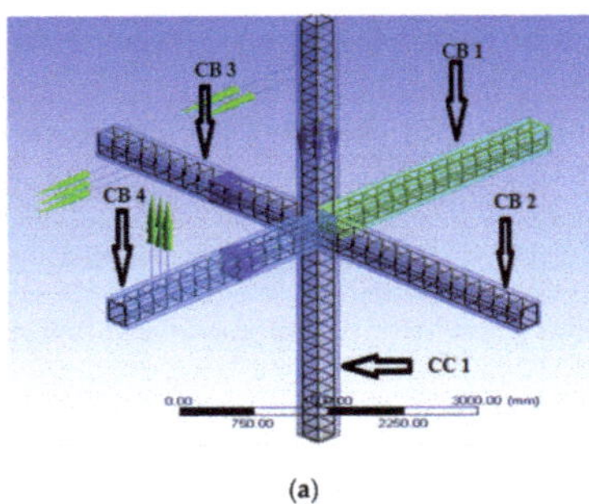

(a)

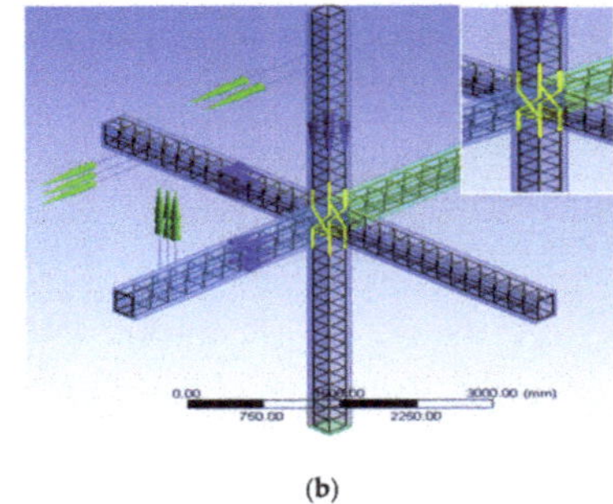

(b)

Figure 3. Geometric model and detail of reinforcement for interior joint; (**a**) controlled specimen, (**b**) strengthened specimen.

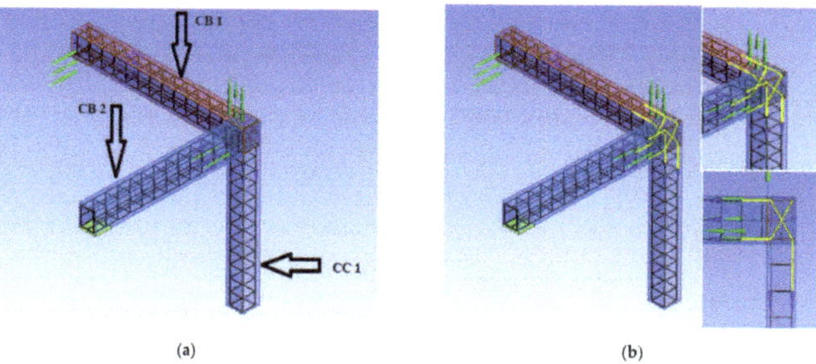

Figure 4. Geometric model and detail of reinforcement for corner joint; (**a**) controlled specimen, (**b**) strengthened specimen.

5. Results Obtained by Numerical Modeling

FE analysis findings for controlled and strengthened specimens were obtained utilizing ANSYS Workbench [37]. The specimens were subjected to cyclic loads ranging between 0–500 kN. For both specimens, maximum strain, shear stress, and total deformation are compared. According to the results, the above three interior and corner beam–column junction parameters are managed with improved reinforcing methods at the joint region (Tables 4 and 5). At the joint, the deformation that occurred in the controlled specimen is reduced in a strengthened specimen (Figure 5b), and the same is also transmitted to the CB1 beam of the corner beam–column junction, according to the controlled specimen's overall deformation model (Figure 5a).

Table 4. Analyzed findings of interior beam–column joint.

Measured Parameter	Highest Value with No MRT	Highest Value Using MRT	Variation in%
Overall deformation (mm)	0.87369	0.09106	89.5
Maximum Shear stress (MPa)	19.92	9.0418	79.3
Maximum Shear strain (mm/mm)	0.0065	0.00062	90.4

Table 5. Post analysis findings of corner beam–column joint.

Measured Parameter	Highest Value with No MRT	Highest Value Using MRT	Variation in%
Overall deformation (mm)	5.7922	0.13358	97.7
Maximum shear stress (MPa)	52.112	10.808	79.3
Maximum shear strain (mm/mm)	0.0009396	0.0003444	63.3

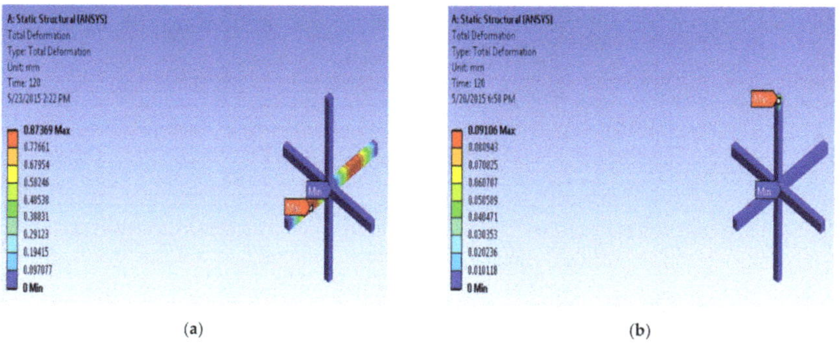

Figure 5. Total deformation for (**a**) control specimen with four beams (ICS) and (**b**) strengthened specimen with four beams (ISS).

From the total deformation model of the control specimen of an interior joint (ICS) (Figure 5a), it is observed that the minimum deformation is in the column and the maximum deformation is in the beam with no modified reinforcement techniques. Furthermore, the total deformation after strengthening with the modified reinforcement technique (ISS) was controlled, as shown in Figure 5b.

From Figure 6a, the maximum shear stress for the control specimen (i.e., without MRT) is observed to be equal in all sections of the interior joint. On the other hand, for the strengthened specimen, as shown in Figure 6b, it is found that there is minimum stress in the column and maximum stress in the beam; the stress in the beam–column joint was also controlled using the MRT technique.

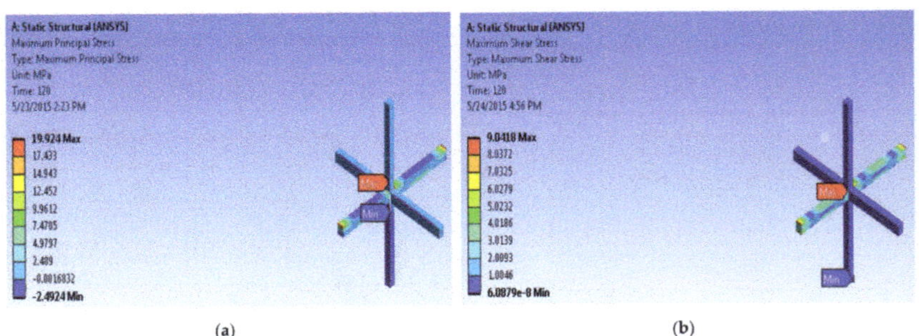

Figure 6. Maximum Principal stress in the interior joint; (**a**) control specimen with four beams (ICS) and (**b**) strengthened specimen with four beams (ISS).

From Figure 7a, the maximum principal strain for the control specimen on the interior beam–column junction (without MRT) is observed at the center of the column; while the minimum strain at the bottom support of the column is the strengthened specimen from Figure 7b, it is observed that the maximum principal strain for strengthening the specimen (with MRT) is at the support of the beam and the minimum principal strain is in the column. The strain in the joint is controlled by introducing MRT at the joint.

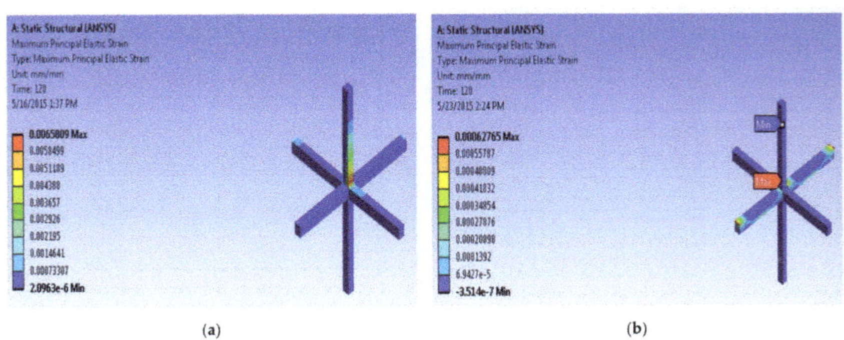

Figure 7. Maximum principal elastic strain in the interior joint for (**a**) the control specimen with four beams (ICS) and (**b**) the strengthened specimen with four beams (ISS).

Figure 8a shows that the highest deformation is at the corner joint and the minimum deformation is at the bottom support of the column for the controlled specimen. Figure 8b shows that the total deformation in the corner joint is controlled, compared to the controlled specimen after being strengthened with the modified reinforcement technique.

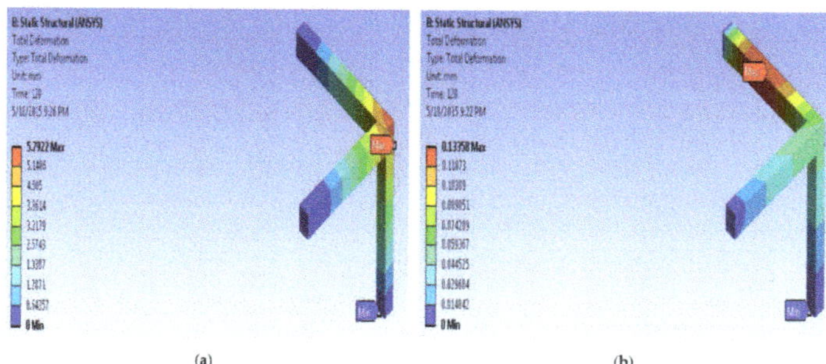

(a) (b)

Figure 8. Total deformation in the corner joint for (**a**) the control specimen (CCS) and (**b**) the strengthened specimen (CSS).

From Figure 9a, it is observed that the maximum principal stress for the control specimen in the corner joint (without MRT) is at the support of the beam, and for the strengthened specimen from Figure 9b, the maximum and minimum stress is found at the support of the beam. It can also be observed that the stress in the beam–column junction is controlled.

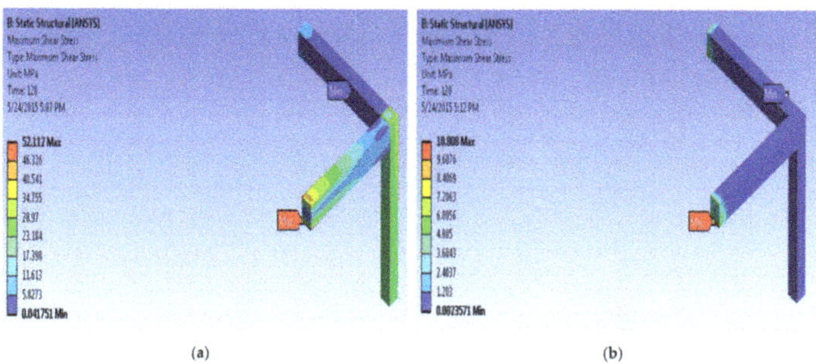

(a) (b)

Figure 9. Maximum shear stress in the corner joint for (**a**) the control specimen (CCS) and (**b**) the strengthened specimen (CSS).

In Figure 10a, the maximum shear strain for the control specimen of the corner beam–column junction (without MRT) is observed at the support of the beam, while the minimum strain is at the corner joint. For the strengthened specimen, it is found that the strain in the corner joint is controlled by introducing MRT at the joint.

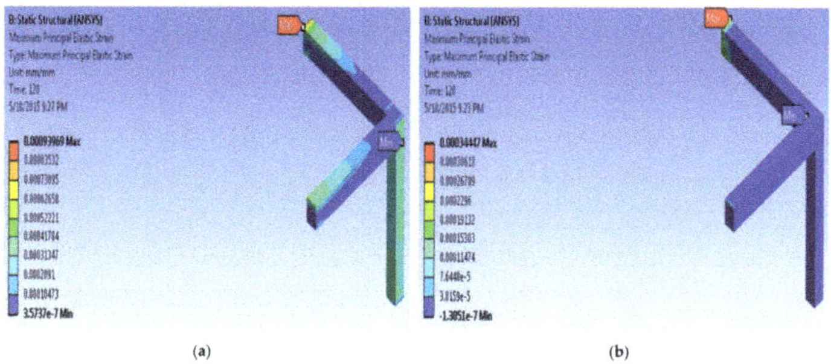

(a) (b)

Figure 10. Maximum principal strain in the corner joint for (**a**) the control specimen (CCS) and (**b**) the strengthened specimen (CSS).

5.1. Validation of Results

The results obtained through finite-element analysis for interior and corner beam–column joints were validated with experimental results in load-deformation behavior, as shown in Figures 11a,b and 12a,b, for the control and strengthened specimens. The load-deformation behavior found in the simulation was very similar to the findings of the experimental studies, with the variation of load ranging from 3% to 5%.

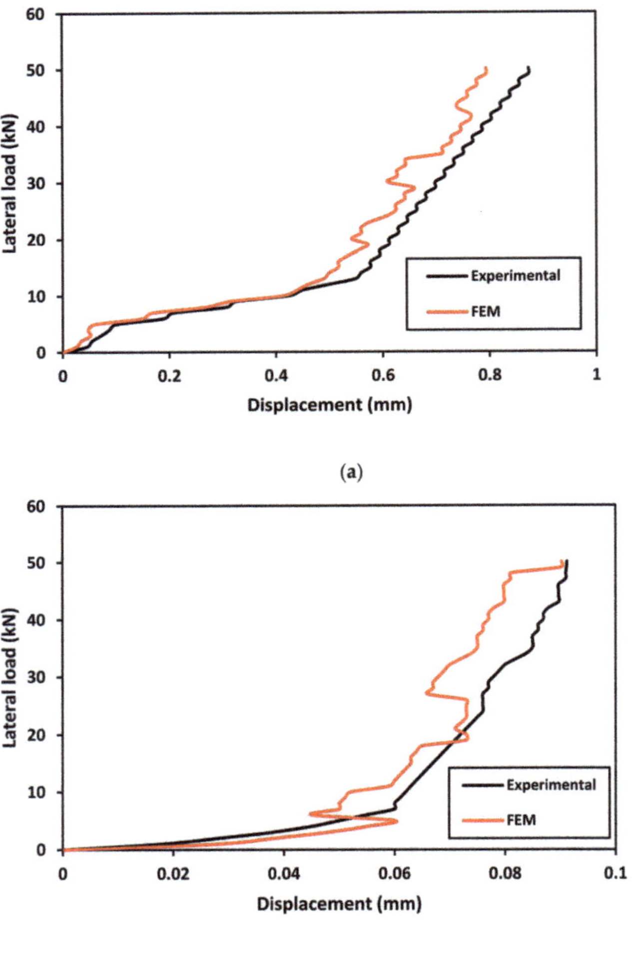

(a)

(b)

Figure 11. (**a**) Load-displacement response of controlled specimen (ICS). (**b**) Load-displacement response of strengthened specimens (ISS).

5.2. Load-Displacement Behaviour for Beam–Column Joints

In Figures 11a and 12a the load-displacement behavior curves are used to compare the results obtained through experimental and finite-element analysis of controlled specimens. The comparison of strengthened specimens of interior and corner beam–column joints are shown in Figures 11b and 12b, respectively. The results obtained through finite-element analysis were in great concurrence with the experimental results. The load-deformation behavior shown in the simulation was extremely close to that observed in experimental studies, with load variations ranging from 3% to 5%. Compared to the controlled specimens, the strengthened specimens displayed elastic behavior at the beginning stage in both the cases (interior and corner joint). Analysis determines the load-displacement characteristics indicated the better performance of strengthened specimens featuring cross-inclined reinforcement at the junction, which resulted in overall managed deformation and raised the ultimate loading capability compared to the controlled specimen in both types of joints, whether corner or inner.

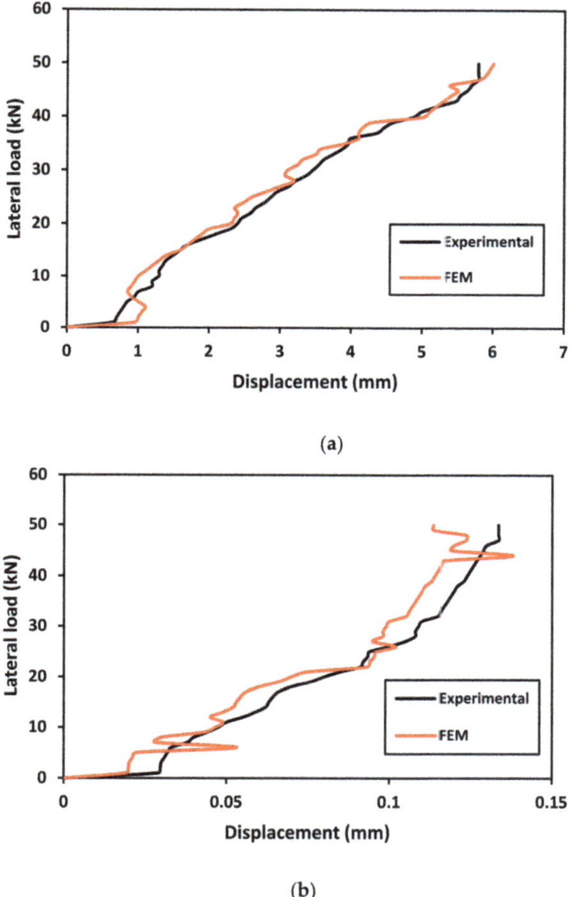

(a)

(b)

Figure 12. (**a**) Load-displacement response of controlled specimen (CCS). (**b**) Load-displacement response of strengthened specimens (CSS).

Thus, considering the ultimate load-carrying capacities from numerical studies, the specimens with diagonal confining bars (modified reinforcement technique) performed well for both cases of column axial loads. Furthermore, it can be observed that the displacement is more controlled for the ISS and CSS specimens by using cross-inclined bars at the joint than that of the ICS and CCS specimens for both the column axial load cases.

6. Results and Discussions

Though the reinforcement detailing of structures conform to the general construction code of practice, it may not adhere to modern seismic provisions. Current seismic code specifications for reinforced concrete-framed constructions are frequently deemed unrealistic by structural experts. They lose structural efficiency when a beam–column junction is subjected to significant lateral stresses, such as strong winds, earthquakes, or explosions. To satisfy the requirements of strength, stiffness, and ductility under cyclic loading, significant percentages of transverse hoops in the cores of joints are required in these locations. Provisions with a high percentage of hoops generate steel congestion, which causes construction difficulties.

On three fronts, researchers are looking into both kinds of beam–column joints (corner and interior).

The factors influencing the behavior of cyclically loaded corner and interior beam–column junctions are examined in the first approach. IS 456:2000 [36] was used to detail the joints. This method measures the maximum principal elastic strain and shear stress and overall deformation of controlled specimens (without MRT) under cyclic loading.

The controlled specimens are strengthened at the joint area in the second approach by using a modified reinforcement method (MRT) on both sides of the column, having a 12-mm diameter crossbar of length 450 mm (as per IS 456:2000) [36] installed. Testing of the joints under the same cyclic loads as the controlled specimens was performed. According to the findings, the improved reinforcing approach boosted the joint's shear resistance capability while simultaneously limiting overall deformation.

Comparison of all the FEM findings of both the control and strengthened specimens of corner and interior beam–column junctions is completed in the third approach. The cyclic response of the corner, as well as the inner beam–column connection, is found to be improved by utilizing updated reinforcing techniques concerning the maximum principal elastic strain and stress as well as overall deformation. The findings are compared using the lateral-loading vs. lateral-displacement curve, loading vs. deflection hysteretic curve, and deflection time history curve along with shear stress vs. load-cycle curve. The findings of the examination of the corner and interior joints are displayed in Table 6.

Table 6. Comparison of specimens with and without MRT.

Specimen ID	Overall Deformation (mm)	Maximum Shear Stress (Mpa)	Maximum Principal Elastic Strain (mm/mm)
CS4	0.873	19.92	0.00062
SS4	0.091	9.924	0.00018
CS5	5.7922	52.11	0.00093
SS5	0.1336	10.81	0.00034

6.1. Hysteretic Behavior of Corner and Interior Beam–Column Junctions

In consideration of shear capacity and deformation capability, the stress-strain behavior of both beam–column junctions, i.e., corner and interior, is investigated. Hysteretic curves in Figure 13a–d depicts load-displacement equations for fixed and strengthened specimens. The overall deformation in controlled specimens of both the interior and corner junctions (ICS and CCS) is higher than strengthened specimens (ISS and CSS). The loading capacities of ISS and CSS are significantly higher than ICS and CCS, as per the findings of the hysteretic analysis. The ductility is increased without compromising the stiffness. In general, specimens with diagonal crossbars perform better than their conventionally detailed counterparts.

6.2. Shear Stress vs. Loading Cycle Behavior of Joints

The ductility of specimens reinforced using the cross-inclined bar, as per IS 456:2000, at the corner and interior beam–column junctions outperforms the regulated specimen with no cross-inclined reinforcement. The addition of cross-inclined reinforcement boosted both the ultimate load-bearing capacity and ductility of the interior as well as of the corner junction in both load circumstances (downward and upward), according to the numerical investigation. The inclusion of slanted bars creates a new shear-transmission mechanism. The corner and interior beam–column junctions using the modified reinforcement method (MRT) have better strength, as shown in Figure 14a,b.

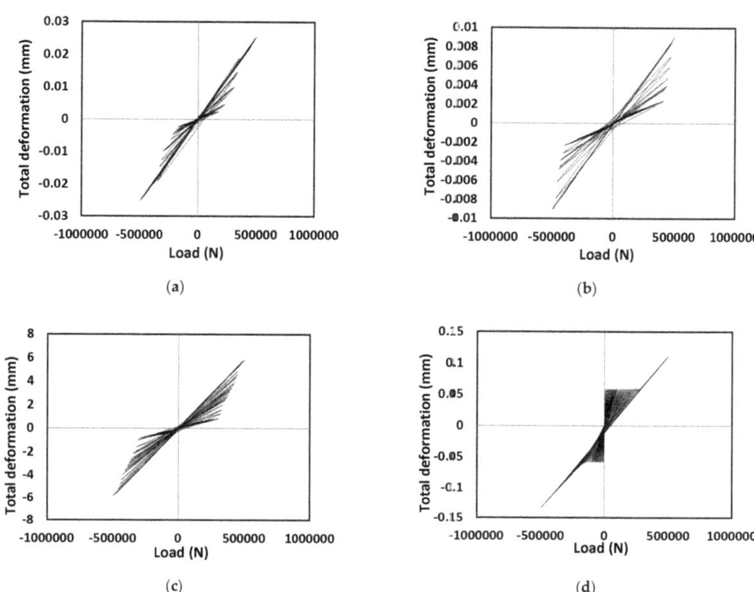

Figure 13. (**a**) Load vs. total deformation hysteretic graph for the controlled specimen (ICS) (without MRT) and (**b**) the strengthened specimen (ISS) (with MRT). (**c**) Load vs. total deformation graph for the controlled specimen (CCS) (without MRT) and (**d**) the strengthened specimen (CSS) (with MRT).

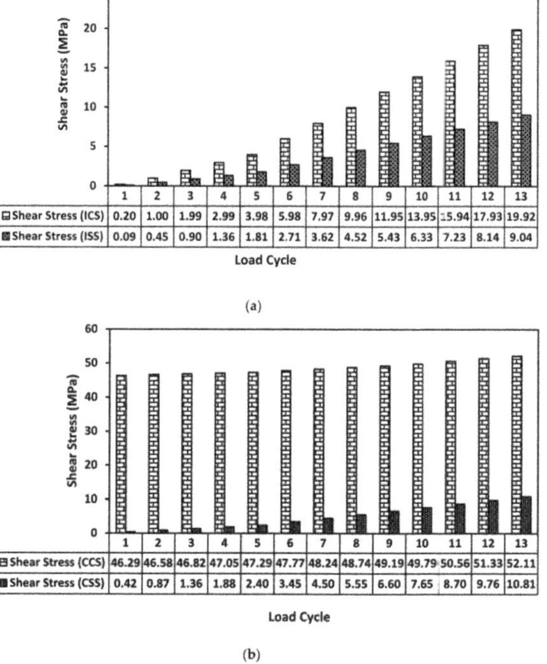

Figure 14. (**a**) Shear stress vs. load cycle for the controlled specimen (ICS) and the strengthened specimen (ISS). (**b**) Shear stress vs. load cycle for the controlled specimen (CCS) and the strengthened specimen (CSS).

6.3. Displacement Time History Curve for Beam–Column Joints

Figures 15a and 16a illustrate the lateral load-displacement time histories curve obtained through numerical analysis for the controlled specimens. Figures 15b and 16b represent the numerical findings that strengthened the specimen's lateral load-displacement time histories. All of the cycles progressed to the push motion after being started with the pull motion. Adopting cross-inclined bars at the joint location to reinforce beam–column joints offers more strength than the controlled specimen in both cases.

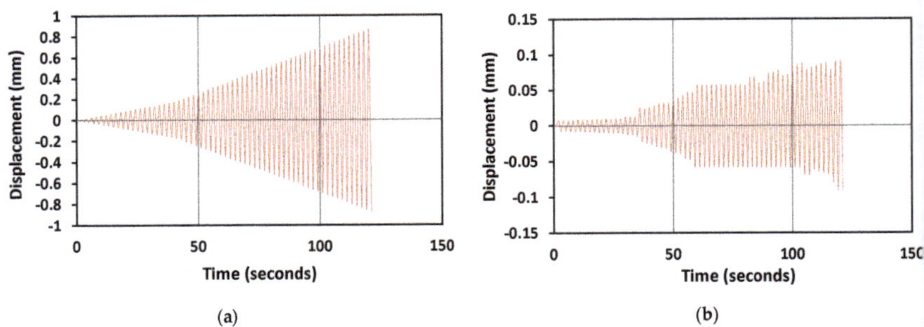

Figure 15. (**a**) Displacement time history of the controlled specimen (ICS) and of (**b**) the strengthened specimen (ISS).

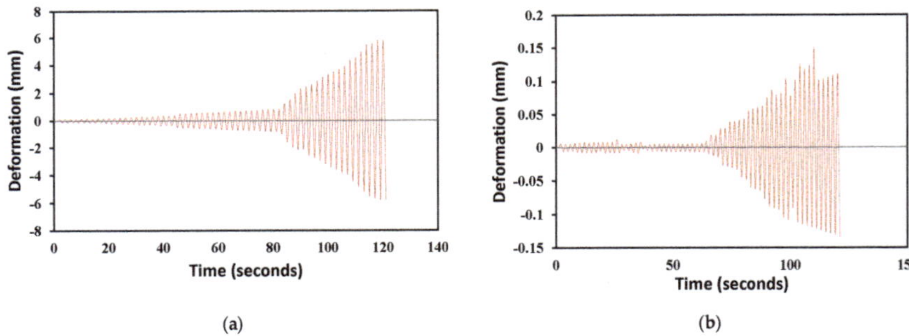

Figure 16. (**a**) Displacement time history of the controlled specimen (CCS) and of (**b**) the strengthened specimen (CSS).

7. Conclusions

The behavior of beam–column joints in RC structures is of great importance for the seismic behavior of a whole structure. Hence, the investigation and research in the field is beneficial, and new methods, techniques, and procedures can help achieve a better understanding of the complex behavior of the joints themselves. Strengthening and upgrading such elements can be completed, e.g., by high-strength steel bars [39], steel jacketing [40], cementitious composites [41], FRP ropes [42], self-centering friction haunches [43], etc. The RC beam–column joints can also be predicted by using modern techniques such as machine learning [44]. In this study, the performance of interior and corner beam–column joints were analyzed through an experimental program, and the results obtained through tests were validated using finite-element software ANSYS. Similar studies were performed by Santarsiero [45]. The following findings may be derived:

i. Based on the present research, the most critical parameters influencing joint shear capacity are the stirrups quantity, the aspect ratio of the joint, the beam longitudinal reinforcement anchorage, and the compressive strength of concrete;

ii.   The results obtained through the experimental studies were validated with numerical analysis in terms of load-deformation behavior, and the numerical results were in great concurrence with the experimental data;

iii. The findings of the finite-element model are compared to the controlled and strengthened specimens, and it is discovered that adding diagonal cross bars (modified reinforcing techniques) to beam–column joints exposed to cyclic loads enhances their performance more than using a controlled specimen in both interior and corner beam–column joints;

iv.   The corner beam–column joint models for the controlled and strengthened specimens are analyzed for similar loadings with different reinforcement arrangements. The larger deformations and stresses, which are reported in the controlled specimen, are reduced in the strengthened specimens after employing modified reinforcement techniques;

v.   When the controlled and strengthened specimens for the interior beam–column joint are analyzed, it is found that the maximum stress and deformation caused in the joint are controlled by using additional diagonal cross bars at the joint region;

vi.   Modified reinforcement techniques with the diagonal cross bar at the joint region is a viable option for enhancing the shear capacity of beam–column joints. The diagonal cross bars help to create an extra shear-transfer mechanism;

vii. A beam–column junction loses structural efficiency when it is exposed to large lateral stresses, such as high winds. Therefore, against such stresses, the specimens with a diagonal crossbar at the junction work best;

viii. In both upward- and downward-load situations, the introduction of cross-inclined bars at the junction area of a strengthened corner and an interior beam–column junction maximizes the joint's stiffness, enhances its load-carrying capacity, as well as its ductility, according to an improved reinforcing approach.

Author Contributions: Conceptualization: A.K.T., S.S. (Sandeep Singh) and S.S. (Shubham Sharma); methodology: A.K.T., S.S. (Sandeep Singh), J.S.C., R.K., S.S. (Shubham Sharma) and M.S.; formal analysis: A.K.T., S.S. (Sandeep Singh), J.S.C., R.K., S.S. (Shubham Sharma) and M.S.; investigation: A.K.T., S.S. (Sandeep Singh), S.S. (Shubham Sharma) and M.S.; resources: S.S. (Shubham Sharma), S.C., F.A. and M.S.; writing—original draft preparation: A.K.T., S.S. (Sandeep Singh), J.S.C., R.K. and S.S. (Shubham Sharma); writing—review and editing: J.S.C., R.K., S.S. (Shubham Sharma), S.C., F.A. and M.S.; supervision: S.S. (Sandeep Singh), J.S.C., R.K., S.S. (Shubham Sharma) and M.S.; funding acquisition: M.S. and S.S. (Shubham Sharma). All authors have read and agreed to the published version of the manuscript.

Funding: This work receives no external funding.

Institutional Review Board Statement: Not applicable.

Informed Consent Statement: Not applicable.

Data Availability Statement: The data presented in this study are available on request from the corresponding author.

Conflicts of Interest: The authors declare no conflict of interest.

Abbreviations

MRT	Modified reinforcement technique
ICS	Controlled specimen of interior joint without MRT
CCS	Controlled specimen of corner joint without MRT
ISS	Strengthened specimen of interior joint with MRT
CSS	Strengthened specimen of corner joint with MRT

References

1. Bossio, A.; Fabbrocino, F.; Lignola, G.P.; Prota, A.; Manfredi, G. Simplified Model for Strengthening Design of Beam–Column Internal Joints in Reinforced Concrete Frames. *Polymers* **2015**, *7*, 1732–1754. [CrossRef]
2. Calvi, G.M.; Magenes, G.; Pampanin, S. Relevance of beam-column joint damage and collapse in rc frame assessment. *J. Earthq. Eng.* **2002**, *6*, 75–100. [CrossRef]
3. Park, R. A summary of results of simulated seismic load tests on reinforced concrete beam-column joints, beams and columns with substandard reinforcing details. *J. Earthq. Eng.* **2002**, *6*, 147–174. [CrossRef]
4. Alsayed, S.H.; Al-Salloum, Y.A.; Almusallam, T.H.; Siddiqui, N.A. Seismic Response of FRP-Upgraded Exterior RC Beam-Column Joints. *J. Compos. Constr.* **2010**, *14*, 195–208. [CrossRef]
5. Alhaddad, M.S.; Siddiqui, N.A.; Abadel, A.A.; Alsayed, S.H. Numerical Investigations on the seismic Behavior of FRP and TRM Upgraded RC Exterior Beam-column Joints. *J. Compos. Constr.* **2012**, *16*, 308–321. [CrossRef]
6. Rossetto, T.; Elnashai, A. Derivation of vulnerability functions for European-type RC structures based on observational data. *Eng. Struct.* **2003**, *25*, 1241–1263. [CrossRef]
7. Ricci, P.; De Luca, F.; Verderame, G.M. 6th April 2009 L'Aquila earthquake, Italy: Reinforced concrete building performance. *Bull. Earthq. Eng.* **2010**, *9*, 285–305. [CrossRef]
8. Pejovic, J.; Stepinac, M.; Serdar, N.; Jevric, M. Improvement of Eurocode 8 Seismic Design Envelope for Bending Moments in RC Walls of High-rise Buildings. *J. Earthq. Eng.* **2020**, 1–25. [CrossRef]
9. Kišiček, T.; Stepinac, M.; Renić, T.; Hafner, I.; Lulić, L. Strengthening of masonry walls with FRP or TRM. *J. Croat. Assoc. Civ. Eng.* **2020**, *72*, 937–953. [CrossRef]
10. Bindhu, K.R.; Jaya, K.P. Strength and Behavior of Exterior Beam Column Joints with Diagonal Cross Bracing Bars. *Asian J. Civil. Eng.* **2010**, *11*, 397–410.
11. Ramírez-Herrera, M.-T.; Romero, D.; Corona, N.; Nava, H.; Torija, H.; Maguey, F.H. The 23 June 2020 MW 7.4 la crucecita, Oaxaca, Mexico earthquake and tsunami: A rapid response field survey during COVID-19 crisis. *Seismol. Res. Lett.* **2020**, *92*, 26–37. [CrossRef]
12. Cetin, K.O.; Mylonakis, G.; Sextos, A.; Stewart, J.P. Reconnaissance of 2020 M 7.0 Samos Island (Aegean Sea) earthquake. *Bull. Earthq. Eng.* **2021**, 1–6. [CrossRef]
13. Stepinac, M.; Lourenço, P.B.; Atalić, J.; Kišiček, T.; Uroš, M.; Baniček, M.; Novak, M.Š. Damage classification of residential buildings in historical downtown after the ML5.5 earthquake in Zagreb, Croatia in 2020. *Int. J. Disaster Risk Reduct.* **2021**, *56*, 102140. [CrossRef]
14. Almusallam, T.H.; Al-Salloum, Y.A. Seismic response of interior RC beam-column joints upgraded with FRP sheets. II: Analy-sis and Parametric Study. *J. Compos. Constr.* **2007**, *11*, 590–600. [CrossRef]
15. Dabiri, H.; Kaviani, A.; Kheyroddin, A. Influence of reinforcement on the performance of non-seismically detailed RC beam-column joints. *J. Build. Eng.* **2020**, *31*, 101333. [CrossRef]
16. Dabiri, H.; Kheyroddin, A.; Kaviani, A. A Numerical Study on the Seismic Response of RC Wide Column–Beam Joints. *Int. J. Civ. Eng.* **2018**, *17*, 377–395. [CrossRef]
17. Tsonos, A.-D.; Kalogeropoulos, G.; Iakovidis, P.; Bezas, M.-Z.; Koumtzis, M. Seismic Performance of RC Beam–Column Joints Designed According to Older and Modern Codes: An Attempt to Reduce Conventional Reinforcement Using Steel Fiber Reinforced Concrete. *Fibers* **2021**, *9*, 45. [CrossRef]
18. Hanif, F.; Kanakubo, T. Shear Performance of Fiber-Reinforced Cementitious Composites Beam-Column Joint Using Various Fibers. *J. Civ. Eng. Forum* **2017**, *3*, 383. [CrossRef]
19. Li, X.; Li, Y.; Yan, M.; Meng, W.; Lu, X.; Chen, K.; Bao, Y. Cyclic behavior of joints assembled using prefabricated beams and columns with Engineered Cementitious Composite (ECC). *Eng. Struct.* **2021**, *247*, 113115. [CrossRef]
20. Oinam, R.M.; Kumar, P.C.A.; Sahoo, D.R. Cyclic performance of steel fiber-reinforced concrete exterior beam-column joints. *Earthq. Struct.* **2019**, *16*, 533–546.
21. Annadurai, A.; Ravichandran, A. Seismic Behavior of Beam–Column Joint Using Hybrid Fiber Reinforced High-Strength Concrete. *Iran. J. Sci. Technol. Trans. Civ. Eng.* **2018**, *42*, 275–286. [CrossRef]
22. *IS 13920:1993*; Indian Standard Code on Criteria for Earthquake Resistant Design of Structures. Bureau of Indian Standards: New Delhi, India, 2002.
23. *ACI Committee 318*; Building Code Requirements for Structural Concrete. American Concrete Institute: Farmington Hills, MI, USA, 2005.
24. Zainal, S.M.I.S.; Hejazi, F.; Rashid, R.S.M. Enhancing the Performance of Knee Beam–Column Joint Using Hybrid Fibers Reinforced Concrete. *Int. J. Concr. Struct. Mater.* **2021**, *15*, 1–28. [CrossRef]
25. Dehghani, A.; Mozafari, A.R.; Aslani, F. Evaluation of the efficacy of using engineered cementitious composites in RC beam-column joints. *Structures* **2020**, *27*, 151–162. [CrossRef]
26. Yuan, F.; Pan, J.; Xu, Z.; Leung, C.K.Y. A comparison of engineered cementitious composites versus normal concrete in beam-column joints under reversed cyclic loading. *Mater. Struct.* **2012**, *46*, 145–159. [CrossRef]
27. Alwash, N.A.; Kadhum, M.M.; Mahdi, A.M. Rehabilitation of Corrosion-Defected RC Beam-Column Members Using Patch Repair Technique. *Buildings* **2019**, *9*, 120. [CrossRef]

23. Bossio, A.; Fabbrocino, F.; Monetta, T.; Lignola, G.P.; Prota, A.; Manfredi, G.; Bellucci, F. Corrosion effects on seismic capacity of reinforced concrete structures. *Corros. Rev.* **2019**, *37*, 45–56. [CrossRef]
29. Santarsiero, G.; Manfredi, V.; Masi, A. Numerical Evaluation of the Steel Plate Energy Absorption Device (SPEAD) for Seismic Strengthening of RC Frame Structures. *Int. J. Civ. Eng.* **2020**, *18*, 835–850. [CrossRef]
30. Lowes, L.N.; Altoontash, A. Modeling Reinforced-Concrete Beam-Column Joints Subjected to Cyclic Loading. *J. Struct. Eng.* **2003**, *129*, 1686–1697. [CrossRef]
31. Brooke, N.J.; Ingham, J.M. Seismic Design Criteria for Reinforcement Anchorages at Interior RC Beam-Column Joints. *ASCE* **2013**, *139*, 1895–1905. [CrossRef]
32. Palermo, D.; Vecchio, F.J. Compression field modeling of reinforced concrete subjected to reversed loading: Formulation. *ACI Struct. J.* **2003**, *100*, 2–234.
33. Rajagopal, P.S.S.; Thomas, H. Seismic behaviour evaluation of exterior beam-column joints with headed or hooked bars using nonlinear finite element analysis. *Earthq. Struct.* **2014**, *7*, 861–875. [CrossRef]
34. Park, S.; Mosalam, K.M. Experimental Investigation of Nonductile RC Corner Beam-Column Joints with Floor Slabs. *J. Struct. Eng.* **2013**, *139*, 1–14. [CrossRef]
35. Ramin, K.; Fereidoonfar, M. Finite Element Modeling and Nonlinear Analysis for Seismic Assessment of Off-Diagonal Steel Braced RC Frame. *Int. J. Concr. Struct. Mater.* **2015**, *9*, 89–118. [CrossRef]
36. *IS 456:2000*; Indian Standard Code on Plain and Reinforced Concrete. Bureau of Indian Standards: New Delhi, India, 2000.
37. ANSYS. *ANSYS Manual*; ANSYS INC: Canonsburg, PA, USA, 2009.
38. Alsayed, S.H.; Almusallam, T.H.; Al-Salloum, Y.A.; Siddiqui, N.A. Seismic Rehabilitation of Corner RC Beam-Column Joints Using CFRP Composites. *J. Compos. Constr.* **2010**, *14*, 681–692. [CrossRef]
39. Feng, J.; Wang, S.; Meloni, M.; Zhang, Q.; Yang, J.; Cai, J. Seismic Behavior of RC Beam Column Joints with 600 MPa High Strength Steel Bars. *Appl. Sci.* **2020**, *10*, 4684. [CrossRef]
40. Santarsiero, G.; Masi, A. Seismic Upgrading of RC Wide Beam–Column Joints Using Steel Jackets. *Buildings* **2020**, *10*, 203. [CrossRef]
41. Shang, X.-Y.; Yu, J.-T.; Li, L.-Z.; Lu, Z.-D. Strengthening of RC Structures by Using Engineered Cementitious Composites: A Review. *Sustainability* **2019**, *11*, 3384. [CrossRef]
42. Golias, E.; Zapris, A.G.; Kytinou, V.K.; Osman, M.; Koumtzis, M.; Siapera, D.; Chalioris, C.E.; Karayannis, C.G. Application of X-Shaped CFRP Ropes for Structural Upgrading of Reinforced Concrete Beam–Column Joints under Cyclic Loading–Experimental Study. *Fibers* **2021**, *9*, 42. [CrossRef]
43. Veismoradi, S.; Yousef-Beik, S.M.M.; Zarnani, P.; Quenneville, P. Seismic strengthening of deficient RC frames using self-centering friction haunches. *Eng. Struct.* **2021**, *248*, 113261. [CrossRef]
44. Ganasan, R.; Tan, C.G.; Ibrahim, Z.; Nazri, F.M.; Sherif, M.M.; El-Shafie, A. Development of Crack Width Prediction Models for RC Beam-Column Joint Subjected to Lateral Cyclic Loading Using Machine Learning. *Appl. Sci.* **2021**, *11*, 7700. [CrossRef]
45. Santarsiero, G. FE Modelling of the Seismic Behavior of Wide Beam-Column Joints Strengthened with CFRP Systems. *Buildings* **2018**, *8*, 31. [CrossRef]

sustainability

MDPI

Article

Influence of Friction on the Behavior and Performance of Prefabricated Timber–Bearing Glass Composite Systems

Vlatka Rajčić, Nikola Perković *, Domagoj Damjanović and Jure Barbalić

Civil Engineering, Structural Department, University of Zagreb, 10000 Zagreb, Croatia;
vlatka.rajcic@grad.unizg.hr (V.R.); domagoj.damjanovic@grad.unizg.hr (D.D.); jure.barbalic@grad.unizg.hr (J.B.)
* Correspondence: nikola.perkovic@grad.unizg.hr

Abstract: The basic concept of seismic building design is to ensure the ductility and sufficient energy dissipation of the entire system. The combination of wood and bearing glass represents a design in which each material transmits the load, and with the mutual and simultaneous interaction of the constituent elements, it is also earthquake resistant. Such a system has been developed so that the glass directly relies on the wooden frame, which allows the load to be transferred by contact and the friction force between the two of materials. Within the seismic load, friction between glass and wood is an important factor that affects both the behavior and performance of a wood–glass composite system. The set-up system consists of a single specimen of laminated or insulating glass embedded between two CLT elements. The friction force was determined at the CLT–glass contact surface for a certain lateral pressure, i.e., normal force. Friction depends on the way the elements (especially glass) are processed, as well as on the lateral load introduced into the system. Conducted experimental research was accompanied by numerical analyses. Experimental research was confirmed by numerical simulations.

Keywords: composites; timber; CLT; load-bearing glass; earthquake; friction; FEM analysis

Citation: Rajčić, V.; Perković, N.; Damjanović, D.; Barbalić, J. Influence of Friction on the Behavior and Performance of Prefabricated Timber–Bearing Glass Composite Systems. *Sustainability* **2022**, *14*, 1102. https://doi.org/10.3390/su14031102

Academic Editor: Syed Minhaj Saleem Kazmi

Received: 15 December 2021
Accepted: 13 January 2022
Published: 19 January 2022

Publisher's Note: MDPI stays neutral with regard to jurisdictional claims in published maps and institutional affiliations.

Copyright: © 2022 by the authors. Licensee MDPI, Basel, Switzerland. This article is an open access article distributed under the terms and conditions of the Creative Commons Attribution (CC BY) license (https://creativecommons.org/licenses/by/4.0/).

1. Introduction

In the current situation of increasingly acknowledging climate change as a threat to our environment and human society, binding agreements have been made during the COP26, taking place in Glasgow in 2021. The building sector has a huge impact and must provide answers on how to tackle climate change, develop a circular economy, and provide a sustainable environment. The building sector should base future technologies on environmentally friendly materials and construction processes. Timber is the leading bio-based material and, through newly designed engineered wood-based materials, the material of the future. One innovative engineering wood product, known as cross-laminated timber (CLT), was introduced in the early 1990s in Austria and Germany. Due to its good mechanical properties, good fire resistance as well as advanced durability, and rheological properties, it has been seen as a potential material to replace reinforced concrete in low- and high-rise buildings. On the other hand, there has been significant development and increase in the use of glass as a load-bearing material. Rajčić et al. concluded that load-bearing glass combined with a timber frame represents a load-bearing composite element, which has very good potential for excellent behavior under normal and seismic loads; it is cost-effective, energy-efficient, and aesthetically acceptable [1–3]. The lack of experience and scientific research as well as non-existing standards and codes covering the design of structural components made as composites from laminated glass and laminated timber limit the implementation of such structural elements in practice. Admittedly, there is a national guideline for the design of glass elements [4]. The purpose of these instructions is to seek to provide an overview that is as complete as possible for the various aspects that must be considered in the design, construction, and control of glass elements with regard to verifying their mechanical strength and stability. There are some existing harmonized standards for glass products, for instance [5,6], necessary for their CE marking, but there are no European harmonized standards that can serve as codes needed for the design of glass structures. Therefore, the European Committee for Standardization (CEN) started the preparation of a new code in the series of Eurocodes in order to clarify the design of safe glass-based structures [7]. Currently, they are in the form of technical specifications (CEN/TS 19100-1–CEN/TS 19100-3). Regarding composite elements of cross-laminated timber and laminated glass, extensive research work should be carried out, which will be a basis for implementation in code. Although there is a limited amount of research dealing with timber-glass composites, the need for large transparent surfaces led architects to use such elements. Adding the aesthetic value of timber and glass and the environmental friendliness of materials that can be fully recycled at the end of their life cycles, a new type of structural element was introduced as timber–glass composite structural systems. These types of structures are also built-in earthquake-prone areas (south Europe, Japan, China, USA). There is a significant concern that should be overcome. Generally, the opinion is that the glass has brittle behavior and cannot be used as a structural element in seismic zones. Antolinc, with his research [8], has contributed to the understanding of the behavior of hybrid structural components based on laminated glass and cross-laminated timber frames. Such a structural component, made of a cross-laminated timber frame infilled with load-bearing laminated glass, has a high potential for various applications. It may be used as for façade element timber houses, as a bracing element for newly built or existing frame structures, as a strengthening structural component in existing timber buildings, or as a supporting structural component in historic buildings during and after their retrofitting and restoration.

During the project financed by the Croatian scientific fund "VETROLIGNUM", led by Prof. Vlatka Rajčić, the system was analyzed in terms of load-bearing capacity, stability, seismic performance, energy efficiency, water tightness, and airtightness. Building with wood is very fast and completely suitable for prefabrication in factories. The LCA (cradle to cradle) shows extremely good results in terms of cost-effectiveness and sustainable construction and a reduced CO_2 footprint [9]. Considering the complexity of the wood-bearing glass composite system and the intentions of presenting the most realistic performance and characteristics of such systems, the research is divided into two sections: laboratory testing and research on numerical models.

The contact between glass and wood, as well as a type of connection in the angles of the wooden frame, are the details that need to be given the utmost attention since it is precisely the manner of joining these elements that greatly determines the behavior of the entire composite system [10]. The most usual way of joining load-bearing glass and wooden structures is by adhesives and different types of mechanical fasteners [3]. Using steel mechanical fasteners results in complicated design solutions and details that damage the edge of the glass, which is the most sensitive part of the glass element. Using adhesives can provide a good connection, but that poses the question of the durability of such systems. In addition to said problems, the seismic load causes damage to the structure at the joint positions, and consequently, the failure of the entire load-bearing system [2,3]. In order to solve that problem and maximize energy dissipation during earthquake loading, a system was developed where the load-bearing glass was inserted into the wooden frame without additional mechanical fasteners and adhesives. The system is designed to allow the glass to move freely in a wooden frame while securing glass stability with additional wooden slats. Therefore, the load is transmitted by direct contact, i.e., friction between two elements [3,8,10,11].

During the experimental campaign, 45 cyclings (racking tests) of the composite panels were performed with four different types of connectors in the laminated timber frame. Tests have shown that failure of a composite panel occurs in a corner of a wooden frame [3]. Due to the partly free movement, the load-bearing glass panels remain intact, which is very important in such a composite system since the wooden elements are easily replaceable, while the mechanical characteristics and properties of the load-bearing glass are retained.

The problem of friction was investigated and discussed in the already mentioned project "Vetrolignum" led by Prof. Vlatka Rajčić Ph.D. The results of the research are presented here.

2. Timber–glass Hybrid Elements: A Brief Literature Overview

When designing hybrid timber–glass structural systems, special attention should be paid to the appropriate type of connectors to use. The main goal when choosing a connector and type of connection is to avoid the concentration of stress, such as the local crushing of glass on the edges as well as the occurrence of tensile stresses in the glass, which can lead to sudden failure. The usual way to deal with this problem is through the use of soft coating materials on the metal connectors. An overview of possibly used systems for joining laminated glass structural components is contained in Stepinac et al.'s research [12]. Hamm [13] discussed possible solutions for connecting timber with glass in a composite structural component as well as variants of possible practical application in buildings. Niedermaier, in his research [14], presented the connection of timber frame and glass sheets in a hybrid structural panel using two types of adhesives. In the first case, the glass panels were glued to the timber frame with polyurethane and silicone adhesive. In the second case, the bonding was achieved with an epoxy adhesive. The results of the panel racking test are also presented. The test results show that the distribution of tensile stresses and strains depends on the type of adhesive as well as the geometry of the specimens that are tested. Wellersho et al. [15] presented methods to stabilize the building envelope using glazing. A hinged steel frame was used in which a stabilizing glass sheet was inserted in the first case. The second model used the same steel frame but the glass sheet, in this case, was glued with acrylate and polyurethane adhesives. Weller et al. [16], along with Mocibob [17], further continued to examine the behavior of structural components composed of glass sheets inserted in steel frames using different types of adhesives. It was concluded that the thickness of the glass panel is very important because it determines the lateral in-plane stiffness of the hybrid structural component. Different authors (Hochhauser et.al, Neubauer, and Winter et al.) [18–20] examined a hybrid panel in which the main timber frame is connected with a timber subframe by screws and the glass is glued to the subframe. An analysis of the in-plane loaded hybrid systems by using mechanical modeling was carried out and described by Cruz et al. in [21].

It was discussed that glass sheets significantly participate in the transfer of horizontal and vertical loads. Additionally, they participate in the prevention of excessive deformations and may substitute the usual type of bracings of steel and timber frames. Blyberg et al. [22], along with Nicklish et al. [23], presented the test results of timber-laminated glass panels where the connection was made by gluing. The authors present the characteristics of adhesives that may be used for structural bonds. A special focus on the analysis of failure mechanisms of timber–glass glued composite wall panels was presented by Ber et al. [24]. Amadio et al. [25] discussed the problem of glass panel buckling. It was analyzed using extended finite-element (FE) investigations and analytical methods for the effect of circumferential sealant joints and metal supporting frames. In [26] Štrukelj et al. presented results of the racking experimental tests of hybrid walls consisting of a timber frame and glass infill connected using polyurethane sealing. Ber et al. [27] used a parametric numerical analysis in their research. The racking stiffness of timber–glass walls is affected by different parameters, and their influence was reported in this study. In [28] Santarsiero et al. analyzed the potential use of glass in earthquake-prone areas as well as the lack of design codes and standards for the design of earthquake-resistant structures designed with glass. This paper concludes that during the design of the earthquake-resistant structures from glass it is necessary to ensure high ductility and dissipation capacity to glass components in buildings. In [29] Bedon et al. reported and demonstrated how the optimal design of glass components can be efficient and beneficial for multiple design configurations.

Special mechanical joints were introduced to enhance the dynamic performance of the glass façade. It was reported that well-designed fasteners can introduce additional flexibility and damping capacities when using hybrid panels in a strengthening traditional building. The first published paper with results from the design and analysis of the innovative laminated timber frames infilled by the laminated glass, which is the main subject of this paper, was presented at the WCTE 2012, or the World Conference on Timber Engineering 2012 [1]. The innovation in this hybrid element is the contact connection between the timber frame and the glass panels without an additional adhesive layer. The research started in 2007, and it was a collaborative research project between the University of Zagreb and the University of Ljubljana. Žarnić et al., in their research, followed the conclusions of the EU JRC ELSA Italy feasibility study [30]. The cooperation was established within the former CEN TC250/WG3 and the current TC250/SC11 and is still ongoing. Stepinac et al. recently introduced glued-in steel rods as a standard connector because of their wide use all around the world [31]. Since the innovative element showed very good performance, further cooperation on new parts of structural glass codes and the new parts of the timber structure design will be necessary to upgrade Eurocode 8 to introduce such a new type of hybrid structure for retrofitting and strengthening the existing structures made from various materials (masonry, concrete, etc.). Generally, Neugebauer et al. emphasized that emerging laminated materials and hybrid structures are not sufficiently covered in the Eurocodes [32].

3. Prototype of a Multifunctional Wood–Bearing Glass Composite System

The main objective of the research was to develop, design, and construct a new composite system that will be used as an independent prefabricated structural component for construction in seismically active areas. The solution, in this case, is simplicity, where the desired behavior of the system is achieved by friction, and therefore, without the use of mechanical connectors or adhesives. The purpose of the research was to design composites and construct details of joints that do not adversely affect the load-bearing glass and to develop systems with a high degree of energy dissipation exclusively by friction between wood and glass. Preliminary research shows that certain composite systems can be used in seismically active areas (such as Croatia) [2,3,10,33–35], but system optimization and parameter analysis have not yet been carried out.

In recent years, thanks to the collaboration of the University of Zagreb and the University of Ljubljana, preliminary testing of the wood–bearing glass composite system at monotonous static and cyclic loading has been carried out. The research and development of energy-efficient composite systems are planned in a three-year project entitled VETROLIGNUM (prototype of a multifunctional wood–bearing glass composite system) funded by the Croatian Science Foundation. This project will build on structure dimensioning knowledge and explore new ways to connect load-bearing elements and prepare a study on optimizing certain parts of the panel to maximize energy efficiency. Žarnić et al. [36] concluded that it is required to build a prototype of the wood–bearing glass composite system, which could be installed in a real building. Additionally, this type of hybrid element can be used as an independent element in the construction of wooden structures, as a temporary or permanent reinforcement, or to stabilize the elements of existing facilities and cultural heritage sites, and as an element for the construction of multi-purpose and adaptive façade systems (Figure 1).

One of the most important parameters when using the modal analysis is the horizontal stiffness of a building. Stiffness and mass determine the structure's vibration periods and hence the influence of an earthquake's frequency content on a structure's response. If a low-rise (only ground floor) building's vibration periods are overestimated (too long) the resulting seismic forces are too conservative. If the periods are underestimated (too short), the results are on the non-conservative side. The situation is just the opposite for higher buildings where the overestimated periods are on the non-conservative side and vice versa. Hence, great care must be taken in assigning the wall's correct stiffness. In the case of a timber–glass panel, the stiffness is predominantly dependent on the shear and hold-down behavior of joints in frame corners. However, glass infill greatly increases panel stiffness, whereby the influence of friction between timber and glass also needs to be considered. For the static calculation of frame building reinforced with a timber glass panel, the whole system could be replaced with only one element.

(a)   (b)

Figure 1. 3D—the prototype of a composite system: (**a**) timber frame; (**b**) timber—glass system.

The basic principle of panel installation is to connect the beams of timber frame with horizontal structural elements of the building. A connection derived by angular brackets has a higher bearing capacity than connections in frame angles (single glued-in rod). By such a solution horizontal force is transmitted directly to the panel without compromising the link between panel and frame structure. According to the analysis of the different types of frames with infill, such as concrete or steel frames with different types of infill (masonry infill as well as concrete, steel, and timber panels as infill), there are not many similarities to describe this system. However, certain similarities between the behavior of CLT panels and timber frame composite systems were found, where one of the important parameters is shear stiffness. The shear stiffness $k_{c,\,shear}$ of timber frame connections can be expressed with Equation (1) from [37]:

$$k_{c,shear} = 4 \cdot K_{c,shear} + \frac{0.6 \cdot q_{vert} \cdot L_{glass}^2 \cdot d_{glass} \cdot c}{u_{slip,Rd}} \tag{1}$$

where $K_{c,\,shear}$ is shear stiffness of a glued-in rod, $q_{vert,}$ is the vertical line load at the top of the panel, $u_{slip,\,Rd}$ is the slip of the weakest connector at the design strength, c is the dynamic friction coefficient of timber–glass contact, L_{glass} is the length of the glass panel, and d_{glass} is total glass panel thickness. In order to confirm this hypothesis, it is necessary to know each of the above parameters. Because there is a lack of data in the literature on the value of these stiffnesses, as well as friction coefficients, the main goal of further studies is to determine them experimentally [37].

Based on reverse-cyclic lateral loading tests on structural timber–glass panels [3], the data show a great way of spending seismic energy which contributes to the development of the forces of friction between wood and glass (Figure 2). The failure occurred in the timber frame corner, followed by the friction force between timber and glass, taking over a considerable amount of horizontal load, i.e., the seismic energy was dissipated through the sliding of glass on timber and activation of the joint in the corner of the timber frame.

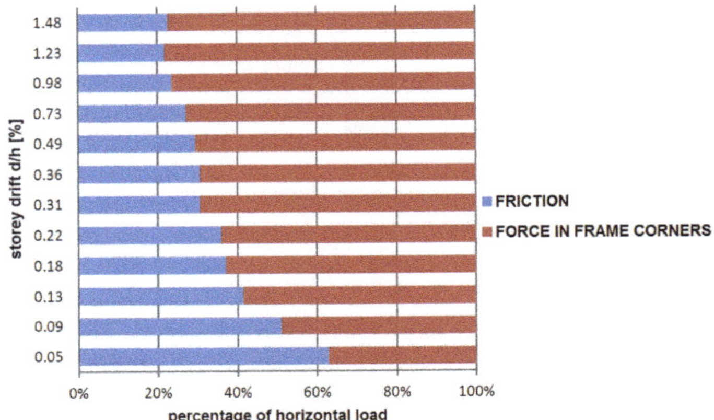

Figure 2. The relationship between the friction force and the bearing capacity of frame angles in horizontal loading (for a double-glazed panel). Reprinted with permission from Ref. [3]. Copyright 2015 Ph.D. thesis "Spojevi kompozitnih sustava drvo-nosivo staklo u potresnom okruženju" by Asst. Prof. Mislav Stepinac, Ph.D.

4. Materials and Methods

Examining the friction between wood and glass is crucial to understanding the operation of the entire timber–bearing glass composite system, in which the glass panel can slide in a wooden frame. It is the sliding, that is, the friction between glass and wood, that is one of the factors that transfers part of the horizontal load [10]. The set-up system consists of a single specimen of laminated or insulating glass embedded between two CLT elements. The positioning of the glass was achieved by making additional wooden slats that prevent the lateral displacement of the glass but do not press it laterally, and therefore, do not affect the force of friction. Based on the test, the friction force was determined at the wood–glass contact surface for a certain lateral pressure, i.e., the normal force). As a result, a coefficient of friction was obtained, which could be used to numerically model the contact between wood and glass in a calculation model. Numerical analysis was carried out with "Ansys" software support.

Description and Preparation of Specimens

The friction testing system consists of CLT elements, glass specimens, and steel elements (lateral force introduction). In order to optimize the system, glass specimens of various types and thicknesses were prepared.

Laminated safety glass is a "sandwich" of two or more glass surfaces that are glued together. "Glue" is a special transparent layer (PVB—polyvinyl butyral, EVA—ethylene vinyl acetate) with a thickness of 1–2 or sometimes more millimeters. In the event of glass breakage, shards and pieces of glass do not scatter but remain retained in the frame thanks to the plastic interlayer. This glass, too, absorbs wide-range sound vibrations and provides better sound insulation than float glass with the same thickness. It is most often single-laminated glass, which does not exclude the possibility of multiple laminating, i.e., joining several glass surfaces between which there is a transparent PVB (or some other) foil. Lamination is performed with PVB (polyvinyl butyral), EVA (ethylene vinyl acetate—transparent or opal), and TPU (thermoplastic polyurethane) foils. In our case, it was lamination with PVB foil.

Insulating glass consists of two or more glass panels that are interconnected at the edge (spacing 6 mm, 9 mm, 12 mm . . .). The connection allows for a flawless and long-lasting seal, and the interspace is filled with dry air or gas. The distance between the glass plates is provided by aluminum holders that are filled with drying agents. Insulating glass can be produced in combination with tempered or laminated glass. The properties of insulating glass are that it reduces heat exchange, reduces energy consumption, and does not allow drafts or condensation, so larger glass surfaces can be used for a given room temperature without increasing energy costs.

For this research, 21 glass specimens measuring 200 mm × 400 mm were prepared (Table 1). The specimens were as follows: 3× laminated glass 2 mm × 6 mm, 3× laminated glass 2 mm × 10 mm (Figure 3), 3× insulating (IZO) glass with double-laminated glazing of 6 mm and a cavity width of 12 mm (Figure 4), 3× insulating (IZO) glass with double-laminated glazing of 10 mm, and a cavity width of 12 mm, 3×—Laminated glass and wooden slat (2 mm × 6 mm) × 2, 3×—Laminated glass and wooden slat (2 mm × 10 mm) × 2, and 3× laminated glass 2 mm × 10 mm smooth ground edges.

Table 1. Specimens.

Specimen Type	Dimensions (mm)	Edge Processing Acc. To DIN 1249-11	Number of Specimens
Laminated glass—2 mm × 6 mm	200 × 400	Bordered edge	3
Laminated glass—2 mm × 10 mm	200 × 400	Bordered edge	3
Insulated (IZO) glass—4 mm × 6 mm	200 × 400	Bordered edge	3
Insulated (IZO) glass—4 mm × 10 mm	200 × 400	Bordered edge	3
(2 mm × 6 mm) × 2—Laminated glass and wooden slat	200 × 400	Bordered edge	3
(2 mm × 10 mm) × 2—Laminated glass and wooden slat	200 × 400	Bordered edge	3
Laminated glass—2 mm × 10 mm smooth ground edges	200 × 400	Smooth ground edge	3

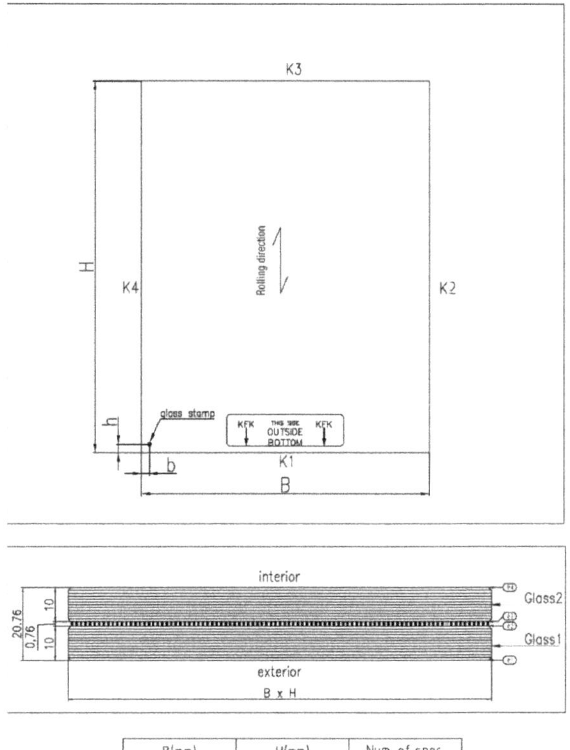

B (mm)	H (mm)	Num. of spec.
200	400	3

Figure 3. Laminated glass panel.

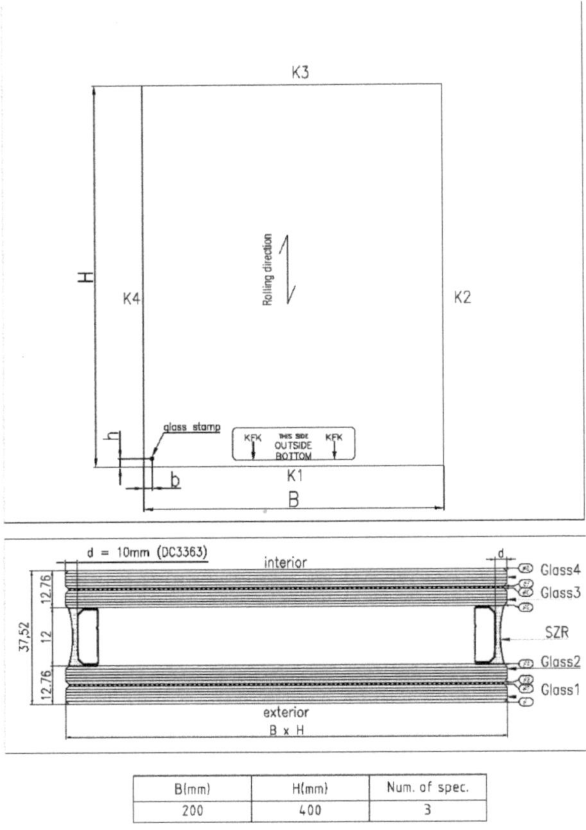

B(mm)	H(mm)	Num. of spec.
200	400	3

Figure 4. Insulating glass panel.

All specimens were ESG-toughened glass according to the EN 12150-1 standard [38]. The manufacturing tolerance was within the permissible limits according to the EN 14179-8 standard [39]. The edges of the specimens were roughly sanded [40]. Laminated glass panes were bonded with a 0.76 mm thick PVB membrane. A total of 90% of the cavity in insulating glass was filled with argon. The spacer was 12 mm wide and made of aluminum with respective DC 3363 butyl and silicone layers. Mechanical characteristics of the glass panels can be seen in Table 2.

Table 2. Mechanical characteristics of glass.

Properties	Value
E—Young's elasticity modulus	70,000 N/mm^2
G—Shear modulus	28.689 N/mm^2
μ—Poisson's ratio	0.22
α—thermal expansion coefficient	8.8×10^{-6}
ρ—density	2.5 g/cm^3
Compressive strength	700–1000 N/mm^2
Tensile strength	30–45 N/mm^2

The timber CLT elements (Figure 5) were processed in the structural testing laboratory of the Faculty of Civil Engineering, University of Zagreb. The CLT consisted of 3 layers, and each layer was 30 mm thick. The timber was class CL24h according to [41]. The additional wooden beams that support the glass were 30 mm × 30 mm. The material and mechanical properties (acc. to [42,43]) of CL24h timber are shown in Tables 3 and 4.

(a)   (b)

Figure 5. CLT specimen: (**a**) wooden slats; (**b**) assembled CLT sample.

Table 3. Material properties of CL24h timber.

Properties	Index	Value
Density	ρ	420 kg/m^3
Young's modulus of elasticity	E_x	11,000 Mpa
	$E_y E_z$	600 Mpa
		580 Mpa
Shear modulus	G_{xy}	600 Mpa
	G_{xz}	690 Mpa
	G_{yz}	580 Mpa
Poisson's ratio	ν_{xy}	0.3
	ν_{xz}	0.25
	ν_{yz}	0.6

Table 4. Mechanical properties of CL24h timber.

Strength	Index	Value
Bending	$f_{m,k}$	24 Mpa
Tension (parallel to the grain)	$f_{t,0,k}$	14 Mpa
Tension (perpendicular to the grain)	$f_{t,90,k}$	0.5 Mpa
Compression (parallel to the grain)	$f_{t,0,k}$	21 Mpa
Tension (perpendicular to the grain)	$f_{t,90,k}$	2.5 Mpa
Shear	$f_{v,k}$	2.5 Mpa

5. Experimental Work

The experiment was carried out by inserting a laminated or insulating glass specimen between two wooden elements. Before starting the experiment and introducing the vertical force, i.e., the force acting in line with the glass pane, it was necessary to secure certain lateral pressure between the glass and the wood. This way we could directly determine the contact point between wood and glass. The lateral force introduction system consists of six steel plates, four threaded rods with nuts, and four springs (Figure 6).

(a)   (b)

Figure 6. Specimen set-up: (**a**) laboratory; (**b**) 3D model.

The system, dimensions, and positions of the elements are shown in Figure 7.

The introduction of lateral pressure of the desired amount (1 kN, 2 kN, or 3 kN) was achieved over a certain amount of spring displacement. Springs were positioned between metal plates. The spring displacement itself was achieved through the displacement of the metal plates that push the springs, that is, by controlled tightening and releasing of the nuts on the threaded rod. Such a system allows constant lateral pressure. In order to determine and control the lateral force that was introduced, a preliminary test was conducted to determine the spring stiffness, i.e., to obtain a force-displacement diagram. The diagram represents the force-displacement ratio for all four springs. The spring stiffness can be seen in Figure 8. The stiffness of one spring was determined in such a way that 25% of the amount of force in the diagram was read. The advantage of this system is its simplicity and accuracy. The distance between the two metal plates, that is, the length of the spring, determines the lateral force. The distance was controlled using a sliding caliper with an expanded measurement uncertainty of 20 μm. The simplicity was manifested in the ability to make spacers in desired dimensions that we could place between the two metal plates and then tighten the bolts.

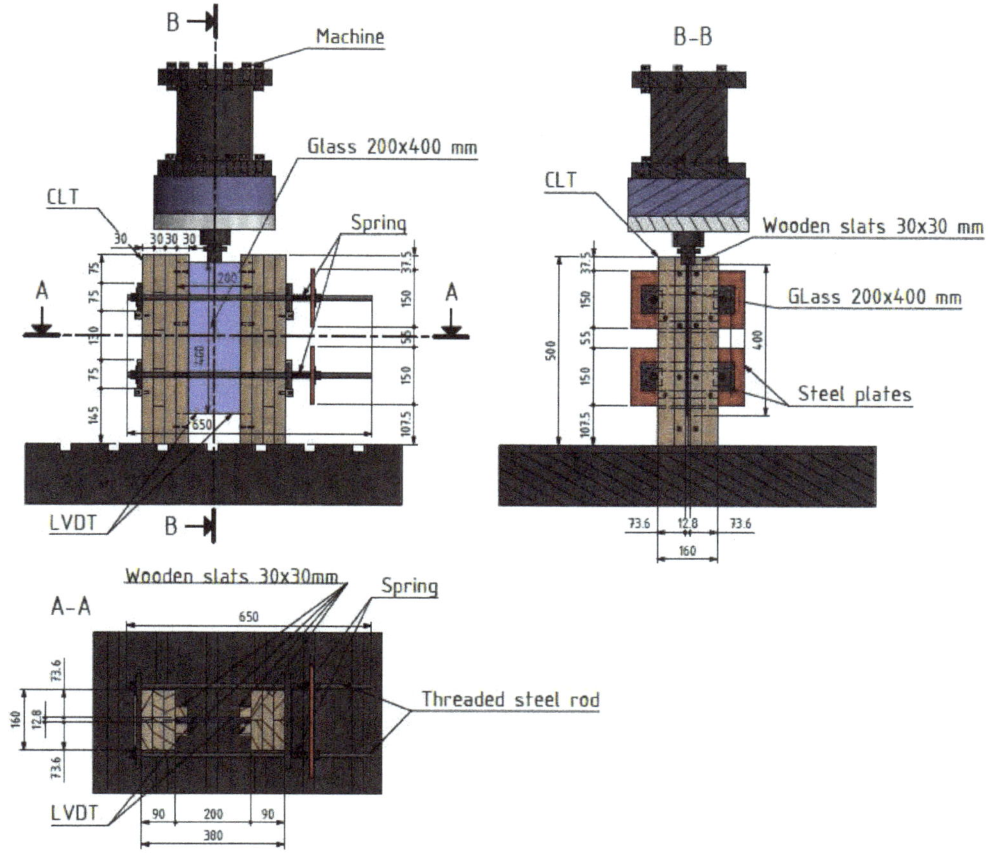

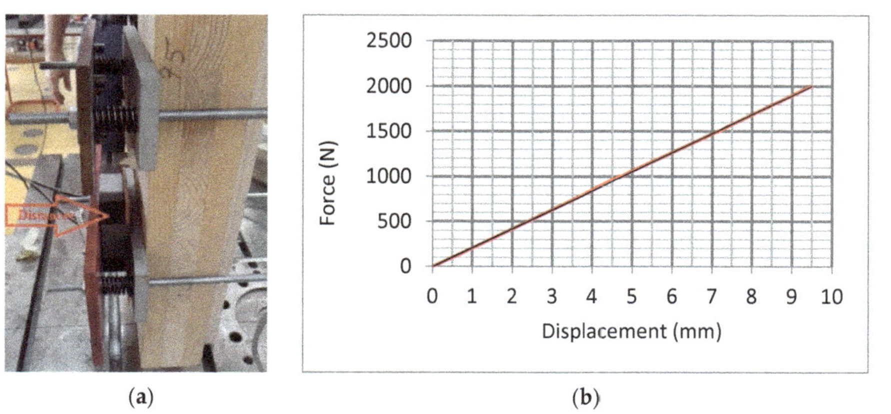

Figure 7. Dimension and positions of the test set-up.

(**a**)

(**b**)

Figure 8. (**a**) Spacer position; (**b**) spring stiffness.

After achieving the desired lateral force (F_n) and centering the specimen, the force was introduced to the glass panel. In order to prevent direct contact between the press (steel) and the glass, a thin rubber washer was placed on the edge of the glass, i.e., at the point of load introduction. The load was applied using a universal electromechanical Zwick/Roell testing machine equipped with force sensor class 0.5 in the range from 1 kN to 50 kN according to EN ISO 7500-1:2018 [44] and displacement sensor class 1 according to EN ISO 9513:2012 [45]. The load was applied by displacement control at a speed of 1 mm/min.

The specimen differed in the thickness and type of glass elements (Table 1). Eighteen samples had rough edges, while three samples had smooth ground edges. The sample with smooth ground edges was tested subsequently to see the impact of the glass treatment itself. Each of the samples was tested with a lateral compressive load of 1 kN, 2 kN and 3 kN. Figure 7 schematically shows the dimensions of the sample and the place of load input. During the experiment, the relative displacement of the glass panels was measured, regarding the fixed CLT elements. Displacement was measured using two LVDTs (Figure 6a) with an expanded measurement uncertainty of 5 µm. The load on the glass panel tangentially to the contact between CLT and glass was measured for a certain normal force F_n. In all experiments, unloading (and then re-loading) was performed in order to eliminate local defects, irregularities, and gaps in the timber material, until the samples fit on the machine surface perfectly, in order to avoid the noise in the data results. The result can be graphically represented as a ratio between the friction force F_t and the longitudinal displacement at a certain normal force (F_n), as shown in Figures 9–15. The friction force (F_t) is half of the force F required to move the glass panels. The coefficient of friction µ was obtained as the ratio of normal (lateral) force (F_n) and frictional force (F_t).

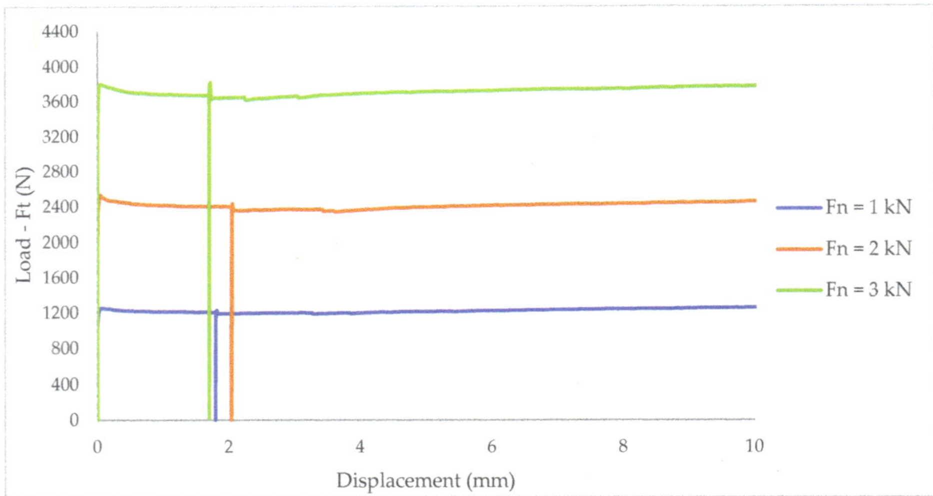

Figure 9. Laminated glass 2 mm × 6 mm.

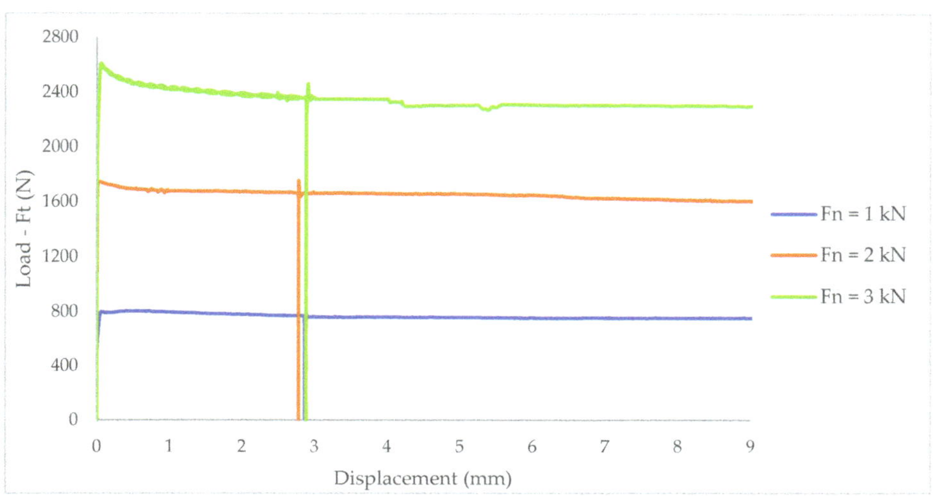

Figure 10. Laminated glass 2 mm × 10 mm.

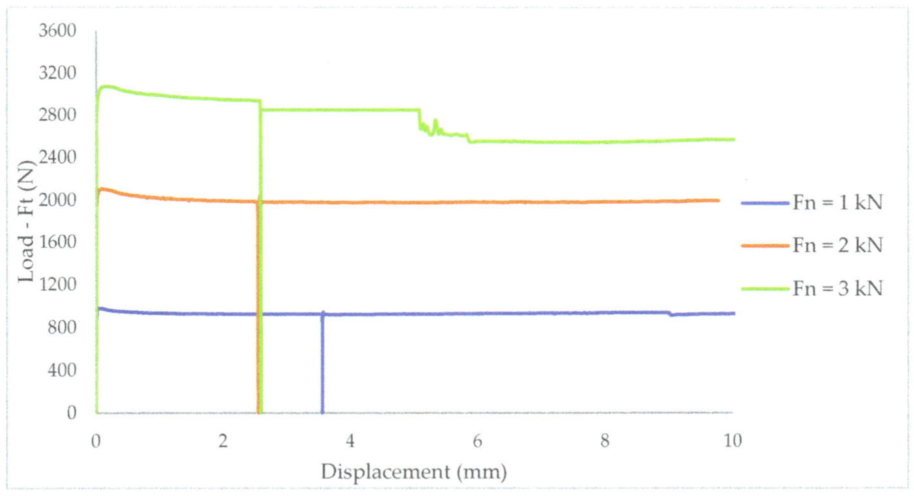

Figure 11. IZO glass 2 mm × 10 mm.

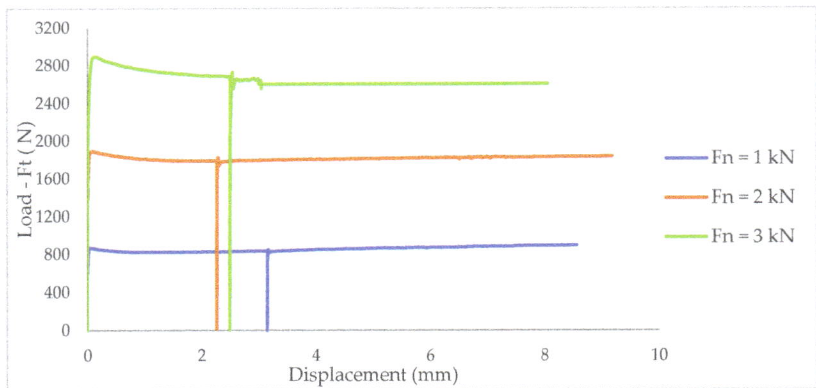

Figure 12. IZO glass 4 mm × 10 mm.

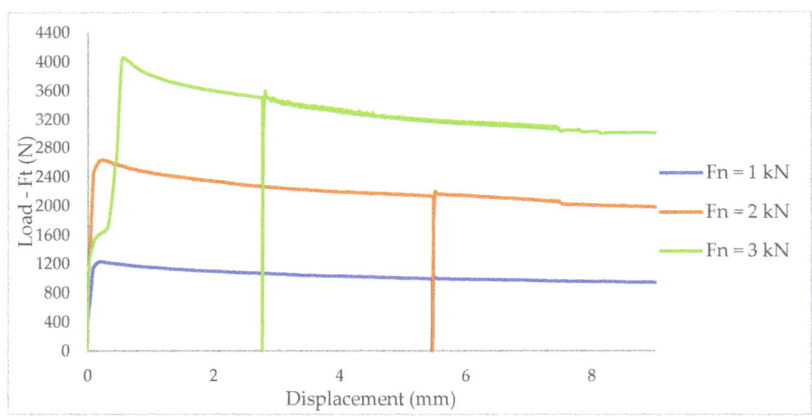

Figure 13. (2 mm × 6 mm) ×2—Laminated glass and wooden slat.

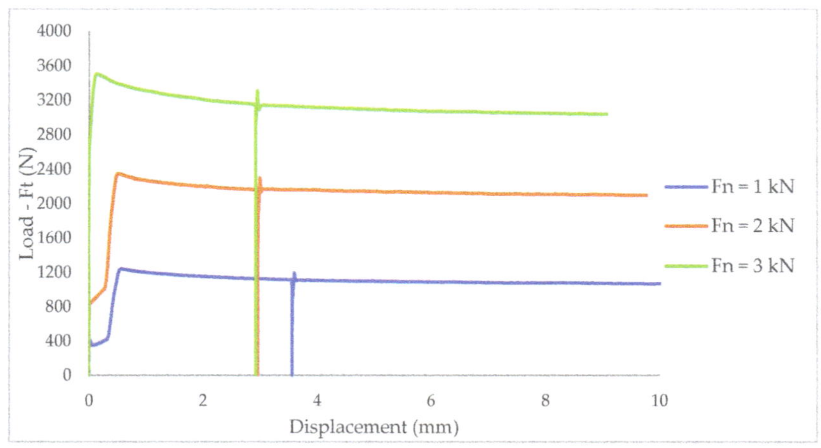

Figure 14. (2 mm × 10 mm) ×2—Laminated glass and wooden slat.

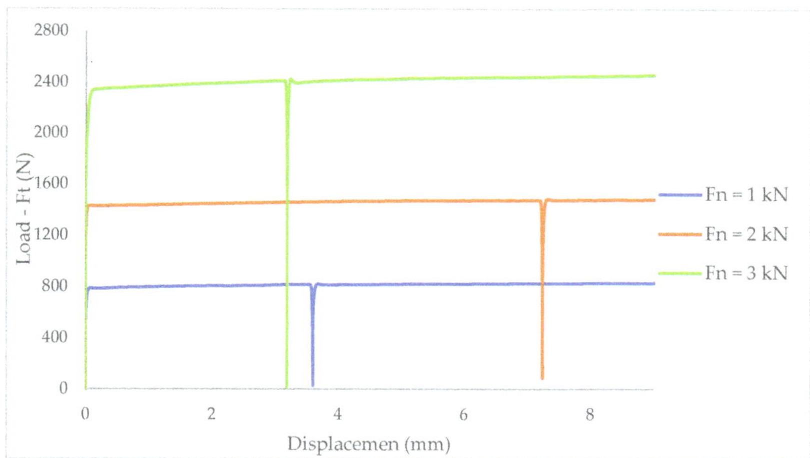

Figure 15. Laminated glass 2 mm × 10 mm—smooth ground edges.

6. FEM Research

The experimental studies carried out were accompanied by numerical analyses. The numerical analyses aimed to extend the knowledge of the behavior of the experimental research. Furthermore, the numerical simulations were performed to confirm and complement the experimental results. The analysis was conducted by Ansys software [46,47]. The entire geometry of the model was drawn by the software "Autodesk Inventor" and imported into "Ansys", where a finite element mesh was formed and in which further simulations were carried out (Figure 16). Element geometry, boundary conditions, loads as well as material characteristics were defined following the experiment.

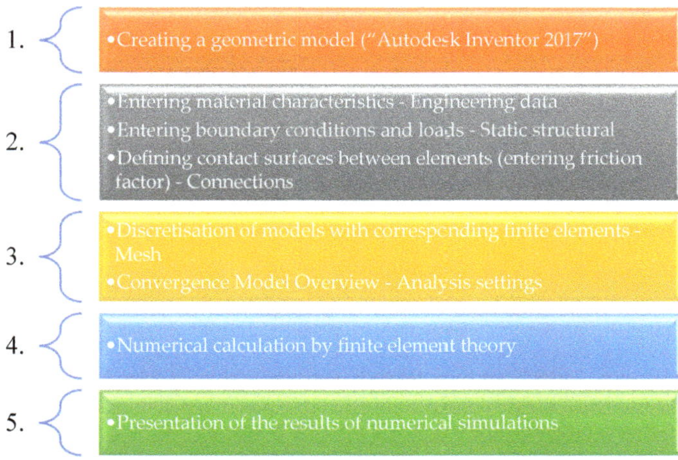

Figure 16. Schematic of numerical analysis procedures.

The model itself is composed of three different materials, namely CLT, glass, and PVB. Boundary conditions and lateral pressure were defined as can be seen in Figure 17.

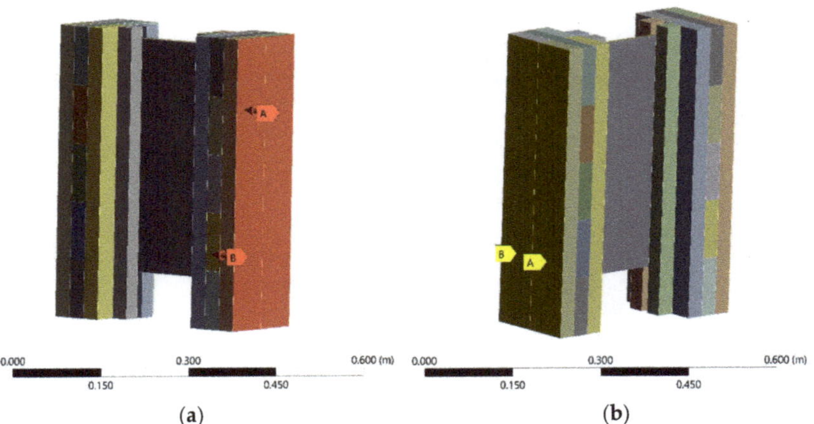

(a)   (b)

Figure 17. Numerical model: (**a**) Normal force (lateral pressure-1 kN, 2 kN or 3 kN); (**b**) displacement (0 mm-fixed).

To discretize the model, the following Ansys mesh tools [48] were used: "edge sizing" and "sphere of influence". The methods MultiZone (allows the creation of models with a denser grid on the contacts) and Hex Dominant (allows the creation of models where the finite element mesh consists mostly of hexahedrons). Both methods were used to model the glass element (Figure 18).

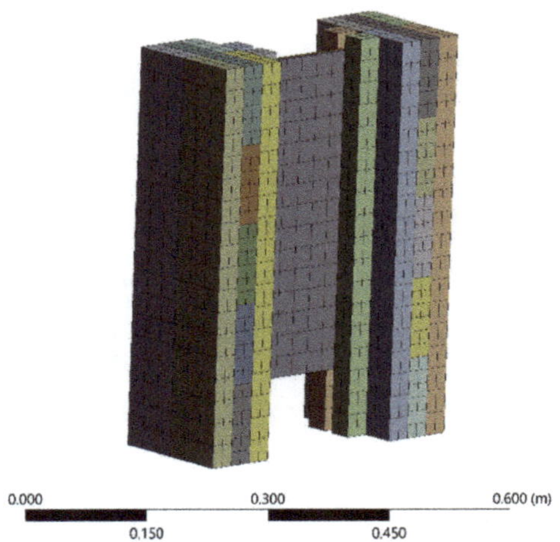

Figure 18. Finite element mesh.

For modeling contact surfaces, absolute stiffness for normal stresses was used, as well as the possibility of the tangential sliding of two surfaces (CLT and glass) with the corresponding coefficient of friction (Figure 19).

The load was introduced by displacement of the glass panel, according to the steps and data obtained from the laboratory. The result of the experiment, i.e., numerical analysis, is the friction stress that occurs on the contact surface.

FEM Results

The numerical analysis aimed to obtain a model and certain behavior legality, which would help the prediction of the behavior of such a system during a seismic event. The main parameter for the control and comparison of numerical simulations and experimental work is the frictional stress that occurs on the contact surfaces. The result obtained from the laboratory was the frictional force required to shift the glass element. In order to compare and evaluate the results of the FEM analysis, the frictional stresses occurring at the contact surfaces were calculated manually, based on the frictional force obtained from the conducted laboratory test. The friction force F_t is expressed as half the force F required for moving the glass element, as the frictional force occurs on the two surfaces where the glass and the timber connect. In addition to the results in the form of frictional stresses, the behavior of the sample (sliding) was obtained by numerical simulations as shown in Figure 20.

Figure 19. Defining contact surfaces.

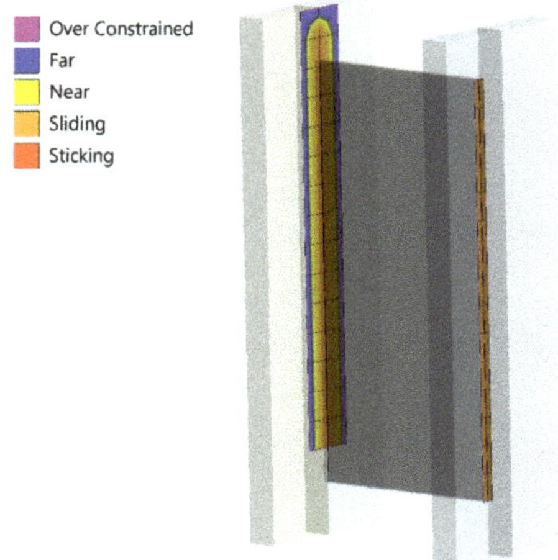

Figure 20. Sample behavior.

For the 2 mm × 6 mm laminated glass sample with a lateral force of 2 kN, the mean frictional stress calculated from the experiment data was 0.25 MPa, while the mean frictional stress obtained by numerical simulations was also 0.25 MPa (Figure 21a). The results of the numerical analysis in form of frictional stress can be seen in Figure 21a. For the same type of specimen, but with a lateral force of 3 kN, the maximum frictional stress was 0.38 MPa, and the same value was obtained by numerical simulation (Figure 21b).

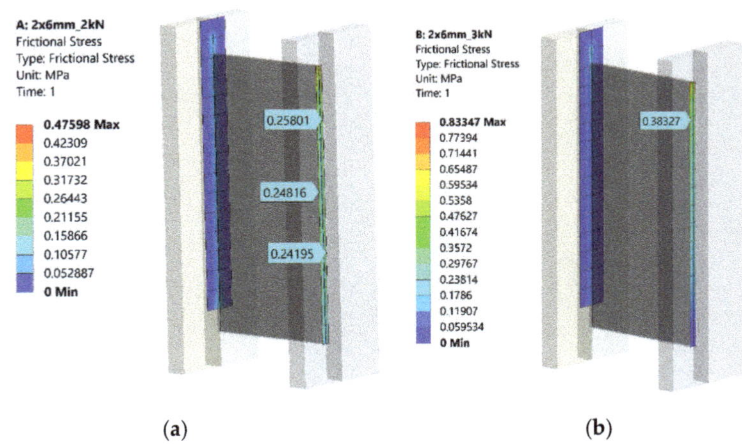

(**a**)   (**b**)

Figure 21. Stresses on contact surfaces of laminated glass 2 mm × 6 mm: (**a**) lateral load 2 kN; (**b**) lateral load 3 kN.

In order to confirm the FEM analysis, other types of specimens were subjected to numerical modeling as follows; for the laminated glass specimen 2 mm × 10 mm and a lateral force of 2 kN, the mean frictional stress calculated from the experiment data was 0.1 MPa, while the mean frictional stress obtained by numerical analysis was also 0.1 MPa. The results of the numerical analysis can be seen in Figure 22. For the same specimen, but with a lateral force of 3 kN, the mean frictional stress obtained by the experiment was 0.155 MPa, and the same value of frictional stress (0.155 MPa) was obtained by numerical simulation in Ansys.

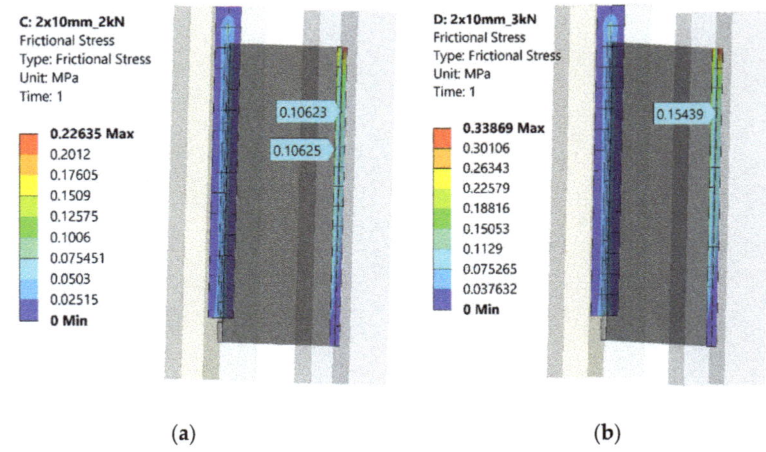

(**a**)   (**b**)

Figure 22. Stresses on contact surfaces of laminated glass 2 mm × 10 mm: (**a**) lateral load 2 kN; (**b**) lateral load 3 kN.

For all specimen types (lateral force of 2 kN), a comparative analysis is presented. Frictional stresses calculated in Ansys were compared with those obtained from the experimental tests. The maximum deviation in the results was 3.3% (Table 5). Furthermore, at different values of lateral pressure on specimens, an analogy can be established, which confirms the validity of the FEM analysis.

Table 5. Comparison of normal stresses between experimental tests and ANSYS.

Specimen Type	Frictional Stress—Lateral Load 2 kN (N/mm²)		Deviation (%)
	Experimental Work	ANSYS	
Laminated glass—2 mm × 6 mm	0.25	0.25	0
Laminated glass—2 mm × 10 mm	0.10	0.10	0
IZO glass—4 mm × 6 mm	0.07	0.072	2.8
IZO glass—4 mm × 10 mm	0.04	0.041	2.5
(2 mm × 6 mm) × 2—Laminated glass and wooden slat	0.07	0.072	2.8
(2 mm × 10 mm) × 2—Laminated glass and wooden slat	0.05	0.051	2
Laminated glass—2 mm × 10 mm- smooth ground edges	0.09	0.093	3.3

7. Discussion

A comparison of the results of all samples is shown in Figures 23 and 24.

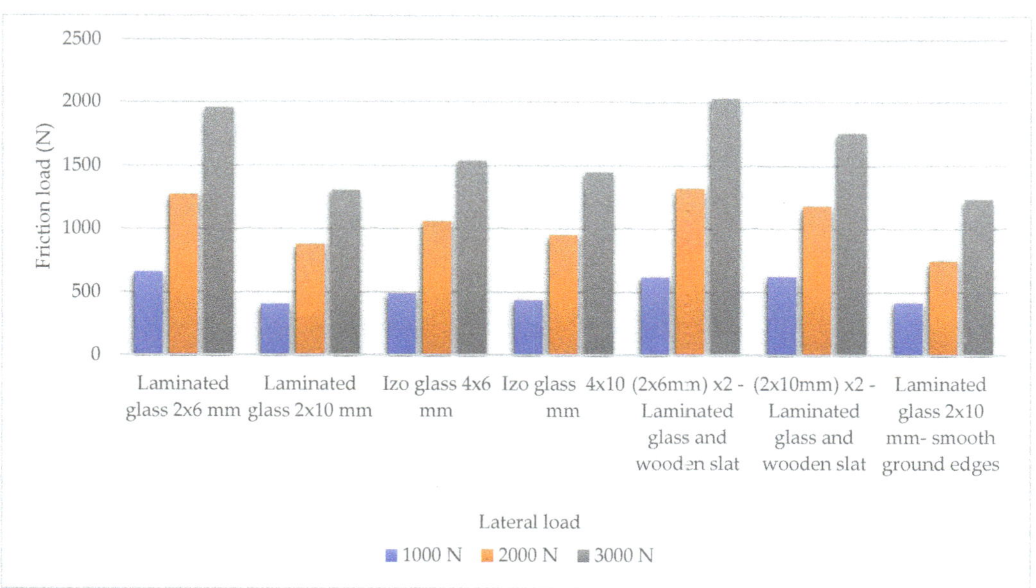

Figure 23. Friction load—comparison.

Based on the presented charts, it was concluded that the friction force increases linearly with increasing lateral force, as expected. Once the legality of the behavior has been determined, a coefficient of friction can be determined for each of the samples. However, to achieve the ultimate goal of the research, it is necessary to highlight and discuss the following findings related to the global behavior of the final product.

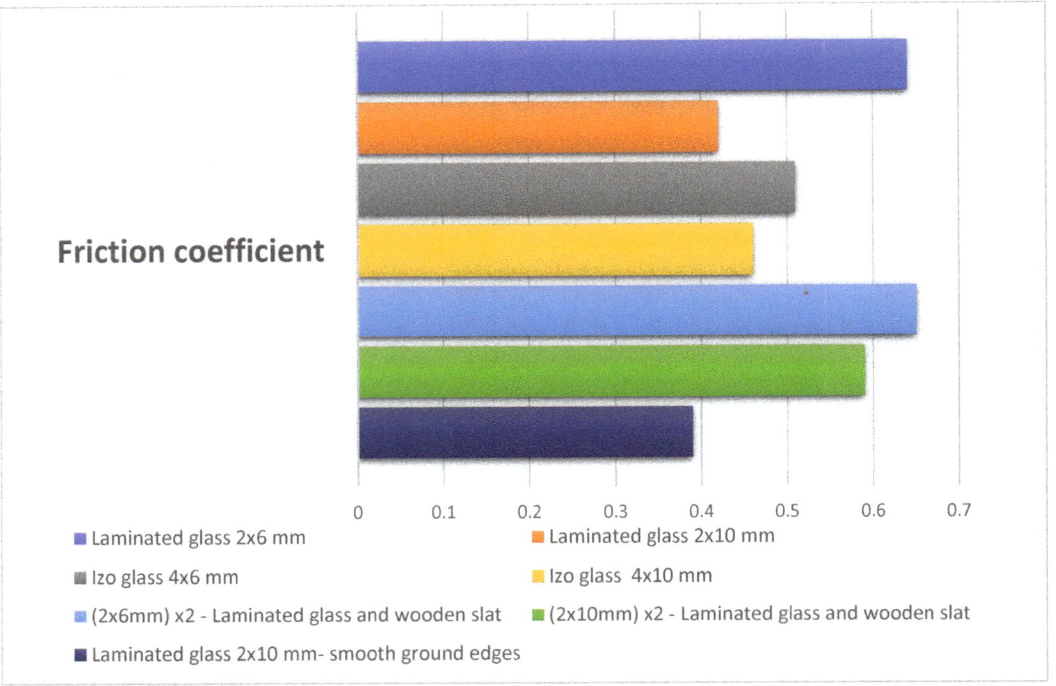

Figure 24. Friction coefficient—comparison.

The previous research [36] showed that the influence of glass infills on the lateral load-bearing capacity is significant. Recommendations for further research should be based on the following facts, taking into account the friction between the timber and the glass:

- Due to the vertical support of the timber frame lintel enabled by glass infill, frame joints are loaded in pure shear for which they have the biggest load-bearing capacity [36].
- Vertical load positively influences the lateral strength of the specimens, by 40%, due to the activation of friction between frame lintels and glass sheets.
- The number of glass sheets (single vs. double glazing) does not influence the lateral strength. The reason is that the friction force acting along the horizontal edges of the glass panel is almost the same.
- The intensity of vertical load influences strength degradation. In the case of specimens with low vertical load, the strength degradation was on average twice as high as in the cases of specimens with a high vertical load. The stiffness degradation was not influenced either by the intensity of vertical load or by the number of glazing panels.

It is possible to formulate this phenomenon with a common equation, which is needed for the definition of the future mathematical model of the tested type of structural hybrid panel components.

Energy dissipation is possible through friction and ductility of the timber frame angle joints. Ductility of the joints in timber structures is a prerequisite, especially in the seismic zones.

8. Conclusions

Insight into the existing literature and the current state of the art reveals a gap in the study of composite systems with load-bearing glass, especially on loads of horizontal forces of variable amounts and directions that occur during seismic loading. In the range of larger story drifts, the effect of glass-to-timber friction plays a major role in energy dissipation.

During horizontal loading, friction between glass and timber is a factor that affects the behavior of the timber—load-bearing composite system. Coefficients of friction were determined for CLT on glass surfaces; in particular, the effects of different lateral pressure levels were investigated. Friction depends on the way the elements (especially glass) are processed, as well as on the load introduced into the system. The difference between the coefficient of friction at rough and smooth ground edges is negligible. There are differences in the coefficient of friction when insulating glass or glass with wooden slats is installed instead of laminated glass, but it is not significant. The reason lies in the fact that samples with wooden slats have a higher friction surface, and in addition, do not act as a singular system, as is the case of insulated glass.

The investigation provided the necessary data for the development of design procedures and computational model design guidance for the new design codes.

In the future, glass elements with polished edges could be investigated, thus expanding knowledge about the behavior and interaction of these two materials. During load transfer of such a composite system, the contact surface on the wooden element changes and "disappears" over time. Furthermore, future considerations should include how atmospheric factors affect changes in wood surfaces (swelling and shrinkage) and the eventual deterioration of the wood surface, which would cause changes in the contact zone between the two materials, and consequently friction between them. Analysis and research of changes in the coefficient of friction over time and at cyclic loading would be of great importance.

Obtaining realistic values of friction coefficients for different types of glass elements is extremely important for numerical simulations. The use of extreme and theoretical values of friction coefficients in numerical simulations often does not represent a real situation and can lead to wrong conclusions and misinterpretation of results. This research emphasized that the effects of friction should not be neglected. Consequently, neglecting the effects of friction does not unavoidably produce a more conservative design situation by magnifying the stresses. Experimental tests have been confirmed by numerical simulations, but there is the possibility for a more detailed analysis of the system. The numerical analysis should be extended to the whole composite framework and realistic conditions, and thus evaluate all components and factors involved in load transfer

Author Contributions: Conceptualization, N.P. and D.D.; methodology, N.P., V.R. and D.D.; software, N.P.; validation, V.R., N.P., D.D. and J.B.; formal analysis, N.P.; investigation, V.R. and N.P.; resources, N.P.; data curation, N.P.; writing—original draft preparation, N.P., V.R., D.D. and J.B.; writing—review and editing, N.P., V.R., D.D. and J.B.; visualization, N.P.; supervision, V.R. and D.D.; project administration, V.R.; funding acquisition, V.R. All authors have read and agreed to the published version of the manuscript.

Funding: The herein reported research work was funded by The Croatian Science Foundation-Project no.: IP-2016-06-3811 "VETROLIGNUM: Prototype of multipurpose timber-structural glass composite panel".

Institutional Review Board Statement: Not applicable.

Informed Consent Statement: Not applicable.

Data Availability Statement: Data available on request due to restrictions, e.g., privacy or ethics. The data presented in this study are available on request from the corresponding author.

Conflicts of Interest: The authors declare no conflict of interest. The funders had no role in the design of the study; in the collection, analyses, or interpretation of data; in the writing of the manuscript; or in the decision to publish the results.

References

1. Rajcic, V.; Zarnic, R. Racking Performance of Wood-Framed Glass Panels. In Proceedings of the World Conference on Timber Engineering, Auckland, New Zealand, 15–19 July 2012; pp. 56–57.
2. Rajcic, V.; Zarnic, R. Seismic Response of Timber Frames with Laminated Glass Infill. In *CIB-W18/45-15-4, Proceedings of the 45th CIB-W18 Meeting, Växjo, Sweden, 27–30 August 2012*; Ingenieurholzbau und Baukonstruktionen Karlsruhe Institute of Technology Germany: Växjo, Sweden, 2012; p. 11.
3. Stepinac, M. Spojevi Kompozitnih Sustava Drvo-Nosivo Staklo u Potresnom Okruženju. Ph.D. Thesis, University of Zagreb, Zagreb, Croatia, 2015.
4. CNR-DT 210/2013; Guide for the Design, Construction and Control of Buildings with Structural Glass Elements. CNR-Advisory Committee on Technical Recommendations for Construction: Roma, Italy, 2013.
5. CEN-EN 572-1; Glass in Building—Basic Soda-Lime Silicate Glass Products—Part 1: Definitions and General Physical and Mechanical Properties. CEN: Brussels, Belgium, 2012.
6. EN 14449:2005; Glass in Building—Laminated Glass and Laminated Safety Glass—Evaluation of Conformity/Product Standard. CEN: Brussels, Belgium, 2005.
7. Feldmann, M.; Kasper, R.; Abeln, B.; Cruz, P.; Belis, J.; Beyer, J.; Colvin, F.E.; Eliášová, M.; Galuppi, L.; Geßler, A.; et al. *Guidance for European Structural Design of Glass Components; Support to the Implementation, Harmonization and Further Development of the Eurocodes*; Publications Office of the European Union: Luxembourg, 2014.
8. Antolinc, D.; Žarnic, R.; Cepon, F.; Rajcic, V.; Stepinac, M. Laminated Glass Panels in Combination with Timber Frame as a Shear Wall in Earthquake Resistant Building Design. In Proceedings of the Challenging Glass 3: Conference on Architectural and Structural Applications of Glass, Delft, The Netherlands, 28–29 June 2012; pp. 623–631.
9. Pascha, V.S. Architectural Application and Ecological Impact Studies of Timber-Glass Composites Structures. In Proceedings of the Engineered Transparency International Conference at Glasstec, Düsseldorf, Germany, 21–22 October 2014.
10. Antolinc, D. Uporaba Steklenih Panelov Za Potresno Varno Gradnjo Objektov: Doktorska Disertacija. Ph.D. Thesis, Univerza v Ljubljani, Ljubljana, Slovenia, 2014.
11. Antolinc, D.; Rajčić, V.; Žarnić, R. Analysis of Hysteretic Response of Glass Infilled Wooden Frames. *Vilnius Gedim. Tech. Univ.* **2014**, *20*, 600–608. [CrossRef]
12. Stepinac, M.; Rajčić, V.; Žarnić, R. Timber-Structural Glass Composite Systems in Earthquake Environment. *Gradjevinar* **2016**, *68*, 211–219.
13. Hamm, J. Development of Timber-Glass Prefabricated Structural Elements. *Proc. IABSE Conf. Innov. Wooden Struct. Bridges* **2001**, *85*, 119–124.
14. Niedermaier, P. Shear-Strength of Glass Panel Elements in Combination with Timber Frame Constructions. In Proceedings of the 8th International Conference on Architectural and Automotive Glass (GPD), Tampere, Finland, 15–18 June 2003; pp. 262–264.
15. Wellersho, F.; Sedlacek, G. Stabilization of Building Envelopes with the Use of the Glazing. In Proceedings of the Glass Processing Days, Tampere, Finland, 17–20 June 2005; pp. 281–283.
16. Weller, B.; Schadow, T. Designing of Bonded Joints in Glass Structures. In Proceedings of the 10th International Conference on Architectural and Automotive Glass (GPD), Tampere, Finland, 15–18 June 2007.
17. Mocibob, D. Glass Panel under Shear Loading-Use of Glass Envelopes in Building Stabilization. Ph.D. Thesis, Ecole Polytechnique Fédérale de Lausanne, EPFL, Lausanne, Switzerland, 2008.
18. Hochhauser, W. A Contribution to the Calculation and Sizing of Glued and Embedded Timber-Glass Composite Panes. Ph.D. Thesis, Vienna University of Technology, Vienna, Austria, 2011.
19. Neubauer, G. Holz-Glas-Verbundkonstruktionen. Weiterentwicklung und Herstellung von Holz-Glas-Verbundkonstruktionen Durch Statisch Wirksames Verkleben von Holz und Glas Zum Praxiseinsatz im Holzhausbau. Available online: https://glassolutions.at/sites/glassolutions.eu/files/2017-12/hgv_holzforschung_austria_als_gebaeudeaussteifung.pdf (accessed on 6 December 2021).
20. Winter, W.; Hochauser, W.; Kreher, K. Load Bearing and Stiening Timber-Glass Composites. In Proceedings of the World Conference on Timber Engineering, Trento, Italy, 20–24 June 2010.
21. Cruz, P.J.S.; Pequeno, J.; Lebet, J.-P.; Mocibob, D. Mechanical Modelling of In-Plane Loaded Glass Panes. In Proceedings of the Challenging Glass 2—Conference on Architectural and Structural Applications of Glass, Delft, The Netherlands, 20–21 May 2010; pp. 309–318.
22. Blyberg, L.; Serrano, E.; Enquist, B.; Sterley, M. Adhesive Joints for Structural Timber/Glass Applications: Experimental Testing and Evaluation Methods. *Int. J. Adhes. Adhes.* **2012**, *35*, 76–87. [CrossRef]
23. Nicklisch, F.; Dorn, M.; Weller, B.; Serrano, E. Joint Study on Material Properties of Adhesives to Be Used in Load-Bearing Timber-Glass Composite Elements. In Proceedings of the Engineered Transparency—International Conference at Glasstec, Düsseldorf, Germany, 21–22 October 2014.
24. Ber, B.; Šušteršič, I.; Premrov, M.; Štrukelj, A.; Dujič, B. Testing of Timber–Glass Composite Walls. In *Proceedings of the Institution of Civil Engineers-Structures and Buildings*; Thomas Telford Services: London, UK, 2015; Volume 168, pp. 500–513. Available online https://www.icevirtuallibrary.com/doi/10.1680/stbu.13.00105 (accessed on 7 January 2022).
25. Amadio, C.; Bedon, C. Effect of Circumferential Sealant Joints and Metal Supporting Frames on the Buckling Behavior of Glass Panels Subjected to In-Plane Shear Loads. *Glass Struct. Eng.* **2016**, *1*, 353–373. [CrossRef]

26. Štrukelj, A.; Ber, B.; Premrov, M. Racking Resistance of Timber-Glass Wall Elements Using Different Types of Adhesives. *Constr. Build. Mater.* **2015**, *93*, 130–143. [CrossRef]
27. Ber, B.; Finžgar, G.; Premrov, M.; Štrukelj, A. On Parameters Affecting the Racking Stiffness of Timber-Glass Walls. *Glass Struct. Eng.* **2018**, *4*, 69–82. [CrossRef]
28. Santarsiero, M.; Bedon, C.; Moupagitsoglou, K. Energy-Based Considerations for the Seismic Design of Ductile and Dissipative Glass Frames. *Soil Dyn. Earthq. Eng.* **2019**, *125*, 105710. [CrossRef]
29. Bedon, C.; Amadio, C. Numerical Assessment of Vibration Control Systems for Multi-Hazard Design and Mitigation of Glass Curtain Walls. *J. Build. Eng.* **2018**, *15*, 1–13. [CrossRef]
30. Žarnić, R.; Tsionis, G.; Gutierrez, E.; Pinto, A.; Geradin, M.; Dimova, S. Purpose and Justification for New Design Standards Regarding the Use of Glass Products in Civil Engineering Works-Publications Office of the EU. Available online: https://op.europa.eu/hr/publication-detail/-/publication/9ae98636-ec59-4560-8e14-a650a2fec188 (accessed on 7 December 2021).
31. Stepinac, M.; Rajcic, V.; Tomasi, R.; Hunger, F.; van de Kuilen, J.-W.G.; Serrano, E. Comparison of Design Rules for Glued-in Rods and Design Rule Proposal for Implementation in European Standards. In Proceedings of the CIB-W18 Timber Structures, Meeting 46, Vancouver, BC, Canada, 26–29 August 2013.
32. Neugebauer, J.; Larcher, M.; Rajcic, V.; Zarnic, R. Activity Report of COST Action TU0905 Working Group 1-Predicting Complex Loads on Glass Structures. In Proceedings of the Challenging Glass 4 & COST Action TU0905 Final Conference, Lausanne, Switzerland, 6–7 February 2014; pp. 11–20.
33. Bedon, C.; Amadio, C. Flexural–Torsional Buckling: Experimental Analysis of Laminated Glass Elements. *Eng. Struct.* **2014**, *73*, 85–99. [CrossRef]
34. *EN 1998-1*; Eurocode 8: Design of Structures for Earthquake Resistance–Part 1: General Rules, Seismic Actions and Rules for Buildings. CEN: Brussels, Belgium, 2004.
35. Krstevska, L.; Tashkov, L.; Rajcic, V.; Zarnic, R. Shaking Table Test of Innovative Composite Panel Composed of Glued Laminated Wood and Bearing Glass. In Proceedings of the 15th World Conference on Earthquake Engineering, Lisbon, Portugal, 24–28 September 2012; pp. 1–10.
36. Žarnić, R.; Rajčić, V.; Kržan, M. Response of Laminated Glass-CLT Structural Components to Reverse-Cyclic Lateral Loading. *Constr. Build. Mater.* **2020**, *235*, 117509. [CrossRef]
37. Barbalić, J.; Rajčić, V.; Žarnic, R. Numerical Evaluation of Seismic Capacity of Structures with Hybrid Timber-Glass Panels. In Proceedings of the World Conference on Timber Engineering (WCTE), Vienna, Austria, 22–25 August 2016.
38. EN 12150-1:2015+A1:2019-Glass in Building-Thermally Toughened Soda Lime Silicate Safety Glass. Available online: https://standards.iteh.ai/catalog/standards/cen/d6f942f1-f173-4a45-bcb5-ccb6e0df280d/en-12150-1-2015a1-2019 (accessed on 7 January 2022).
39. EN 14179-1:2016-Glass in Building-Heat Soaked Thermally Toughened Soda Lime Silicate Safety. Available online: https://standards.iteh.ai/catalog/standards/cen/f97d0a10-fe69-491b-aa45-7e8ff3d11a9e/en-14179-1-2016 (accessed on 7 January 2022).
40. DIN 1249-11 -Glass in Building-Part 11: Glass Edges-Terms and Definitions, Characteristics of Edge Types and Finishes. Available online: https://www.en-standard.eu/din-1249-11-glass-in-building-part-11-glass-edges-terms-and-definitions-characteristics-of-edge-types-and-finishes/ (accessed on 7 December 2021).
41. EN 16351:2015-Timber Structures-Cross Laminated Timber-Requirements. Available online: https://standards.iteh.ai/catalog/standards/cen/cbfe13a0-c6c2-4c1c-9e7f-d6c46155e3e9/en-16351-2015 (accessed on 7 January 2022).
42. EN 14080:2013-Timber Structures-Glued Laminated Timber and Glued Solid Timber-Requirements. Available online: https://standards.iteh.ai/catalog/standards/cen/f5a36e99-021d-44cb-8dec-9e41df42d7a5/en-14080-2013 (accessed on 7 January 2022).
43. EN 14081-1:2016+A1:2019-Timber Structures-Strength Graded Structural Timber with Rectangular. Available online: https://standards.iteh.ai/catalog/standards/cen/f0f4bbff-b863-4571-98f6-364add96ae79/en-14081-1-2016a1-2019 (accessed on 7 January 2022).
44. ISO 7500-1:2018-Metallic Materials—Calibration and Verification of Static Uniaxial Testing Machines—Part 1: Tension/Compression Testing Machines—Calibration and Verification of the Force-Measuring System. Available online: https://www.iso.org/standard/72572.html (accessed on 9 December 2021).
45. ISO 9513:2012-Metallic Materials—Calibration of Extensometer Systems Used in Uniaxial Testing. Available online: https://www.iso.org/standard/41619.html (accessed on 9 December 2021).
46. ANSYS®. *Academic Teaching Mechanical. ANSYS Workbench User's Guide*; ANSYS, Inc.: Canonsburg, PA, USA, 2009.
47. ANSYS®. *ANSYS Theory Reference*; SAS IP, Inc.: Canonsburg, PA, USA, 1999.
48. Ansys, Inc. ANSYS Meshing User's Guide. *Renew. Sustain. Energy Rev.* **2011**, *80*, 724–746.

 *sustainability*

Article

On the Use of Cloud Analysis for Structural Glass Members under Seismic Events

Silvana Mattei , Marco Fasan and Chiara Bedon *

Department of Engineering and Architecture, University of Trieste, 34127 Trieste, Italy;
silvana.mattei@phd.units.it (S.M.); mfasan@units.it (M.F.)
* Correspondence: chiara.bedon@dia.units.it; Tel.: +39-040-558-3837

Abstract: Current standards for seismic-resistant buildings provide recommendations for various structural systems, but no specific provisions are given for structural glass. As such, the seismic design of joints and members could result in improper sizing and non-efficient solutions, or even non-efficient calculation procedures. An open issue is represented by the lack of reliable and generalized performance limit indicators (or "engineering demand parameters", EDPs) for glass structures, which represent the basic input for seismic analyses or q-factor estimates. In this paper, special care is given to the q-factor assessment for glass frames under in-plane seismic loads. Major advantage is taken from efficient finite element (FE) numerical simulations to support the local/global analysis of mechanical behaviors. From extensive non-linear dynamic parametric calculations, numerical outcomes are discussed based on three different approaches that are deeply consolidated for ordinary structural systems. Among others, the cloud analysis is characterized by high computational efficiency, but requires the definition of specific EDPs, as well as the choice of reliable input seismic signals. In this regard, a comparative parametric study is carried out with the support of the incremental dynamic analysis (IDA) approach for the herein called "dynamic" (M1) and "mixed" (M2) procedures, towards the linear regression of cloud analysis data (M3). Potential and limits of selected calculation methods are hence discussed, with a focus on sample size, computational cost, estimated mechanical phenomena, and predicted q-factor estimates for a case study glass frame.

Keywords: seismic design; structural glass; q-factor; engineering demand parameters (EDPs); finite element (FE) numerical models; non-linear incremental dynamic analyses (IDA); cloud analysis; linear regression

Citation: Mattei, S.; Fasan, M.; Bedon, C. On the Use of Cloud Analysis for Structural Glass Members under Seismic Events. *Sustainability* **2021**, *13*, 9291. https://doi.org/10.3390/su13169291

Academic Editor: Syed Minhaj Saleem Kazmi

Received: 14 June 2021
Accepted: 16 August 2021
Published: 18 August 2021

Publisher's Note: MDPI stays neutral with regard to jurisdictional claims in published maps and institutional affiliations.

Copyright: © 2021 by the authors. Licensee MDPI, Basel, Switzerland. This article is an open access article distributed under the terms and conditions of the Creative Commons Attribution (CC BY) license (https://creativecommons.org/licenses/by/4.0/).

1. Introduction

The large use of glass structures in civil engineering applications represents a challenging issue for designers. In addition to intrinsic mechanical features of the involved load-bearing materials [1,2], careful consideration should be paid in earthquake-prone regions to satisfy rigid resistance and displacement demands. This is the case of primary, stand-alone glass structures, but also secondary glass systems belonging to different primary buildings and constructional assemblies [3–8].

According to various literature studies, the seismic capacity of glass structures can benefit from innovative tools and special fasteners [9–11]. At the component level, refined calculation approaches and investigations of literature have been dedicated to both the pre- and post-cracked analysis of laminated glass (LG) elements [12–14], including considerations of their residual strength [15]. In any case, glass structures are still a rather new domain for several professional designers, and certainly require dedicated methods of analysis [16]. Among others, an open issue is represented by the seismic design of glass structures. Most of the available technical documents do not provide specific recommendations for glass [17,18], but suggest the use of "reliable calculation methods" to verify the seismic resistance/displacement capacity of glass components and restraints. Such a technical difficulty is further enforced by the need for a realistic calibration of the expected q-factor [19].

The main goal of present study, in this regard, is to assess the sensitivity of q-factor for glass structures based on simplified or more advanced calculation approaches. As shown in Section 2, consolidated strategies are common for conventional constructional typologies/materials. Moreover, the q-factor itself (with $q > 1$) is known to represent the intrinsic dissipation capacity of the structure/material to verify. At the same time, established performance indicators (or "engineering demand parameters", EDPs) in support of seismic analysis and design are available in literature for structural members and systems made of steel, reinforced concrete, timber, or masonry, while such a calibration is missing for glass. To summarize the present discussion, the numerical analysis is focused on a case study glass frame that was earlier investigated in [19]. Differing from [19], however, the attention was given to the seismic performance and capacity of the full-size frame, rather than its key base connections only. To this aim, an original finite element (FE) numerical model was developed and optimized to support the local/global analysis of the frame as a whole. Extended sets of non-linear dynamic analyses were in fact carried out for the frame under in-plane seismic lateral loads. In doing so, three selected methods of analysis that are deeply consolidated for ordinary constructions (M1 to M3 in Section 3) were adapted to the examined structural glass frame and assessed for the q-factor prediction. Basic comparative calculations were first carried out with the support of the incremental dynamic analysis (IDA) approach for the herein called "dynamic" (M1) and "mixed" (M2) procedures. The cloud analysis procedure (M3), as shown, is characterized by high efficiency compared to M1 and M2 methods, but requires the calibration of specific EDPs for glass, as well as an accurate selection of input signals for the structural system to verify. FE comparative results are thus discussed in Sections 4–7, showing the potential and limits of selected M1 to M3 calculation methods, in support of a realistic and computationally efficient estimation of seismic behavioral trends for glass structures, thus resulting in their optimized structural design.

2. State-of-Art and Literature Review on q-factor Methods

Following EC8 [17], the seismic design of buildings is today conducted by using the so-called force-based design (FBD) method. The design base shear is conventionally obtained as the ratio between the elastic base shear and the q-factor of the structure to verify (Figure 1). The q-factor introduction, as such, simplifies its complex energy dissipation capacity (by means of plastic deformations) to a linear elastic model. Due to its strategic role, the q-factor definition is thus a topic which has been deeply discussed in the seismic engineering field as a primary focus of several studies since the 1950s.

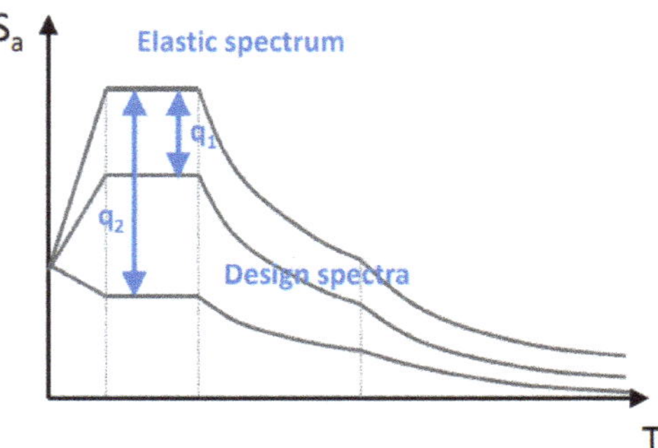

Figure 1. Examples of design spectra calculated for two different q-factor values.

A first simple formulation was proposed for the q-factor in the 1980s [20]. Further, it was first recognized by the modern design strategy that structures able to resist severe earthquakes are expected to experience permanent damage. As a matter of fact, design seismic actions are scaled by taking advantage of an intrinsic plastic capacity that is correlated to irreversible deformations.

Typical q-factor values by EC8 are known to span in the range of 1.5 (inverted pendulum systems), 2.0 (torsionally flexible systems), and 3.0 (frame systems), or even higher. Due to this, these constant EC8 values should be generally treated as an upper bound, and thus moving the decision on the acceptable level of damage becomes a designer responsibility. This decision is directly affected by structural features, details, and material properties. In such a general discussion, the choice of standards to adopt constant q-factor values looks very conservative. Several literature studies proved that the dissipative capacity of a structure is generally greater than the recommended limit values. For example, as concerns steel moment resisting frames (MRFs), the EC8 prescribes different q-factors for medium (DCM, local plastic deformations) or high (DCH, global plastic deformations) ductility classes. Macedo et al. [21] evaluated the consequences of adopting the EC8 recommended q-factor and presented a more rational selection methodology based on the specific structure and the site seismic hazard. Costanzo et al. [22] discussed existing design provision for both MRFs and chevron concentrically braced frames (CCBFs), giving evidence of a large lateral overstrength due to the codified design requirements. Also for reinforced concrete frames, studies by Kappos [23], Borzi and Elnashai [24], and Chryssanthopoulos et al. [25] assessed the reliability of q-factor values by EC8. and emphasized their high conservativity. Although the cited results from [23–25] looked conflicting, the joint EC8 conservatism was jointly justified with either structural overstrength or ductility supply, or both the aspects. In this context, it is thus recognized that the primary goal of standardizing committees is to simplify, on the safe side, the computational burden for designers. Such a strategy makes it possible to avoid performing complex non-linear analyses, but at the same time can severely limit the actual structural plastic capacity of the examined building systems.

For glass structures, to date, legislative and research efforts have not provided a general recommendation about realistic q-factor values that could be adopted in design. Furthermore, it is already required to satisfy global and local verifications for resistance and displacement capacities in seismic conditions [3]. As such, the typical effect often takes the form of fully elastic design ($q = 1$). The present study aimed to investigate further the expected structural behavior trends of seismically loaded glass members, based on the observations of a case study frame. Calibrated parameters are presented to possibly support the adaptation of consolidated general procedures to glass structures. The potentials/issues of available methodologies are assessed towards the q-factor calculation for similar structural typologies.

3. q-factor and Selected Calculation Methods

Different approaches can be used from literature to analytically or numerically predict the q-factor of a given structural system [26]. As far as the computational effort and accuracy of a method increase, and the reference EDPs are well defined, moreover, the q-factor estimation is progressively more robust and reliable. Figure 2 shows a typical push-over (PO) analysis result for a reinforced concrete building, in which the EDPs are qualitatively pointed out, depending on various performance levels and limit states. As usual, the most common EDPs are represented:

- for structural components, by inter-story drift ratios (IDR), with inelastic component deformations and associated forces;
- for non-structural components (and contents), by inter-story drift ratios (IDR) or peak floor accelerations (PFA), see [27–29].

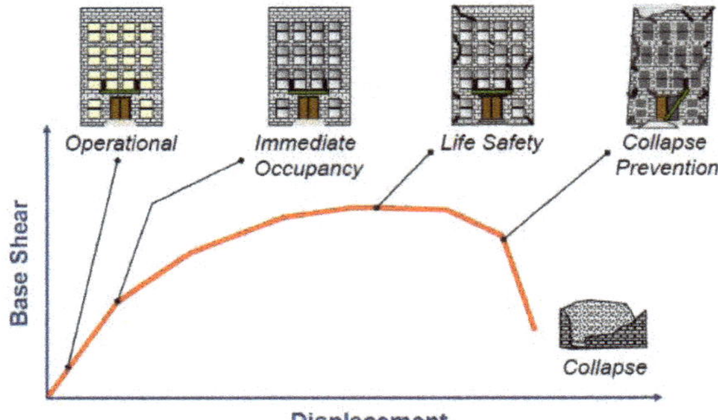

Figure 2. Expected building response and damage under seismic events. In evidence, the reference limit states and EDPs for design.

It is worth noting that recommended EDPs are available for traditional constructional materials and systems, but these parameters cannot be directly transferred to glass structures.

It follows that secondary glass members that take place in a primary building must necessarily accommodate the seismic performance and capacity of the building itself (and thus satisfy the corresponding EDPs). For primary/stand-alone glass structures, otherwise, no recommended parameters are available, and thus the present study tries to provide some research developments in this direction.

The above issue arises for novel structural systems and/or innovative materials (glass included), for which no or indicators are provided by design standards for earthquake resistant buildings. Relevant examples of literature can be found in [30–35].

From a practical point of view, the flowchart in Figure 3 can be adapted to general constructions/materials, once standardized procedures are established and a primary calculation method is chosen. In case of structural glass (as well as other innovative solutions), the critical step takes place in #2, as a direct/major effect of the analysis method choice and its input basic assumptions (first of all, the reference EDPs).

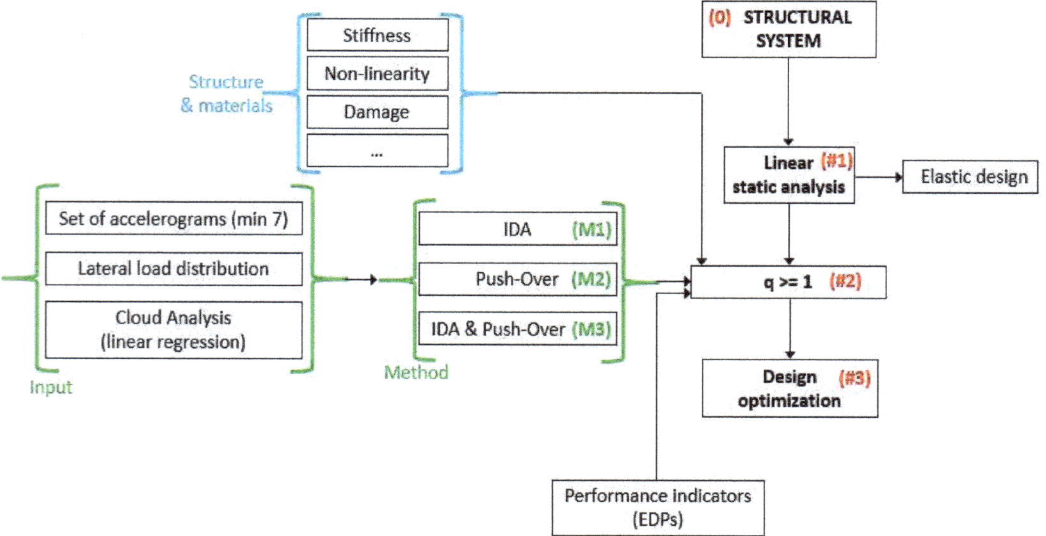

Figure 3. Reference flowchart for the seismic assessment of novel structural systems/materials.

In the present paper, such an issue is further discussed with a focus on the structural glass frame described in Figure 4 and [19]. Three calculation methods (M1, M2, and M3 from Figure 3) are compared in terms of predicted q-factor, computational efficiency, accuracy, and sufficiency of results. In doing so, special care is taken for the detection of reliable EDPs that could be used for design, especially with regard to the key configurations of yielding and collapse. The so-called M1, M2, and M3 methods herein explored find inspiration from literature, but in the current study are specifically adapted to glass frames. Examples for traditional structures can be found in [36], as regards the M1 (Section 3.1) and M2 (Section 3.2) methods whilst, for the M3 one (Section 3.3), the procedure in use for the construction of fragility curves is adapted to glass. As such, the M3 q-factor is derived from linear regression on a cloud of points that is obtained from non-linear time-history analyses.

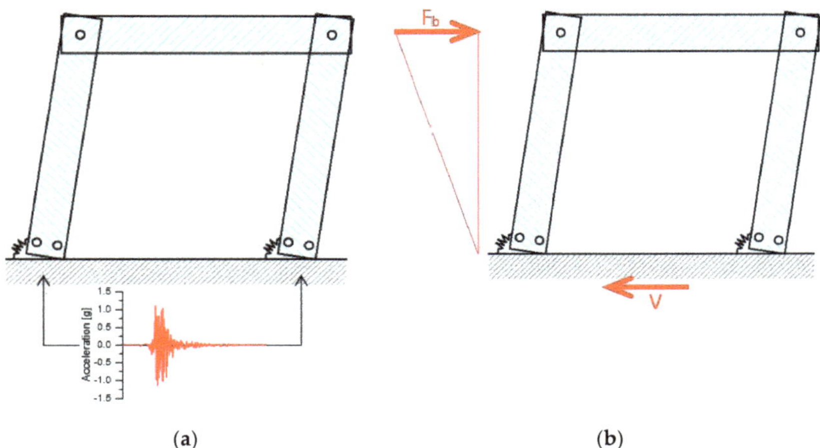

Figure 4. Reference structural glass frame (adapted from [19]) for the q-factor estimation, based on (**a**) non-linear incremental dynamic analyses (IDA) or (**b**) push-over (PO) numerical procedures.

3.1. Dynamic (or PGA) Method (M1)

The dynamic (or PGA) method (M1) conventionally defines the q-factor as the ratio between PGA_u and PGA_y, that is the peak ground acceleration values corresponding to "collapse" or "first yielding" respectively:

$$q = \frac{PGA_u}{PGA_y} \tag{1}$$

In accordance with Equation (1), IDA were thus carried out in this paper. Based on Figure 4a and a set of input accelerograms, sequential non-linear time-history numerical analyses were performed to estimate the PGA_u and PGA_y values of interest. A minimum set of 7 input signals was taken into account [17].

3.2. Mixed Method (M2)

The mixed (M2) method examined in this paper still takes advantage from efficient FE simulations. The q-factor estimation was based in this case on two different contributions, that is:

$$q = \frac{PGA_u}{PGA_y} \cdot \frac{V_y}{V_d} \tag{2}$$

In Equation (2), PGA_u and PGA_y values agree with the definition in Section 3.1, and can be derived from the non-linear IDA for the structural system object of analysis.

At the same time, V_y and V_d in Equation (2) denote the base shear load corresponding to "first significant yield strength" and "allowable design strength". These base shear values were conventionally derived from non-linear PO simulations according to Figure 4b.

3.3. Cloud Analysis (M3) with Linear Regression

The q-factor of the examined frame was finally estimated in this paper by using the inelastic response spectrum, with the support of the so-called cloud analysis and the spectral acceleration (S_a) definitions [36]. Successful cloud analysis applications can be found in [37–39] for various structural typologies and materials.

Differing from Sections 3.1 and 3.2 (IDA procedure), the cloud analysis is carried out with the support of a set of unscaled accelerograms. The set of input signals (60 in the present study, Table A1 [40]) must be established to ensure an appropriate distribution of cloud data [41,42]. It is in fact known that the major issue of IDA method may consist of a significant computational cost, and most often a marked scaling of original records to various intensity levels, before the desired EDPs could be achieved. This effort is not required in cloud analysis. Moreover, the use of unscaled natural accelerograms in cloud analysis allows to keep all the information related to the event, also known as "record-to-record variability". From unscaled signals and non-linear dynamic analyses, the correlation is established between selected EDPs and some intensity measure (IM) values of the imposed signals by taking advantage of linear regression [37]. As in case of IDA, however, the unscaled signals require an accurate definition of reference EDPs and are also expected to cover a useful range of values for identifying the required limit states. These signals are thus sensitive to the fundamental vibration period T_1 (to predict) and the characteristics of the structure to verify (material properties, damage mechanisms, etc.). Moreover, the input signals should be selected to be representative of the seismic hazard of the site under investigation. When appropriate signals are not available, site-specific ground motion modeling techniques can also be used [43–45]. Based on EC8, the q-factor can be finally calculated as:

$$q = \frac{S_a(T_1)_u}{S_a(T_1)_y} \tag{3}$$

where $S_a(T_1)$ is the spectral ordinate corresponding to the characteristic period of the design spectrum. The subscripts "u" and "y" in Equation (3) refer to the "collapse" and "first yielding" configurations.

4. Case-Study Glass Frame

4.1. Geometrical and Mechanical Properties

The current research study follows and extends the analytical and numerical investigation reported in [19]. As such, some geometrical and mechanical features are summarized herein for the system in Figure 5a. Each glass frames followed the layout in Figure 5b, with $H = 6$ m and $L = 8$ m. Both the beam and column sections were composed of heat-strengthened (HS) LG members, with uniform size ($h = 600$ mm high $\times$ $t_{tot} = 66$ mm thick) given by $5 \times t_g = 12$ mm glass layers and $t_{int} = 1.52$ mm thick ionoplast foils. The mechanical connection at each beam–column interception took the form of an ideal pin, see Figure 5b. Possible out-of-plane deformations of the frame were restrained, and the related mechanisms (including lateral-torsional buckling for beams [46,47], or coupled bending-compressive buckling for columns [48]) can be preliminarily disregarded. For the base restraints of columns, stainless steel pins pass through two holes in the glass ($\varphi_g = 32$ mm in diameter, with $\varphi = 24$ mm the nominal diameter of bolts and $D = 500$ mm their distance). Four mild steel brackets (S235 steel) fix the columns to the foundation ($t_s = 15$ mm, $B_s = 200$ mm, $b_s = 165$ mm, $H_s = 300$ mm and $L_s = 200$ mm). The restraint was finally locked by n_b anchoring bolts (Figure 5c).

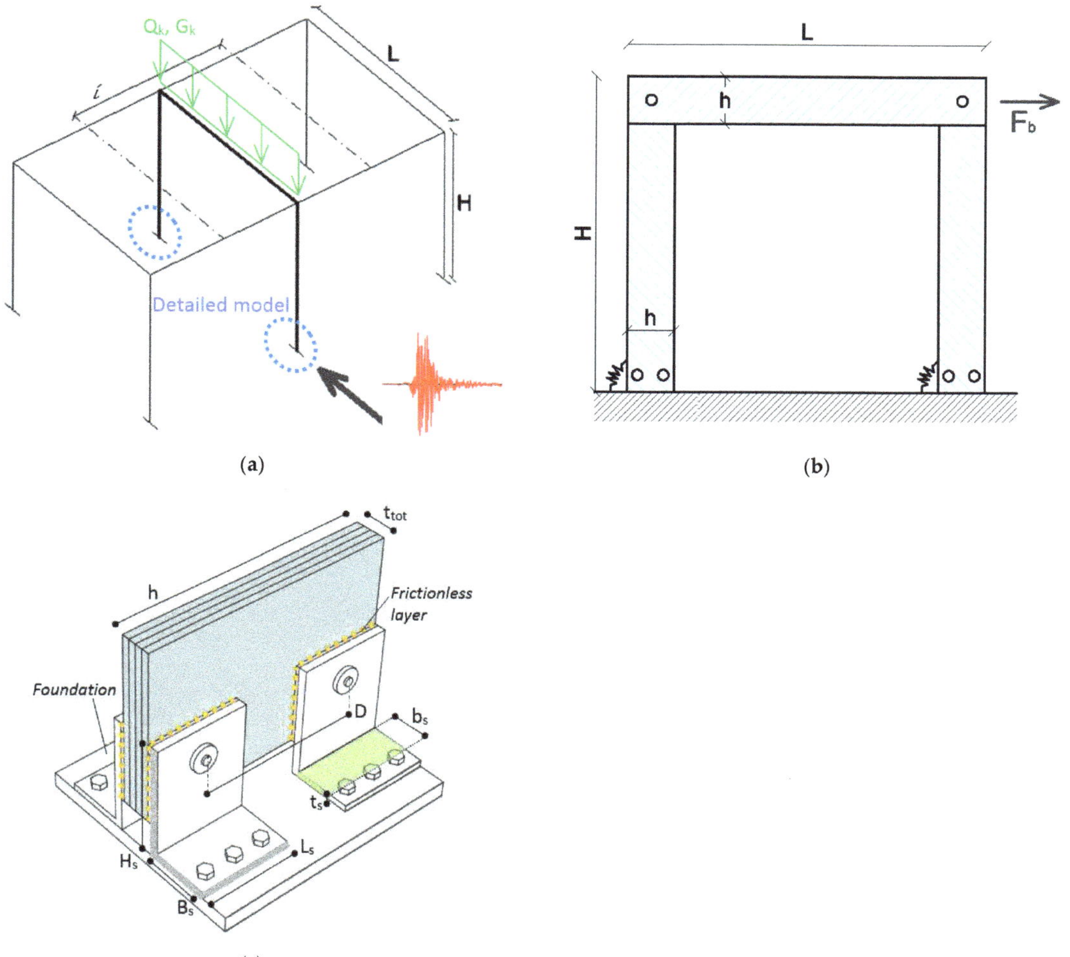

Figure 5. Schematic representation of the case study frame: (**a**) design concept (axonometry), with (**b**) static scheme of the structural glass frame object of analysis and (**c**) detail view of a typical push-pull moment connection at the base of the columns (adapted from [19]).

4.2. Preliminary Elastic Seismic Design of the Frame

For calculation purposes, the glazed assembly of Figure 5 is located in a high seismicity region of Italy. Based on [17,49,50], its strength and stiffness should be verified to resist the most unfavorable expected seismic combination of actions E_d, that is:

$$E_d \leq R_d \tag{4}$$

where R_d is the structural capacity.

A simple design of glass members should properly verify that LG columns and beams are not subjected—due to the imposed in-plane seismic loads—to relevant stress peaks and premature fracture. As for regular structures in plan and elevation, the input seismic force F_i the frame should resist is given by:

$$F_i = F_b \frac{z_i m_i}{\sum_j^n z_j m_j} \tag{5}$$

with $i = 1, \ldots n$, m_i, m_j the story masses and z_i, z_j their height from the foundation, while the base shear F_b is:

$$F_b = \frac{S_d(T_1) \, W \, \lambda}{q} \tag{6}$$

and:

- W the aboveground total mass of the building object of analysis,
- $S_d(T_1)$ the design acceleration from the reference spectrum, as a function of the vibration period T_1, with $S_d = 0.35$ g the peak ground acceleration (high seismic region of Italy),
- $\lambda = 1$ a correction factor for one-story buildings with $T_1 > 2T_C$ (otherwise 0.85), and
- $q \geq 1$ the behavior factor of the system.

Given the lack of more appropriate recommendations, the conventional design of the case study columns suggests the assumption that $q = 1$. Disregarding the vertical loads that the LG members must sustain (as a part of the framed system of Figure 5), the in-plane lateral force affects the region of glass holes at the base connections, that is:

$$\sigma_{t,max} = K_t \cdot \sigma_t = 2.71 \cdot 77 \approx 208.9 MPa \tag{7}$$

with:

$$K_t = 2 + \left(1 - \frac{\phi_g}{h/2}\right)^3 = 2.71 \tag{8}$$

the magnification factor for stresses [51–53], while the tensile stress σ_t in glass is given by:

$$\sigma_t = \frac{F_t}{\left(\frac{h}{2} - \phi_g\right) \cdot t_{tot}} \approx 77 MPa \tag{9}$$

with:

$$F_t = \frac{(0.5 \cdot F_b) \cdot H}{D} = 1236 kN \tag{10}$$

The resistance verification of the LG columns in seismic conditions requires that:

$$\sigma_{t,max} \leq f_{g;d} \tag{11}$$

with:

$$f_{g;d} = \frac{k_{mod} k_{ed} k_{sf} \lambda_{gA} \lambda_{gl} f_{g;k}}{R_M \gamma_M} + \frac{k'_{ed} k_v \left(f_{b;k} - f_{g;k}\right)}{R_{M;v} \gamma_{M;v}} \tag{12}$$

the design resistance [49]. Among the coefficients in Equation (12), the short-term duration of seismic events (conventionally set in 30 s [3]) suggests $k_{mod} = 0.78$. Given that the columns are composed of HS glass, Equation (12) results in $f_{g;d} \approx 75$ MPa, that is $\approx 1/3$rd the maximum stress from Equation (7), due to the seismic shear from Equation (6).

To avoid the improper sizing of load-bearing glass members, the design (with given input parameters) would require the exploitation of a minimum $q_{min} \approx 3$. In other words, the ratio of Equation (7) to Equation (12) and combination with Equation (6) can be used for simple analytical estimates of (minimum) required plastic capacities of the frame, towards the seismic demand. In this regard, it is also worth noting that the analytical model developed in [19] for the ductility estimation of base angle brackets (and properly combined with the stress analysis in Equations (7)–(12)), would result in $q = 4.58$ (collapse governed by stress peaks in the region of glass holes). However, such an analytical prediction is not able to account for complex mechanisms in the frame as a whole, under dynamic seismic accelerations.

5. Finite Element Numerical Investigation

5.1. Numerical Model

The reference numerical analysis was carried out in ABAQUS/Standard [54] on a full three-dimensional model representative of the glass frame object of analysis, inclusive of LG members, and reproducing the geometrical details for base connections (Figure 6). For symmetry, 1/4th the geometry was taken into account.

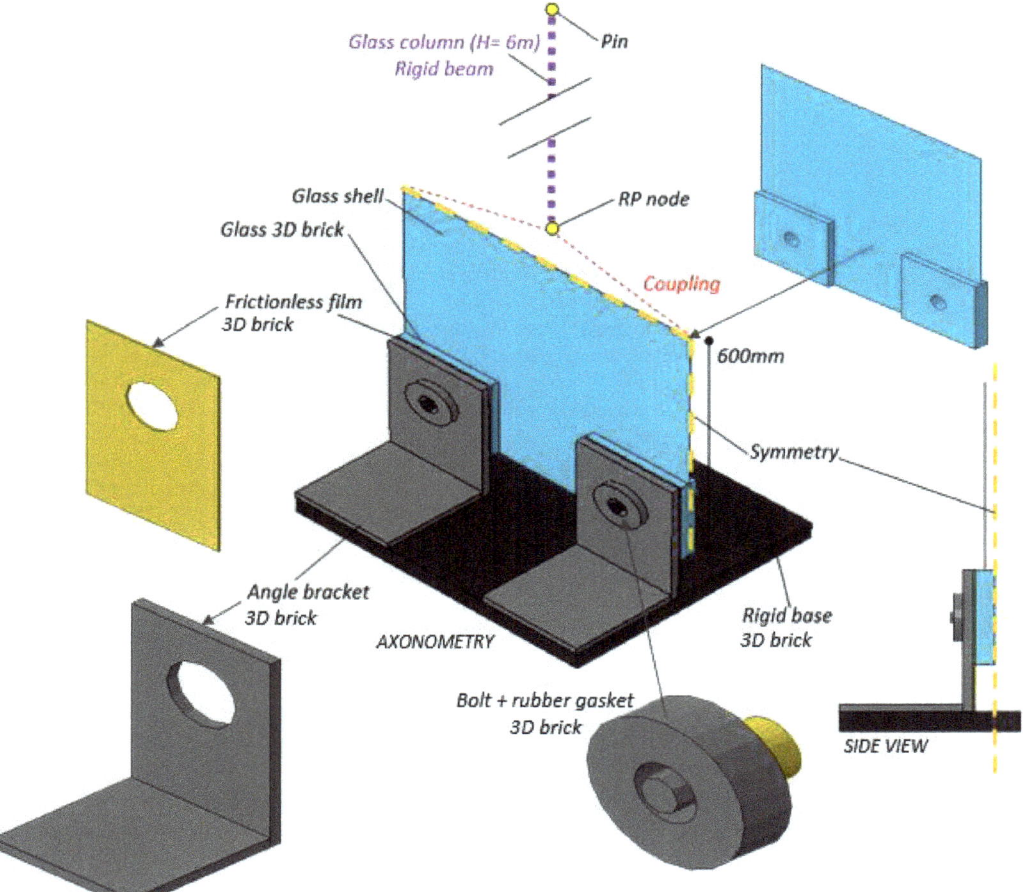

Figure 6. Numerical model of the case study structural glass frame under in-plane seismic loads (detail of the base region and beam/column connection, ABAQUS).

Differing from [19], the seismic response of the frame as a whole was explored for the purpose of this study. To this aim, a novel optimized FE model was developed to maximize its computational efficiency.

Solid brick elements (C3D8R type from ABAQUS library) were used for rigid base support, angular members, frictionless foils, and bolts. For the glass plate, a mix of brick solid elements and shell elements was used to preserve the accuracy of stress distributions in the regions of holes. Finally, the column in elevation (and top beam) were described as pinned rigid beams. The overall symmetry assumption resulted in 18,000 solid/shell elements and 75,000 DOFs for the frame model in Figure 6. After preliminary validation, such a solution was used to replace the FE assembly from [19], in which a total of 45,000 solid elements and 170,000 DOFs were used for half geometry of the base connection only.

5.2. Materials and Contact Interactions

Key mechanical assumptions for glass and steel members were derived from [19]. An elastic-perfectly plastic law was used for mild steel, with E_s = 210 GPa the modulus of elasticity, ν_s = 0.3 the Poisson' ratio, and $\sigma_{s,y} = \sigma_{s,u}$ = 235 MPa the yielding/failure strength, with corresponding strain values equal to ε_y = 0.112% and ε_u = 25%. The ductile damage material option was also accounted for in FE analyses to detect the possible initiation of ductile failure mechanism in angle brackets.

An elasto-plastic law was also used for steel bolts, with $\sigma_{b,y} = \sigma_{b,u}$ = 1000 MPa the yielding/ultimate resistance (8.8 resistance class). Finally, the rubber layers were described in the form of an equivalent elastic-perfectly plastic material, with E_r = 30 GPa and ν_r = 0.3. The yielding/ultimate stress was conventionally set at 2.4 MPa [19].

The tensile brittleness of glass was included with the concrete damaged plasticity (CDP) model from the ABAQUS library [19]. While the CDP model was primarily developed for concrete, literature studies show that the same model can be efficiently used for structural glass members (under specific loading/boundary conditions), as in the present application. From the post-processing of FE results, fracture initiation in glass was in fact assumed as a reference for "failure". This means that the overall post-cracked stage was disregarded but the simulation was prevented from additional uncertainties that are typical of the post-cracked response of glass under cyclic loads. In doing so, the nominal mechanical properties for HS were taken into account (E_g = 70 GPa, ν_g = 0.23 and σ_{tk} =70 MPa). Further, the characteristic compressive strength was set to σ_{ck} = 300 MPa (350–500 MPa [19] the reference strength).

A set of surface-to-surface contacts at the interface of adjacent FE components allowed to reproduce the in-plane lateral response of the frame under seismic loads (with "penalty" tangential characteristic (friction μ = 0.3) and "hard" normal features). "Tie" mechanical constraints were also used to rigidly connect some FE components (i.e., the head of each bolt and the corresponding angle bracket, or the frictionless layer and the adjacent angle bracket).

5.3. Loading Strategy

The frame was investigated by taking into account the presence of in-plane seismic loads and dead loads due to constructional members, plus a vertical accidental load Q_k = 3 kN/m^2 (with i = 1 m). The FE assembly of Figure 6 was used for both the required non-linear static PO and time-history dynamic analyses. As such, two different solving procedures were taken into account, based on two separate steps representative of:

- S1 = an initial stage for introduction of dead and accidental loads (5 s), followed by
- S2 = seismic analysis of the pre-loaded glass frame (60 s).

In case of IDA, the main seismic input consisted in the selected accelerogram in Figure 7 (acceleration-time history at the base of the frame for M1 and M2 methods). The used earthquake records were derived from [55], that is considering a PGA of 0.35 g, with type A soil (rock soil), topographic category T1, and a reference nominal life of 50 years. A maximum lower and upper tolerance of 10% was also considered.

For the PO analyses (M2 approach), otherwise, the FE system of Figure 6 was subjected to a linear increasing in-plane shear force in accordance to Figure 7 (base connection rigidly fixed to ground). Finally, in case of cloud analysis (M3), a total of 60 unscaled signals (Table A1 [40]) was taken into account to replace the 7 input signals of M1 and M2 procedures.

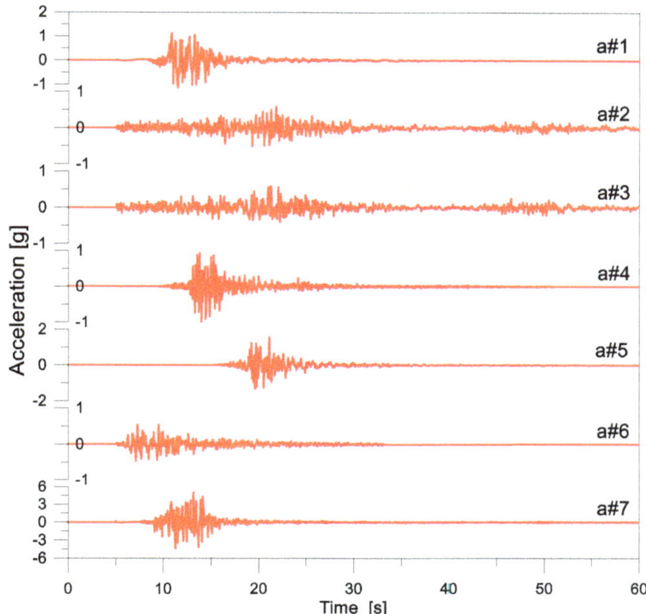

Figure 7. Reference set of time-acceleration histories, as derived from REXEL [55] for the non-linear incremental dynamic analysis (IDA) of the case study glass frame.

6. Discussion of M1 and M2 Results

6.1. Detection of First Yielding, Collapse, and Allowable Strength Parameters

Given the general definitions of calculation approaches in Sections 4.1 and 4.2 and the structural system object of study, special care must be taken for the definition of the key design configurations. Major design challenges derive from the lack of explicit recommendations for glass structures in current design standards for seismic resistant buildings. On the other hand, the in-plane seismic response of the frame is strongly affected by the intrinsic mechanical properties of its basic components, namely:

- brittle elastic glass panels (with holes), and
- flexible angle brackets at the base of the frame.

In other words, the "first yielding" condition of the system of Figure 6 was defined in this project as the first plastic deformation of angle brackets in tension (with $\sigma_{s,y}$ = 235 MPa the reference strength and δ_y = 0.677 mm the corresponding vertical deformation [19]). Regarding the "collapse" damage state for the frame, maximum drift amplitudes (or column rotations, or even vertical deformations of the steel angle brackets) should be checked. Globally, for the parametric investigation herein summarized, the control of local and global critical conditions for the frame was primarily based on local stress and displacement controls, namely representing a potential:

- (C1) Tensile cracking of glass, close to the column base (region of holes),
- (C2) Compressive fracture of glass, close to the column base (region of holes),
- (C3) Ultimate deformation for steel angle brackets (plastic strain and vertical deformation, with δ_u =54.10 mm based on [19]),
- (C4) Possible yielding of steel bolts for the base connection.

In addition, for comparative purposes, conventional deformation limits available in design standards were also taken into account. For the case study frame, as far as the glass holes are properly protected from potential local damage, the seismic analysis could take advantage of the intrinsic flexibility and dissipative capacity of angle brackets. In this sense, the reliability of limit values of Table 1 from FEMA 356 [56], Vision 2000 [57], UBC

1997 [58], EC8 [17], and NTC2018 [50] documents and their applicability to the examined frame were taken into account in this study.

Table 1. Recommended limit configurations for the collapse prevention of steel structures, according to selected international design standards.

		Limit Drift Value (u/H)		Column Rotation (rad)
Structural system	FEMA 356 [56]	Vision 2000 [57]	UBC 1997 [58]	EC8, NTC2018 [17,50]
Steel braced frames	0.02	0.025	0.02	0.03
Steel moment frames	0.05			

Finally, the "allowable strength" condition required by the M2 approach should also be defined. As far as the brackets are assumed responsible of the overall in-plane seismic performance of the frame, the yielding stress of steel suggests that:

$$V_d = f(\sigma_{s,adm}) = \frac{\sigma_{s,y}}{\gamma_M} = \frac{235}{1.05} = 223\ MPa \tag{13}$$

with γ_M the partial safety factor, thus a minimum:

$$\frac{V_y}{V_d} = \gamma_M = 1.05 \tag{14}$$

to account in Equation (2).

Figure 8 shows the base shear-lateral deformation of the frame from PO analysis. It is clear that the above assumption can strongly penalize the frame response, and the critical glass members prove to offer a safety factor in the order of $\approx$2.1 against potential tensile cracks. Moreover, it is possible to see that the compressive limit in glass holes is not achieved, neither under large displacements. This results from the high deformation capacity of the frame, thanks to detailing of base connections explored in [19].

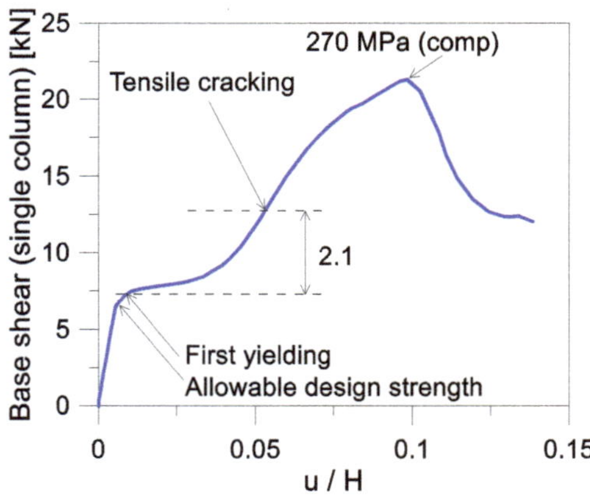

Figure 8. PO analysis of the frame, with evidence of relevant EDPs (ABAQUS).

6.2. Seismic Performance Assessment

The seismic response of the frame was found to agree with Figure 9, where the typical IDA deformed shape (detail) is proposed for brackets under large in-plane lateral displacements.

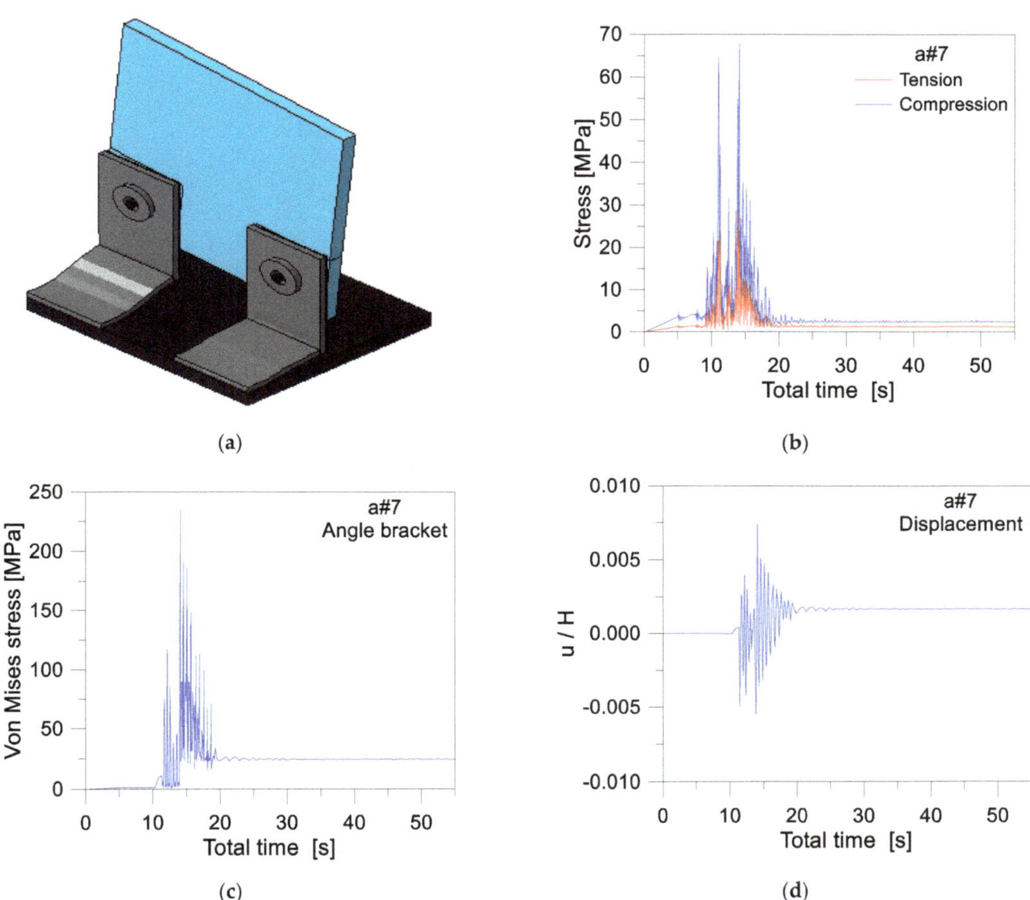

Figure 9. Example of IDA results (ABAQUS): (**a**) deformed shape (extruded axonometric detail, scale factor = 300, at a total time of 15 s); (**b**) stress envelope in the glass holes; (**c**) Von Mises stress in the angle brackets; and (**d**) in-plane lateral deformation. Results for the seismic record a#7 ($\times 1$, PGA = 5 m/s^2).

A total of 140 non-linear analyses was carried out with the imposed scaled accelerograms from Figure 7 (with an average of $\approx$20 differently scaled simulations for each accelerogram). The IDA results still confirmed the close correlation with PO results in Figure 8, with a qualitative agreement of damage phenomena and maximum effects due to the imposed design loads. Figure 9b, in this regard, presents the evolution of maximum stress peaks in the region of holes, while Figure 9c,d focus on the bracket and frame responses, respectively.

A more detailed analysis of IDA results can be found in Figures 10 and 11, in terms of relevant EDPs, as a function of the imposed PGA for each one of the input scaled signals. It is worth noting that the parametric analysis was carried out in the ideal PGA range of 0–50 m/s^2 to address the performance of structural components. In this regard, typical PGA values can be seen as associated to limited stress levels in the structure, as is expected due to the limited structural mass and high flexibility of the system. Key benefits derive also from gaps in the region of glass holes to prevent premature stress peaks at the edges. Such an approach is also in line with other studies on the seismic performance of structural systems with flexible joints (see for example [59]).

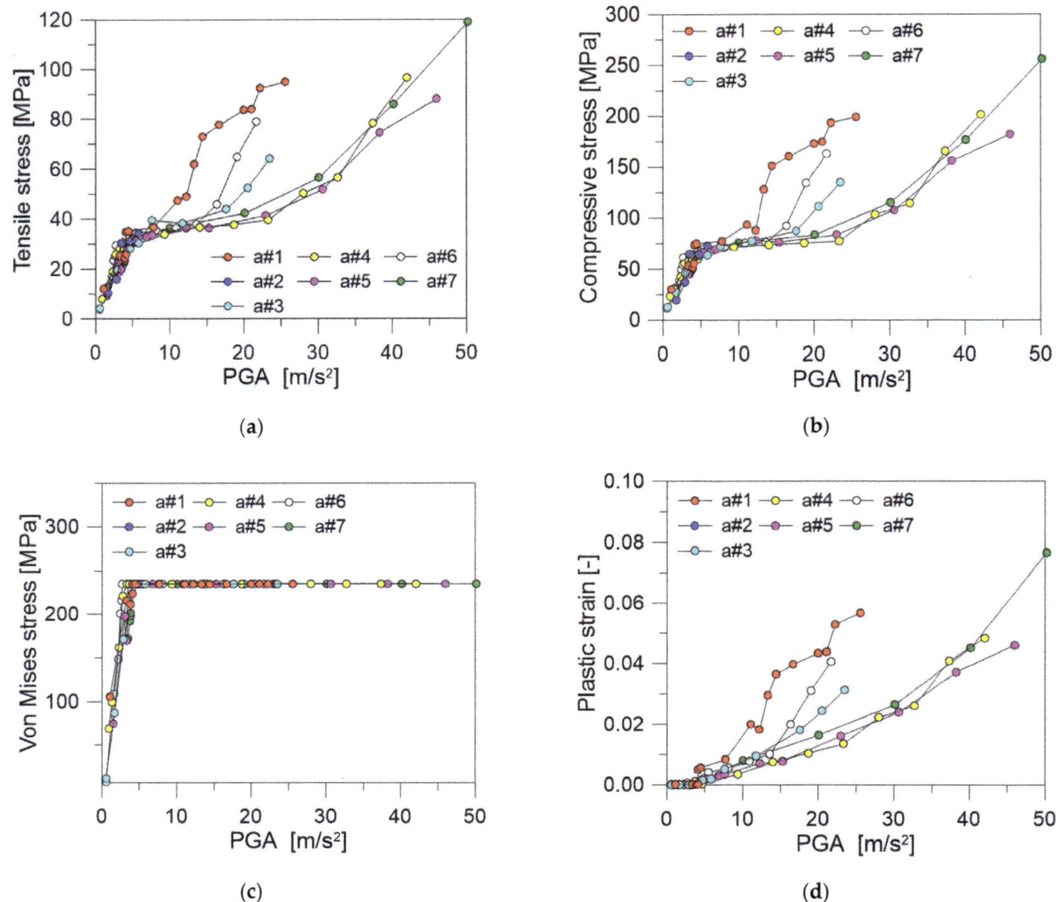

Figure 10. Selected IDA numerical results, as a function of the maximum imposed PGA (ABAQUS): (**a**) tensile and (**b**) compressive stress in glass holes, with (**c**) Von Mises stresses and (**d**) plastic strain in the steel angle brackets.

As shown in Figure 10a,b, the compressive stress peaks were mostly observed to double the corresponding tensile peaks in glass, due to a combination of in-plane lateral and vertical loads. Otherwise, it is also interesting to notice that the LG members can sustain relatively strong earthquake motions, before glass could fracture. A relevant aspect is hence represented, in both figures, by the non-linear evolution of stress peaks with the imposed PGA. A first linear trend of the charts can be observed for PGA up to ≈ 6 m/s^2, and such a slope change coincides with first yielding (and progressive plastic deformation) of angle brackets. This limit condition was generally achieved for PGA in the order of ≈ 4 m/s^2 (Figure 10c,d).

Compared to the stress evolution in the holes region, similar trends can also be observed for the deformations of the frame in Figure 11.

The vertical displacement δ, in-plane lateral drift u / H and base rotation θ are proposed, as obtained from IDA and maximum envelopes of selected EDPs. Under the input assumptions of this study, the collapse condition is never achieved on the side of angle brackets (Figure 11a). The limit drift of 2% or 5% is exceeded for PGA in the order of ≈ 10 m/s^2 and 20 m/s^2 (average value), see Figure 11b. The 2% drift, finally, is mostly in line with the 0.03 rad rotation of the frame (Figure 11c).

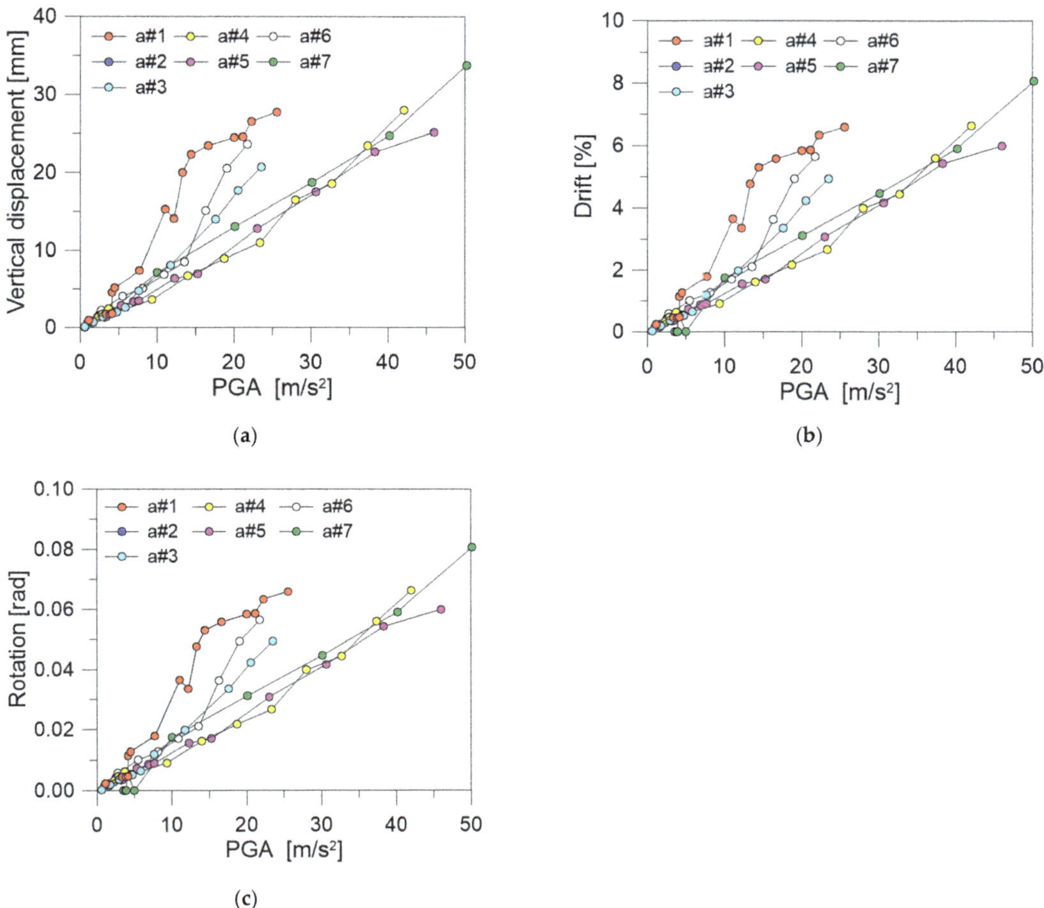

Figure 11. Selected IDA numerical results, as a function of the maximum imposed PGA (ABAQUS): (**a**) vertical displacement; (**b**) lateral drift, and (**c**) base rotation of the frame.

6.3. q-factor Estimates

The analysis was first focused on the so-called M1 method. For the M2 case, the M1 value was adapted with the magnification factor in Equation (14). The so-calculated IDA results are proposed in Figure 12.

Note that the attention was focused on the most unfavorable collapse mechanism for the frame as a whole. This was generally observed to coincide with tensile glass cracking ("glass"), while in two cases, only the deformability of the base joint allowed to reach a lateral drift of 2% ("Drift 2%"). The corresponding q-factor values are presented for the M1 method (Equation (1)), in the range from 1 (a#1) to 5 (a#6). The average value of $q = 2.59$ is compared with the corresponding M2 estimate from Equation (2), $q = 2.72$. The preliminary analytical requirement (Equations (7)–(12) combined with Equation (6)) is also highlighted ($q_{min} = 3$), while the analytical value based on local analysis ($q = 4.58$ from [19]) gives evidence of intrinsic limits due to simple predictions carried out for the base connection only (with collapse governed by stress peaks in the region of glass holes).

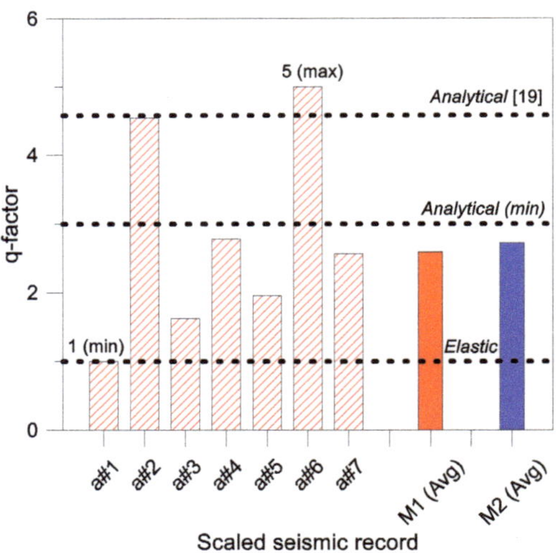

Figure 12. M1 and M2 calculated q-factor from IDA (ABAQUS).

7. Cloud Analysis

7.1. Input Records and EDPs

The maximum inter-story drift (IDR) was chosen as reference EDP for the frame ($T_1 = 0.3$ s), with PGA and pseudo-spectral acceleration $S_a(T_1)$ being selected as IM parameters. While the PGA value is only related to the seismic ground motion, the S_a value depends on the dynamic behavior of the structure; therefore, the performance to different IMs depends on the type of structure and the governing failure mechanism.

The preliminary analysis was focused on the distribution of tensile stress peaks in a glass column (with nominal height H) deprived of the angle brackets (rigid base connection). Given the characteristic tensile resistance of HS glass ($\sigma_{tk} = 70$ MPa), an inter-story displacement $u_u = 0.048$ m was calculated as in Figure 13. The reference EDP corresponds to contour plots in Figure 13a,b), while Figure 13c shows the data trend obtained at the column base and in terms of maximum envelope.

It is worthy of interest that the so-calculated value corresponds to $u/H \approx 0.007$ and is in close correlation with consolidated limit values for constructional materials characterized by typical brittle behavior in tension, such as, for example, masonry [50]. At the same time, the calculated value significantly minimizes the expected seismic capacity of the frame, thus confirming the key role of its base connections.

The total set of 60 unscaled ground motion records in Table A1 were chosen from [40], depending on the possible collapse mechanism of the frame. According to procedures for general buildings, special attention was paid to cover a wide range of spectral accelerations, but also to respect the consistency between the characteristics of selected records and the supposed classification for the site of interest. As a result, the selected accelerograms were characterized by a moment magnitude (M_w) between 5.6 and 7.6, an epicentral distances (R) ranging between 3.5 km and 62.9 km, and a soil class type A or B (EC8 classification).

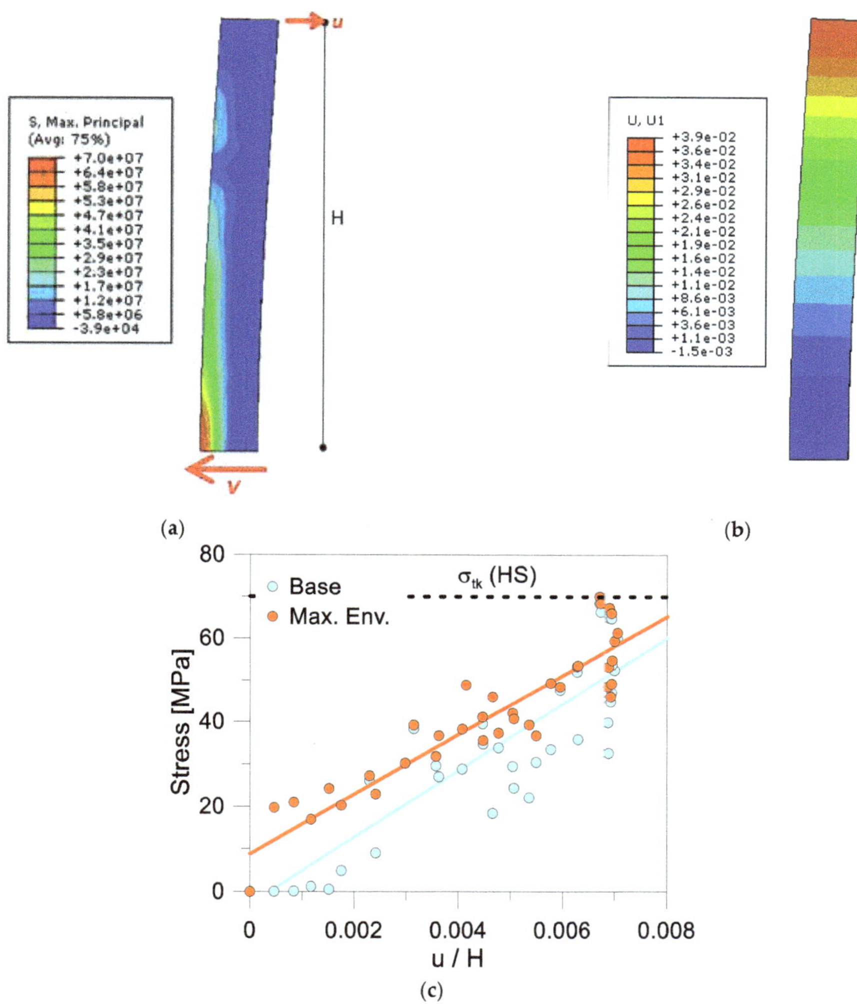

Figure 13. Definition of the reference inter-story drift, based on the distribution of tensile stress peaks in glass (ABAQUS): (**a**) stress analysis at collapse (legend values in Pa) and (**b**) corresponding in-plane lateral deformation (legend values in m), with (**c**) calculated trends at the column base or from maximum envelope data.

7.2. Analysis of M3 Results with Linear Regression

The results of the cloud analysis method related to PGA and $S_a(T_1)$ are separately collected in Figure 14, with attention to the measured inter-story displacement.

Differing from IDA, one of the potential intrinsic limits of the M3 method can manifest in the availability of natural seismic records that possess sufficiently high accelerations to reach the desired EDPs.

In this regard, Figure 15 gives evidence of the typical observed response for the case study frame. As shown, the imposed records are able to lead the angle brackets to yielding (Figure 15a), but still relatively smooth stress peaks are achieved in glass (holes), with tensile and compressive stress peaks in Figure 15b,c. In the same way, the measured in-plane deformations of the frame are still lower than the % limit values earlier discussed for IDA.

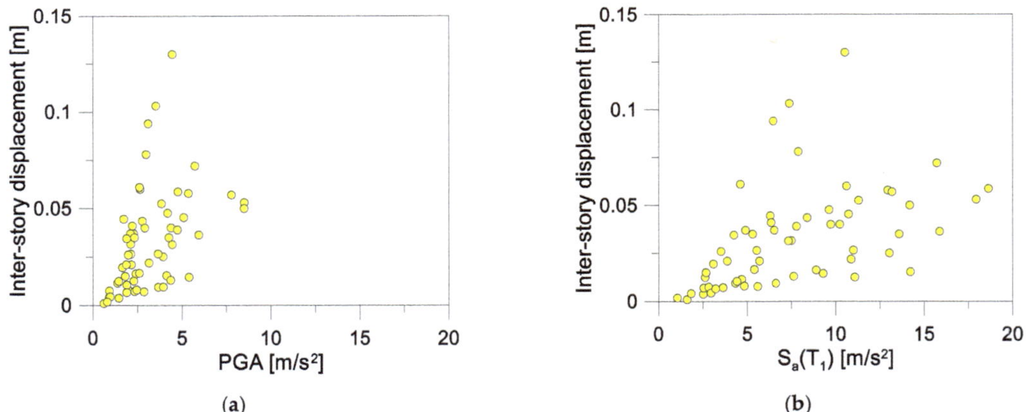

Figure 14. Cloud analysis results (M3) in the form of inter-story lateral displacement, as a function of (**a**) PGA and (**b**) $S_a(T_1)$.

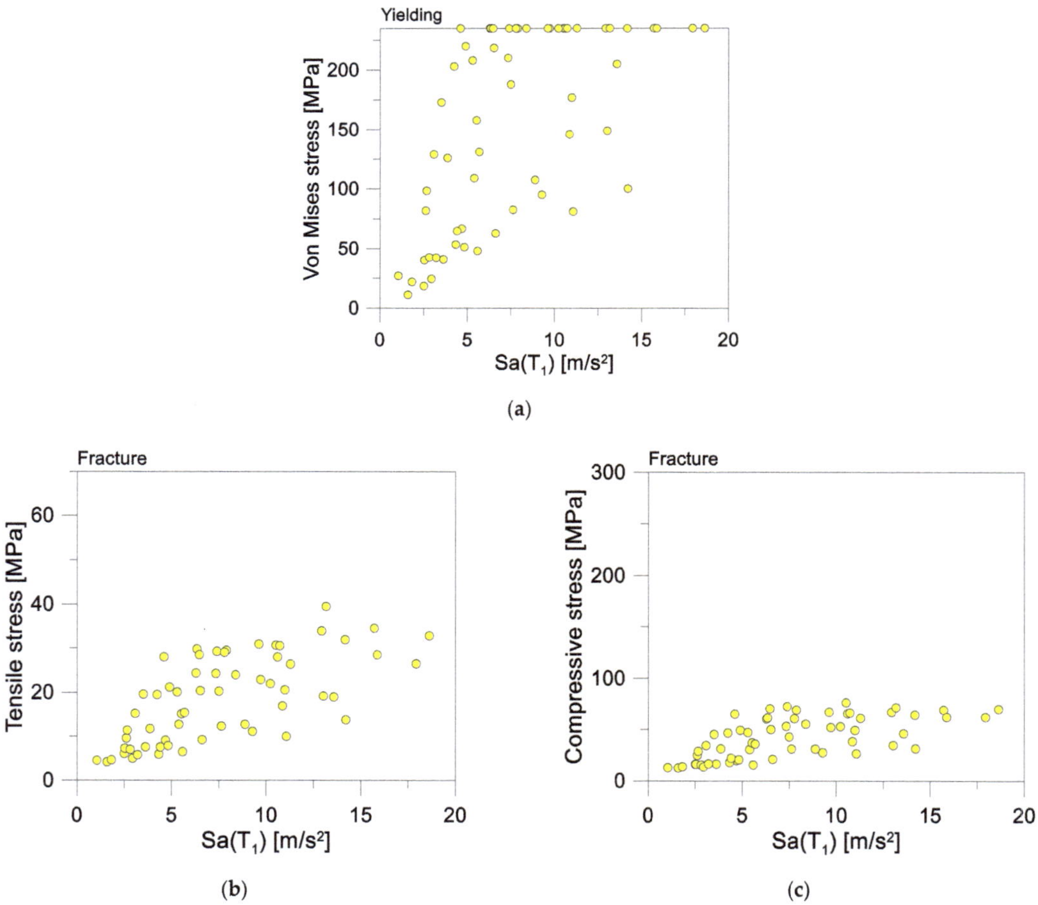

Figure 15. Cloud analysis results (M3) in the form of (**a**) Von Mises stress in the angle brackets, (**b**) tensile stress, and (**c**) compressive stress in glass (hole region), as a function of $S_a(T_1)$.

By considering the entire set of available data from the cloud analysis of the frame, an ordinary least-square linear regression was thus performed. The analysis was carried out in the logarithmic space, given that the use of logarithm of variables improves the fit of the model by transforming the distribution of the features to a more normally shaped bell curve. In order to control the skew and counter problems in heteroskedasticity, both the dependent variable (IM) and the independent variable (EDP) were log-transformed. The final result is proposed in Figure 16, where the linear fits of cloud data are obtained from the least squares method.

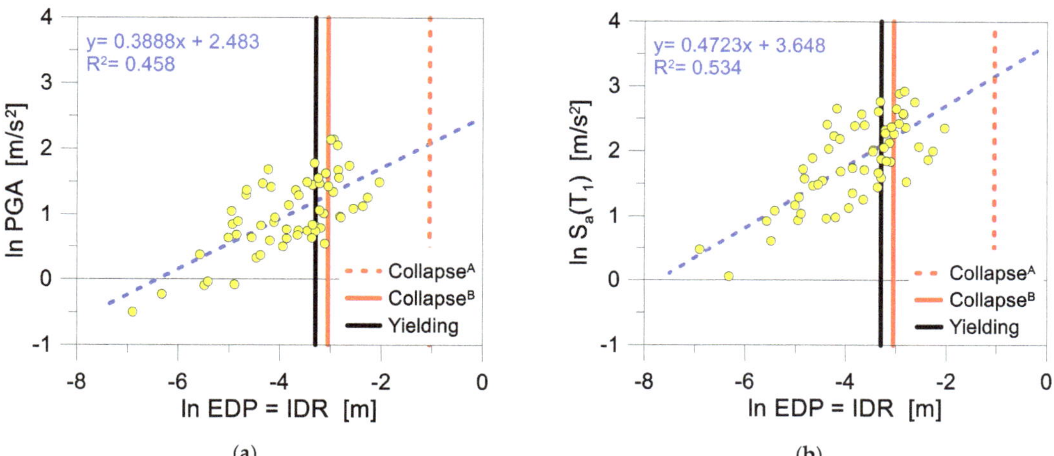

Figure 16. Cloud analysis results (M3) in the form of lateral displacement of the frame, as a function of (a) PGA and (b) $S_a(T_1)$.

The thresholds of "yielding" ($EDP_{Y,50}$ = 0.037 m) and "collapse" ($EDP_{CP,50}$) performance levels are indicated in Figure 16 by vertical lines. Two different thresholds (noted as "A" and "B") were used to identify the collapse prevention limit and to quantify further the influence of the base steel connection in the seismic response of the frame, namely:

- (A) $EDP_{CP,50}$ = 0.048 m (u / $H \approx 0.007$), as calculated by the preliminary PO analysis of the glass column with rigid base connection (Figure 13), and
- (B) $EDP_{CP,50}$ = 0.35 m (u / $H \approx 0.05$), representative of IDR value corresponding to first glass cracking in the PO curve of the frame (Figure 8).

In this regard, it should be noted that the regression line was assumed to be valid for "B" collapse value of displacement even if it is outside the available data cloud. Once the regression line is found, the IM characterizing yielding and collapse prevention were obtained in Figure 16 using the following relations:

$$IM_{CP,50} = \exp(a + b\ln(EDP_{CP,50})) \tag{15}$$

$$IM_{Y,50} = \exp(a + b\ln(EDP_{Y,50})) \tag{16}$$

The q-factor estimation can thus be based on Figure 16, for PGA and $S_a(T_1)$, respectively.

Certainly, the obtained results are affected by base steel joints and thus by $EDP_{CP,50}$. As such, the above outcome should be taken into account as a general approach for basic design considerations of similar structures, given that the resistance and stiffness of joints are strictly responsible for the final ductility of the frame, and thus for the possible fracture initiation in glass.

7.3. Comparative q-factor Predictions

In conclusion, Figure 17 shows the M3 calculated q-factor and a comparison of selected methods (average). As expected, q significantly decreases as far as the M3 approach at collapse disregards the beneficial effect of brackets in the post-yielded stage (collapse "A"). At the same time, as far as the real ultimate inter-story displacement is considered for the tensile fracture of the column (collapse "B"), Figure 17 proves a stable q-factor estimation from M3 or M1–M2 methods.

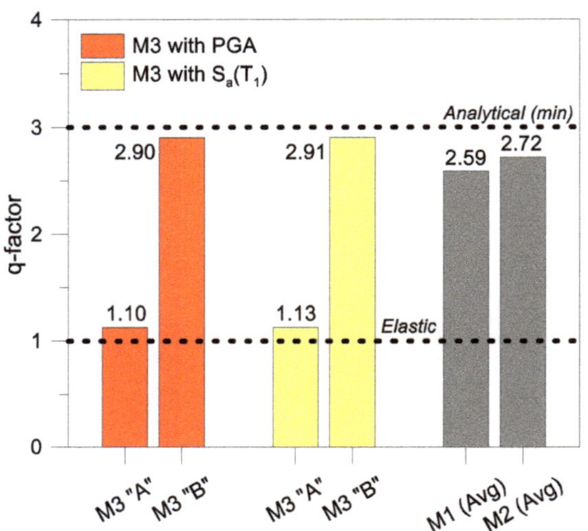

Figure 17. Calculated q-factor for the examined frame, based on M1 to M3 methods and EDPs.

The positive outcome is confirmation of intrinsic ductility and post-yielding capacity for the frame as a whole. This is in line with preliminary results from [19]. Moreover, the local analysis of angle bracket ductility and stress peak estimation in the region of glass holes was quantified in [19] up to $q = 4.58$ for the frame (collapse governed by tensile fracture of glass). The present study, consequently, confirms the need for full-size structural analyses for special structures and joint details.

Comparative data in Figure 17 are also a confirmation of simple analytical expectations about the minimum plastic capacity of the frame from Section 4.2 (with $q_{min} \approx 3$ from Equations (7)–(12) combined with Equation (6)), so as to preserve the glass columns from fracture. At the same time, it is necessary to highlight the relatively stable trend for the q-factor numerical estimates from the M1 to M3 selected approaches. Most importantly, this finding seems to confirm the potential of linear regression method based on cloud analysis, thanks to the computational efficiency of the M3 method. The M1 and M2 procedures, while limited in number of signals, are univocal in EDPs detection but could require major calculation efforts compared to M3 (140 scaled simulations, in the present study). Furthermore, the IDA calculated average q-factor can be highly sensitive to input signals (7 minimum). While the present investigation suggests a very good correlation of M1 to M3 average q-factor predictions, this could not be the case of different structural members, thus requiring even more pronounced calculation efforts from IDA (M1 or M2). Finally, compared to simple analytical estimates that are not able to account for complex mechanical phenomena of the frame as a whole (i.e., $q = 4.58$ from [19]), all the numerical estimates in Figure 17 are on the conservative side, thus confirming the need for refined models and non-linear dynamic procedures in support of seismic design.

8. Conclusions

Available design standards for seismic-resistant buildings provide various recommendations in support of analysis and safe design of several structures subjected to earthquakes, but no specific details are given for glass systems. Among others, major uncertainties derive from the reliable calculation of the seismic performance and dissipation capacity of glass structures, thus their q-factor.

In this paper, attention was focused on the local/global seismic analysis of a structural glass frames under in-plane lateral seismic loads. Careful consideration was paid for the development of efficient finite element (FE) numerical models in support of extended parametric non-linear dynamic analyses that could be used to adapt/assess for glass some consolidated procedures in use for structural systems composed of ordinary materials. For most traditional materials and systems, reference engineering demand parameters (EDPs) are recommended by standards or literature documents. On the other hand, reliable EDPs are still lacking for the methods' adaptation to glass structures.

Three numerical calculation methods were taken into account for q-factor estimates, based on the parametric incremental dynamic analysis (IDA; dynamic "M1" and mixed "M2" methods), and the cloud analysis based on linear regression ("M3"). Numerical calculations were also compared to simple analytical estimates.

From the FE parametric outcomes, more in detail, it was proven that the metal joints in use for structural glass applications were the major source of possible critical failure mechanisms, but also a key source of enhanced ductility performances for glass members. Such a finding was confirmed in line with ductility and flexibility capacities discussed in [19], based on local analysis of the base connection of the frame. In addition, the present study also confirmed the need of full-size FE models and non-linear dynamic procedures.

In terms of calculated q-factor values, more in detail, it was shown that:

- IDA-based approaches (M1 or M2) are univocal in damage detection, thus in the corresponding estimation of reliable EDPs;
- Both M1 and M2 procedures are indeed strongly expensive in computational cost. The present study, for example, was based on a minimum of 7 accelerograms and required up to 140 non-linear dynamic analyses; and
- High sensitivity was observed for the predicted average q-values from M1 or M2, thus recommending a careful selection of input signals, but also the possible use of largest sets of scaled records.

At the same time, the adaptation of M3 method with linear regression to structural glass frames:

- Confirmed the reduced computational cost of the approach, compared to M1 or M2 methods (60 unscaled signals and analyses in total for the present study, compared to 140 simulations); and
- Confirmed that reliable EDPs for special structures should be properly calculated, with the support of refined numerical models or even experimental tests. Existing consolidated EDPs of literature and standards for seismic-resistant structures can hardly adapt to special glass systems and members.
- However, the M3 procedure also gave evidence of some difficulties of dataset interpretation (due to limited stress/deformation levels in the load-bearing members, for some simulations). The reason was found in the set of unscaled input accelerograms that sometimes (when applied to structures characterized by limited self-weight and high flexibility as in the present study) can hardly achieve the desired EDPs at collapse; and
- Furthermore, the FE parametric study proved that—once EDPs and damage mechanisms for relevant limit states are established—the M3 approach can offer rather accurate predictions for glass structures under seismic loads, and thus support as an efficient tool the estimation of q-factor for the seismic design of similar structural systems. For the present case-study frame, the calculated q-factor was in fact in line,

but on the conservative side compared to simple analytical predictions from [19], based on the local analysis of bracket ductility and stress peaks in the region of glass holes. Such a finding also confirms the need for complex numerical models able to capture dynamic mechanical phenomena in similar systems, as a more detailed investigation to combine with simplified analytical procedures.

Author Contributions: Conceptualization, C.B.; methodology, S.M., M.F. and C.B.; validation, C.B.; formal analysis, S.M. and M.F.; data curation, S.M. and M.F.; writing-original, S.M., M.F. and C.B.; writing-review, S.M., M.F. and C.B.; supervision C.B.; project administration, C.B. All authors have read and agreed to the published version of the manuscript.

Funding: This research received no external funding.

Institutional Review Board Statement: Not applicable.

Informed Consent Statement: Not applicable.

Data Availability Statement: Data will be made available upon request.

Conflicts of Interest: The authors declare no conflict of interest.

Appendix A

Table A1. Reference parameters for the selected ground motions records.

Event ID	Date	Soil Type	M_w	R (km)	PGA (m/s^2)	$S_a(T_1)$ (m/s^2)
ME-1979-0003	15/04/1979	B	6.9	6.8	3.53	7.39
ME-1979-0003	15/04/1979	A	6.9	62.9	2.11	7.50
ME-1979-0003	15/04/1979	B	6.9	19.7	2.98	7.89
ME-1979-0003	15/04/1979	B	6.9	19.7	4.45	10.52
ME-1979-0003	15/04/1979	A	6.9	19.7	1.73	6.29
ME-1979-0003	15/04/1979	B	6.9	22	2.77	8.39
GR-1986-0006	13/09/1986	B	5.9	6.6	2.28	6.52
GR-1986-0006	13/09/1986	B	5.9	6.6	2.65	10.61
GR-1986-0006	13/09/1986	B	5.9	5.5	2.91	10.23
EMSC-20161030_0000029	30/10/2016	A	6.5	18.6	4.26	13.57
EMSC-20161030_0000029	30/10/2016	A	6.5	18.6	3.85	11.29
EMSC-20160824_0000006	24/08/2016	B	6	8.5	8.51	17.94
EMSC-20161030_0000029	30/10/2016	B	6.5	26.4	3.94	13.02
IT-2009-0009	06/04/2009	B	6.1	5	4.37	9.72
IT-2009-0009	06/04/2009	B	6.1	4.9	5.35	12.93
EMSC-20161030_0000029	30/10/2016	A	6.5	7.8	4.19	9.63
EMSC-20161030_0000029	30/10/2016	A	6.5	7.8	5.71	15.72
EMSC-20161026_0000095	26/10/2016	B	5.9	14	5.39	9.29
EMSC-20161026_0000095	26/10/2016	B	5.9	39.1	2.40	8.90
EMSC-20161030_0000029	30/10/2016	B	6.5	4.6	4.76	18.63
EMSC-20160824_0000006	24/08/2016	B	6	15.3	3.67	6.61
EMSC-20161030_0000029	30/10/2016	B	6.5	4.6	3.65	11.00
IT-1980-0012	23/11/1980	B	6.9	33.3	3.14	10.87
IT-1980-0012	23/11/1980	B	6.9	33.3	2.21	6.34
EMSC-20161030_0000029	30/10/2016	B	6.5	22.6	4.74	7.79
EMSC-20161030_0000029	30/10/2016	A	6.5	12	7.79	13.17
EMSC-20161030_0000029	30/10/2016	A	6.5	12	8.50	14.17
EMSC-20161030_0000029	30/10/2016	B	6.5	11.4	5.93	15.88
EMSC-20161030_0000029	30/10/2016	B	6.5	11.4	4.13	14.21
EMSC-20161030_0000029	30/10/2016	B	6.5	9.9	2.60	5.39
EMSC-20161030_0000029	30/10/2016	B	6.5	26.1	4.45	7.34
EMSC-20161030_0000029	30/10/2016	B	6.5	26.1	4.36	7.63

Table A1. *Cont.*

Event ID	Date	Soil Type	M_w	R (km)	PGA (m/s²)	$S_a(T_1)$ (m/s²)
TK-2003-0038	01/05/2003	B	6.33	11.8	5.09	10.73
TK-1999-0077	17/08/1999	A	7.6	3.5	2.29	11.07
ME-1979-0012	24/05/1979	B	6.2	8.3	2.61	4.60
ME-1979-0003	15/04/1979	A	6.9	19.7	2.10	4.90
ME-1979-0003	15/04/1979	B	6.9	22	2.32	5.30
GR-1986-0011	15/09/1986	B	-	14.2	1.38	4.68
GR-1993-0027	14/07/1993	B	5.6	4.9	3.95	4.33
GR-1990-0002	17/05/1990	B	-	23	1.98	5.58
GR-1986-0006	13/09/1986	B	5.9	5.5	2.12	5.52
IT-1976-0002	06/05/1976	B	6.4	27.7	3.10	6.47
EMSC-20160903_0000063	03/09/2016	A	4.3	3.6	1.45	2.51
EMSC-20161030_0000029	30/10/2016	B	6.5	20	2.86	2.54
EMSC-20161101_0000060	01/11/2016	A	4.8	18.7	0.61	1.60
EMSC-20161030_0000029	30/10/2016	B	6.5	39.2	0.92	2.82
EMSC-20161030_0000029	30/10/2016	B	6.5	39.2	0.91	1.83
EMSC-20161030_0000029	30/10/2016	B	6.5	39.2	0.96	2.95
EMSC-20161026_0000077	26/10/2016	B	5.4	7.7	2.33	3.63
EMSC-20161026_0000095	26/10/2016	B	5.9	9.2	2.16	3.87
EMSC-20161030_0000029	30/10/2016	B	6.5	17.4	1.89	4.24
ME-1979-0012	24/05/1979	B	6.2	33.3	1.97	3.51
IT-2009-0102	07/04/2009	B	5.5	14.3	1.44	2.62
EMSC-20161026_0000095	26/10/2016	A	5.9	10.8	1.89	4.41
EMSC-20161026_0000095	26/10/2016	A	5.9	16.2	1.65	3.08
EMSC-20161030_0000029	30/10/2016	B	6.5	8.2	2.45	4.82
IT-1977-0008	16/09/1977	B	5.3	7.1	0.80	1.05
IT-1976-0024	11/09/1976	B	5.2	6.1	1.87	5.68
EMSC-20161026_0000077	26/10/2016	B	5.4	8.9	1.81	2.67
EMSC-20161026_0000133	26/10/2016	A	4.5	5.6	1.89	3.21

References

1. Haldimann, M.; Luible, A.; Overend, M. *Structural Use of Glass*; IABSE: Zurich (CH), Switzerland, 2008; ISBN 978-3-85748-119-2.
2. Feldmann, M.; Kasper, R.; Abeln, B.; Cruz, P.; Belis, J.; Beyer, J.; Colvin, J.; Ensslen, F.; Eliasova, M.; Galuppi, L.; et al. *Guidance for European Structural Design of Glass Components—Support to the Imple-Mentation, Harmonization and Further Development of the Eurocodes*; Pinto, D., Denton, F., Eds.; Report EUR 26439-Joint Research Centre-Institute for the Protection and Security of the Citizen: Ispra, Italy, 2014. [CrossRef]
3. Bedon, C.; Amadio, C.; Noé, S. Noé Safety Issues in the Seismic Design of Secondary Frameless Glass Structures. *Safety* **2019**, *5*, 80. [CrossRef]
4. Sucuoglu, H.; Vallabhan, C.V.G. Behaviour of window glass panels during earthquakes. *Eng. Struct.* **1997**, *19*, 685–694. [CrossRef]
5. Lago, A.; Sullivan, T.J. *A Review of Glass Facade Systems and Research into the Seismic Design of Frameless Glass Facades*; ROSE Research Report 2011/11; IUSS Press: Pavia, Italy, 2011; ISBN 978-88-6198-059-4.
6. Baniotopoulos, C.C.; Chatzinikos, K.T. Glass facades of mid-rise steel buildings under seismic excitation. In *Research in Architectural Engineering Series, Volume 1: EU COST C13 Glass and Interactive Building Envelopes*; IOS Press: Amsterdam, The Netherlands, 2007; pp. 239–246. ISBN 978-1-58603-709-3.
7. Sivanerupan, S.; Wilson, J.L.; Gad, E.F.; Lam, N.T.K. Drift performance of point fixed glass façade systems. *Adv. Struct. Eng.* **2014**, *17*, 1481–1495. [CrossRef]
8. Martins, L.; Delgado, R.; Camposinhos, R.; Silva, T. Seismic behaviour of point supported glass panels. In *Proceedings of the Challenging Glass 3. Bos, Delft, The Netherlands, 28–29 June 2012*; Veer Louter, N., Ed.; IOS Press: Amsterdam, The Netherlands, 2012.
9. Casagrande, L.; Bonati, A.; Occhiuzzi, A.; Caterino, N.; Auricchio, F. Numerical investigation on the seismic dissipation of glazed curtain wall equipped on high-rise buildings. *Eng. Struct.* **2019**, *179*, 225–245. [CrossRef]
10. Bellamy, L.; Palermo, A.; Sullivan, T. Developing innovative facades with improved seismic and sustainability performance. In Proceedings of the 12th Conference on Advanced Building Skins, Bern, Switzerland, 2–3 October 2017.
11. Bedon, C.; Amadio, C. Numerical assessment of vibration control systems for multi-hazard design and mitigation of glass curtain walls. *J. Build. Eng.* **2018**, *15*, 1–13. [CrossRef]
12. Bukieda, P.; Engelmann, M.; Stelzer, I.; Weller, B. Examination of Laminated Glass with Stiff Interlayers—Numerical and Experimental Research. *Int. J. Struct. Glas. Adv. Mater. Res.* **2019**, *3*, 1–14. [CrossRef]

13. Zhao, C.; Yang, J.; Wang, X.; Azim, I. Experimental investigation into the post-breakage performance of pre-cracked laminated glass plates. *Constr. Build. Mater.* **2019**, *224*, 996–1006. [CrossRef]

14. Kuntsche, J.; Schuster, M.; Schneider, J. Engineering design of laminated safety glass considering the shear coupling: A review. *Glas. Struct. Eng.* **2019**, *4*, 209–228. [CrossRef]

15. D'Ambrosio, G.; Galuppi, L.; Royer-Carfagni, G. Post-breakage in-plane stiffness of laminated glass: An engineering approach. *Glas. Struct. Eng.* **2019**, *4*, 421–432. [CrossRef]

16. Bedon, C.; Zhang, X.; dos Santos, F.A.; Honfi, D.; Kozłowski, M.; Arrigoni, M.; Figuli, L.; Lange, D. Performance of structural glass facades under extreme loads—Design methods, existing research, current issues and trends. *Constr. Build. Mater.* **2018**, *163*, 921–937. [CrossRef]

17. *Eurocode 8: Design of Structures for Earthquake Resistance—Part 1: General Rules, Seismic Actions and Rules for Buildings*; European Committee for Standardisation: Brussels, Belgium, 1998.

18. *Eurocode 3: Design of Steel Structures—Part 1–5: Plated Structural Elements*; European Committee for Standardisation: Brussels, Belgium, 1993.

19. Santarsiero, M.; Bedon, C.; Moupagitsoglou, K. Energy-based considerations for the seismic design of ductile and dissipative glass frames. *Soil Dyn. Earthq. Eng.* **2019**, *125*, 105710. [CrossRef]

20. ATC3-06. *Tentative Provisions for the Development of Seismic Regulations for Buildings*; Applied Technology Council: Redwood City, CA, USA, 1978.

21. Macedo, L.; Silva, A.; Castro, J.M. A more rational selection of the behaviour factor for seismic design according to Eurocode 8. *Eng. Struct.* **2019**, *188*, 69–86. [CrossRef]

22. Costanzo, S.; Tartaglia, R.; DI Lorenzo, G.; De Martino, A. Seismic Behaviour of EC8-Compliant Moment Resisting and Concentrically Braced Frames. *Buildings* **2019**, *9*, 196. [CrossRef]

23. Kappos, A. Evaluation of behaviour factors on the basis of ductility and overstrength studies. *Eng. Struct.* **1999**, *21*, 823–835. [CrossRef]

24. Borzi, B.; Elnashai, A. Refined force reduction factors for seismic design. *Eng. Struct.* **2000**, *22*, 1244–1260. [CrossRef]

25. Chryssanthopoulos, M.K.; Dymiotis, C.; Kappos, A. Probabilistic evaluation of behaviour factors in EC8-designed R/C frames. *Eng. Struct.* **2000**, *22*, 1028–1041. [CrossRef]

26. Newmark, N.M.; Hall, W. Procedures and criteria for earthquake-resistant design. In *Part of: Selected Papers By Nathan M. Newmark Civil Engineering Classics*; Reprinted from Building Practices for Disaster Mitigation; Building Science Series, No. 46,1973; Natl. Bureau of Standards: Gaithersburg, MD, USA, 1973; Volume 1, pp. 209–236.

27. Megalooikonomou, K.G.; Parolai, S.; Pittore, M. Toward performance-driven seismic risk monitoring for geothermal platforms: Development of ad hoc fragility curves. *Geotherm. Energy* **2018**, *6*, 1–17. [CrossRef]

28. Terrenzi, M.; Spacone, E.; Camata, G. Collapse limit state definition for seismic assessment of code-conforming RC buildings. *Int. J. Adv. Struct. Eng.* **2018**, *10*, 325–337. [CrossRef]

29. D'Aragona, M.G.; Polese, M.; Prota, A. Stick-IT: A simplified model for rapid estimation of IDR and PFA for existing low-rise symmetric infilled RC building typologies. *Eng. Struct.* **2020**, *223*, 111182. [CrossRef]

30. Follesa, M.; Fragiacomo, M.; Casagrande, D.; Tomasi, R.; Piazza, M.; Vassallo, D.; Canetti, D.; Rossi, S. The new provisions for the seismic design of timber buildings in Europe. *Eng. Struct.* **2018**, *168*, 736–747. [CrossRef]

31. Pozza, L.; Scotta, R.; Trutalli, D.; Polastri, A.; Smith, I. Experimentally based q-factor estimation of cross-laminated timber walls. *Struct. Build.* **2016**, *169*, 492–507. [CrossRef]

32. Magenes, G.; Modena, C.; da Porto, F.; Morandi, P. Seismic Behaviour and Design of New Masonry Buildings: Recent Developments and Consequent Effects of Design Codes. In *Proceedings of the Eurocode 8 Perspectives from the Italian Standpoint Workshop*, Edoardo, C., Ed.; CRC Press: Boca Raton, Florida, USA, 2009; pp. 199–212.

33. Zonta, D.; Zanardo, G.; Modena, C. Experimental evaluation of the ductility of a reduced-scale reinforced masonry building. *Mater. Struct.* **2001**, *34*, 636–644. [CrossRef]

34. Casalegno, C.; Russo, S.; Sciarretta, F. Numerical Analysis of a Masonry Panel Reinforced with Pultruded FRP Frames. *Mech. Compos. Mater.* **2018**, *54*, 207–220. [CrossRef]

35. Mazzolani, F.M.; Della Corte, G.; D'Aniello, M. Experimental analysis of steel dissipative bracing systems for seismic upgrading. *J. Civ. Eng. Manag.* **2009**, *15*, 7–19. [CrossRef]

36. D'Ayala, D.; Meslem, A.; Vamvatsikos, D.; Porter, K.; Rossetto, T.; Silva, V. *Guidelines for Analytical Vulnerability Assessment of Low/Mid-Rise Buildings, Vulnerability Global Component Project*; GEM Foundation: Dubai, United Arab Emirates, 2015. [CrossRef]

37. Jalayer, F.; De Risi, R.; Manfredi, G. Bayesian Cloud Analysis: Efficient structural fragility assessment using linear regression. *Bull. Earthq. Eng.* **2015**, *13*, 1183–1203. [CrossRef]

38. Zentner, I.; Gündel, M.; Bonfils, N. Fragility analysis methods: Review of existing approaches and application. *Nucl. Eng. Des.* **2017**, *323*, 245–258. [CrossRef]

39. Bakalis, K.; Vamvatsikos, D. Seismic Fragility Functions via Nonlinear Response History Analysis. *J. Struct. Eng.* **2018**, *144*, 04018181. [CrossRef]

40. Luzi, L.; Lanzano, G.; Felicetta, C.; D'Amico, M.C.; Russo, E.; Sgobba, S.; Pacor, F.; ORFEUSWorking Group, 5. Engineering Strong Motion Database (ESM) (Version 2.0). *Ist. Naz. Geofis. Vulcanol. (INGV)* **2020**. [CrossRef]

41. Kiani, J.; Camp, C.; Pezeshk, S. On the number of required response history analyses. *Bull. Earthq. Eng.* **2018**, *16*, 5195–5226. [CrossRef]

42. Baltzopoulos, G.; Baraschino, R.; Iervolino, I. On the number of records for structural risk estimation in PBEE. *Earthq. Eng. Struct. Dyn.* **2018**, *48*, 489–506. [CrossRef]

43. Bradley, B.A.; Pettinga, D.; Baker, J.W.; Fraser, J. Guidance on the Utilization of Earthquake-Induced Ground Motion Simulations in Engineering Practice. *Earthq. Spectra* **2017**, *33*, 809–835. [CrossRef]

44. Evangelista, L.; Del Gaudio, S.; Smerzini, C.; D'Onofrio, A.; Festa, G.; Iervolino, I.; Landolfi, L.; Paolucci, R.; Santo, A.; Silvestri, F. Physics-based seismic input for engineering applications: A case study in the Aterno river valley, Central Italy. *Bull. Earthq. Eng.* **2017**, *15*, 2645–2671. [CrossRef]

45. Hassan, H.M.; Fasan, M.; Sayed, M.A.; Romanelli, F.; ElGabry, M.N.; Vaccari, F.; Hamed, A. Site-specific ground motion modeling for a historical Cairo site as a step towards computation of seismic input at cultural heritage sites. *Eng. Geol.* **2020**, *268*, 105524. [CrossRef]

46. Bedon, C.; Amadio, C. Design buckling curves for glass columns and beams. *Proc. Inst. Civ. Eng. Struct. Build.* **2015**, *168*, 514–526. [CrossRef]

47. Santo, D.; Mattei, S.; Bedon, C. Elastic Critical Moment for the Lateral–Torsional Buckling (LTB) Analysis of Structural Glass Beams with Discrete Mechanical Lateral Restraints. *Materials* **2020**, *13*, 2492. [CrossRef]

48. Amadio, C.; Bedon, C. A buckling verification approach for monolithic and laminated glass elements under combined in-plane compression and bending. *Eng. Struct.* **2013**, *52*, 220–229. [CrossRef]

49. CNR-DT 210/2013. *Istruzioni per la Progettazione, L'esecuzione ed il Controllo di Costruzioni con Elementi Strutturali di vetro [Guide for the Design, Construction and Control of Buildings with Structural Glass Elements]*; National Research Council of Italy (CNR): Roma, Italy, 2013; (Italian Version). Available online: www.cnr.it/it/node/2630 (accessed on 17 August 2021).

50. NTC2018. *Norme Tecniche Per Le Costruzioni; Design Standard for Buildings*; Ministero delle Infrastrutture e dei Trasporti: Roma, Italy. 17 Gennaio 2018 (In Italian)

51. Pilkey, W.D. *Peterson's Stress Concentration Factors*, 2nd ed.; John Wiley & Sons, Inc.: Hoboken, NJ, USA, 1997.

52. Coker, E.G.; Filon, L.N.G. *Photo-Elasticity*; Cambridge University Press: London, UK, 1931.

53. Frocht, M.M. *Photoelasticity*; Wiley: New York, NY, USA, 1949; Volume 1.

54. Simulia. *ABAQUS Computer Software*; Dassault Systèmes: Providence, RI, USA, 2020.

55. Iervolino, I.; Galasso, C.; Cosenza, E. REXEL: Computer aided record selection for code-based seismic structural analysis. *Bull. Earthq. Eng.* **2010**, *8*, 339–362. [CrossRef]

56. FEMA 356. *Prestandard and Commentary for the Seismic Rehabilitation of Buildings*; FEMA Publication No. 356; American Society of Civil Engineers for the Federal Emergency Management Agency: Washington, DC, USA, 2000.

57. SEAOC. *Vision 2000: Performance Based Seismic Engineering of Buildings*; Structural Engineers Association of California: Sacramento, CA, USA, 1995.

58. UBC. *Uniform Building Code*; International Conference of Building Officials: Whittier, CA, USA, 1997.

59. Bedon, C.; Rinaldin, G.; Fragiacomo, M.; Noè, S. q-factor estimation for 3D log-house timber buildings via Finite Element analyses. *Soil Dyn. Earthq. Eng.* **2019**, *116*, 215–229. [CrossRef]

sustainability

MDPI

Article

Post-Earthquake Damage Assessment—Case Study of the Educational Building after the Zagreb Earthquake

Luka Lulić, Karlo Ožić, Tomislav Kišiček [iD]**, Ivan Hafner** [iD] **and Mislav Stepinac ***[iD]

Faculty of Civil Engineering, University of Zagreb, 10000 Zagreb, Croatia; luka.lulic@grad.unizg.hr (L.L.);
karlo.ozic@grad.unizg.hr (K.O.); tomislav.kisicek@grad.unizg.hr (T.K.); ivan.hafner@grad.unizg.hr (I.H.)
* Correspondence: mislav.stepinac@grad.unizg.hr

Abstract: In the wake of recent strong earthquakes in Croatia, there is a need for a detailed and more comprehensive post-earthquake damage assessment. Given that masonry structures are highly vulnerable to horizontal actions caused by earthquakes and a majority of the Croatian building stock is made of masonry, this field is particularly important for Croatia. In this paper, a complete assessment of an educational building in Zagreb Lower Town is reported. An extensive program of visual inspection and geometrical surveys has been planned and performed. Additionally, an in situ shear strength test is presented. After extensive fieldwork, collected data and results were input in *3Muri* software for structural modeling. Moreover, a non-linear static (pushover) analysis was performed to individuate the possible failure mechanisms and to compare real-life damage to software results.

Keywords: assessment; earthquake; Zagreb; case study; cultural heritage

Citation: Lulić, L.; Ožić, K.; Kišiček, T.; Hafner, I.; Stepinac, M. Post-Earthquake Damage Assessment—Case Study of the Educational Building after the Zagreb Earthquake. *Sustainability* **2021**, *13*, 6353. https://doi.org/10.3390/su13116353

Academic Editor: Giacomo Salvadori

Received: 29 April 2021
Accepted: 27 May 2021
Published: 3 June 2021

Publisher's Note: MDPI stays neutral with regard to jurisdictional claims in published maps and institutional affiliations.

Copyright: © 2021 by the authors. Licensee MDPI, Basel, Switzerland. This article is an open access article distributed under the terms and conditions of the Creative Commons Attribution (CC BY) license (https://creativecommons.org/licenses/by/4.0/).

1. Introduction

On 22 March 2020, at 6 h 22 min, Zagreb Metropolitan area was hit by an earthquake of medium magnitude $M_L = 5.5$, and intensity of VII, in the epicenter, according to the EMS-98 scale [1]. At 7 h 1 min followed the strongest subsequent earthquake of magnitude $M_L = 5.0$ and intensity of VI. The main earthquake damaged most of the buildings in the Lower Town, including residential buildings, universities, schools, kindergartens, hospitals and public buildings. The vast majority of buildings built after the first mandatory earthquake regulations in former Yugoslavia (1964) [2,3] either remained intact or suffered small damage. Nonetheless, the larger part of the city's historical center (Upper and Lower Town) was severely damaged because the buildings in the center were built before any seismic regulations. The damage to historical buildings is enormous. Numerous museums, churches and university buildings have been severely damaged (Figure 1). At the end of the year, Croatia was hit by another devastating earthquake with an epicenter in Petrinja, located approx. 50 km from Zagreb ($M_L = 6.3$). The quake caused subsequent damage to already damaged buildings, but to a lesser extent.

As well as most parts of the European region, many existing buildings in Croatia are built in masonry. Given that most of the so-called "strategic" buildings of cultural significance and high historical importance are built using masonry, such a condition is suggesting that the assessment and rehabilitation of existing masonry structures must be conducted on a very high level [4–8]. An important part of the structural assessment is numerical analysis. When it comes to existing buildings, a more refined non-linear analysis should be adopted. Non-linear static analysis or pushover analysis is important and is recommended in Eurocode 8-3 as a reference method for such situations.

Figure 1. Typical damage to educational buildings after the Zagreb earthquake (photo credit: M. Stepinac).

After the earthquake, the first to respond to a disaster were civil engineers who led and coordinated the entire organization of building assessment and damage detection. Various similar post-earthquake assessment procedures are used worldwide [9–11].

In the first week, a large number of buildings were inspected, with a rapid post-earthquake assessment. The most endangered buildings in the city's center were the ones under cultural heritage protection. The aim of a rapid assessment of buildings is to determine the degree of damage to buildings concerning the protection of life and property, that is, to determine if the buildings are usable, temporarily unusable or unusable. Emerging technological advances allow the usage of artificial intelligence in the post-earthquake assessment process in the form of machine learning methods for more efficient and precise results [6,12–15].

Zagreb's historic urban complex is a protected area regulated by the Law on the Protection and Preservation of Cultural Heritage. The area is divided into two zones, zone A and zone B (Figure 2) [1]. Zone A includes the oldest and most architecturally valuable parts of Zagreb and is characterized by densely-built blocks of buildings made of stone, brick or a combination of materials. Most buildings consist of massive longitudinal and orthogonal walls and masonry ceiling vaults or wooden ceiling beams and wooden roofs (Figure 3) [1]. Many hospitals, schools, business premises, residential and government buildings, cultural institutions, monuments, churches and chapels are located in zone A and are protected either as part of a historic urban complex or as individual heritage buildings per se. A total of 72% [1] of buildings in zone A suffered major damage due to the earthquake; to compare the damage suffered by this area is almost proportional to the value of its cultural heritage. Zone B consists of a variety of urban patterns and a large number of immensely valuable buildings [16]. According to the World Bank report, in the educational sector, 106 buildings intended for preschool education, 214 primary and secondary school buildings and education centers, and 12 pupils' dorms were damaged. In the higher education subsector, the damage was reported to 152 buildings. In addition, the buildings of 29 research institutes were also affected. The total value of damage and losses to the education sector is estimated at EUR 1.8 billion at pre-disaster prices, with 97.9% affecting the City of Zagreb [1].

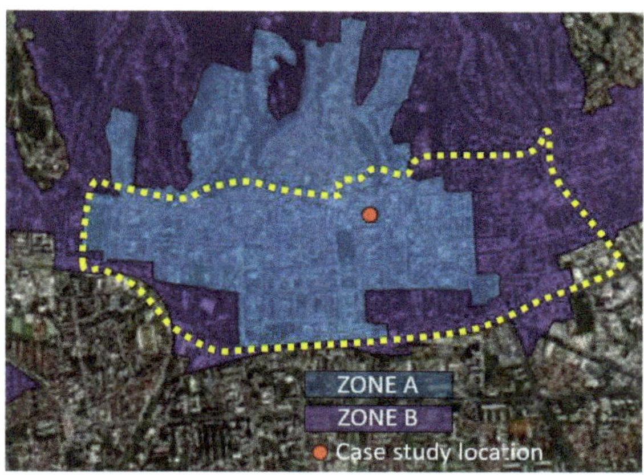

Figure 2. Protected zones A and B with the location of the case study inside the Lower Town of Zagreb (yellow dashed line).

Figure 3. Typical Lower Town masonry building with timber floors and timber roof.

In 2020, there were destructive earthquakes all around the world, causing loss of lives, building collapses, and severe structural and non-structural damage and economic losses (e.g., M7.0 Aegean Sea (Turkey–Greece) [17], M6.7 Elazig (East Turkey) [18], M5.5 and M6.4 Croatia [1]). Identification of vulnerability characteristics and earthquake performance assessment of existing structures are essential steps in reducing earthquake losses, and the topic of seismic assessment of existing masonry structures is actual worldwide. Based on the available state-of-the-art literature on assessment and rehabilitation of existing masonry structures (e.g., [19–23]), this paper presents the Croatian perspective and shows it on an actual case study.

2. The Case Study

The subject of this paper is a building (Figures 4 and 5) located at Vlaška Street 38 as an attached building inside a block. Today's building was built in 1895 by adapting and upgrading two one-story buildings that were built in the early 19th century. The building was upgraded in 1906, while the building's current shape established complete reconstruction in 1997. The building was retrofitted in 1997 for educational purposes and seismic strengthening was not implemented. The building has a rectangular ground plan with the main orientation, which is the longer side of the building in the east-west direction. The building's external dimensions are 12×53 m, with two wings: one, 4.4×7.6 m and the other, 4.2×5.4 m located at the south side of the building. The total floor area of the building is approximately 685 m^2. The building consists of a basement, first, second, third floor and attic. According to the Croatian seismic hazard map [24], the building is located in the area of peak ground acceleration intensity of $a_{gR} = 0.255$ g for a return period of 475 years. The building serves as an educational-scientific institution. The condition of the building before the earthquake, regarding the vertical loads, was satisfactory, and the building was regularly maintained.

Figure 4. Aerial view of the building—north façade (photo credit: M. Stepinac).

Figure 5. Aerial view of the building—south façade (photo credit: M. Stepinac).

The original drawings show foundations that are approximately 1.0 m wide and 1.15 m deep in relation to the surrounding terrain. They were probably built in brick or stone, which is in line with the construction technology of the time. The building is built of a solid brick of a standard format 30 × 15 × 6.5 (unusual for today's standards) used in the late 19th century. The load-bearing wall thicknesses vary throughout the building, reducing with height, and are 51, 43, 28 cm (Figures 6 and 7). Plaster thickness also varies throughout the building from 3 to 6 cm. The ceiling structures before the reconstruction in 1997 were wooden beams, except for the basement ceiling and the first floor where the masonry vaults are located. After the 1997 reconstruction, the attic and 2nd floor ceilings (Figures 8 and 9) were replaced with reinforced concrete slabs, 12 cm and 16 cm, respectively. The 1st floor ceiling is a semi-precast masonry/concrete floor system (Fert ceiling) inserted between the existing wooden beams. In contrast, the ceilings on the ground floor and basement remained masonry vaults. The building also has two auxiliary staircases at the ends of the building made during the 1997 reconstruction and are made of reinforced concrete, and the main central staircase is older and is made of prefabricated stone stairs supported by a wall on one and a beam on the other edge.

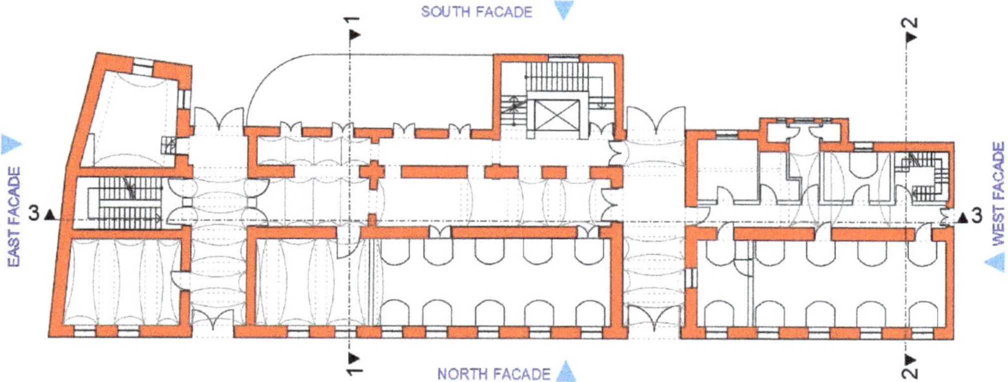

Figure 6. Ground floor plan with load-bearing walls in red.

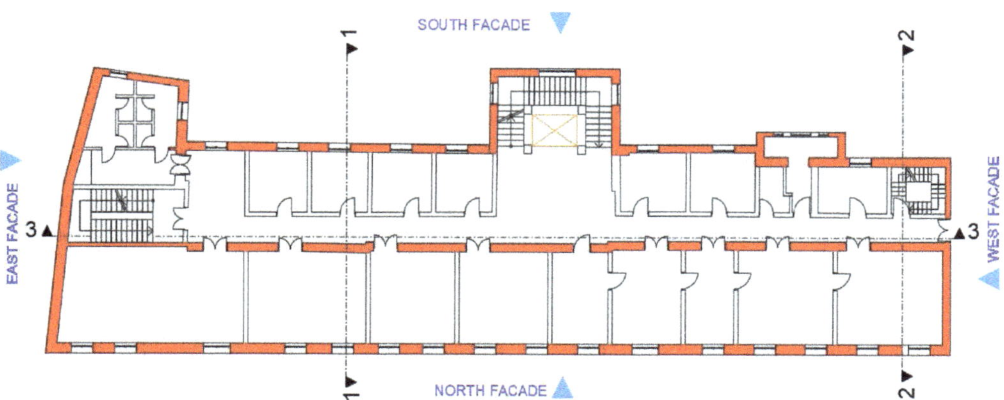

Figure 7. 1st floor plan with load-bearing walls in red. The floor plan of the 2nd floor is identical to the 1st floor plan.

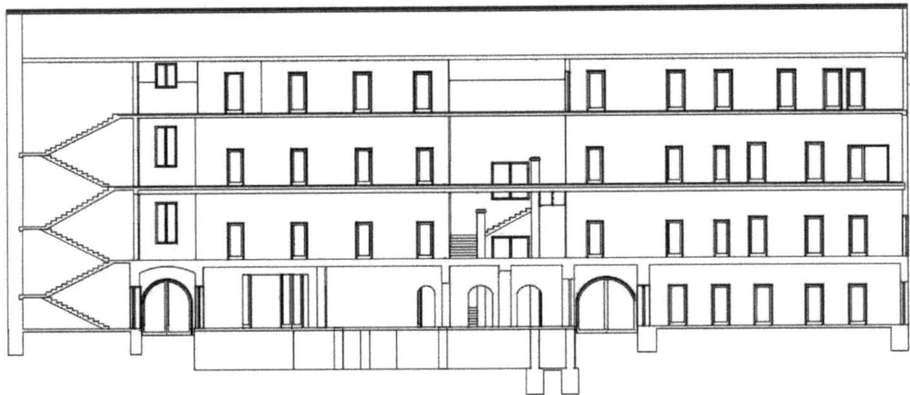

Figure 8. Longitudinal building section.

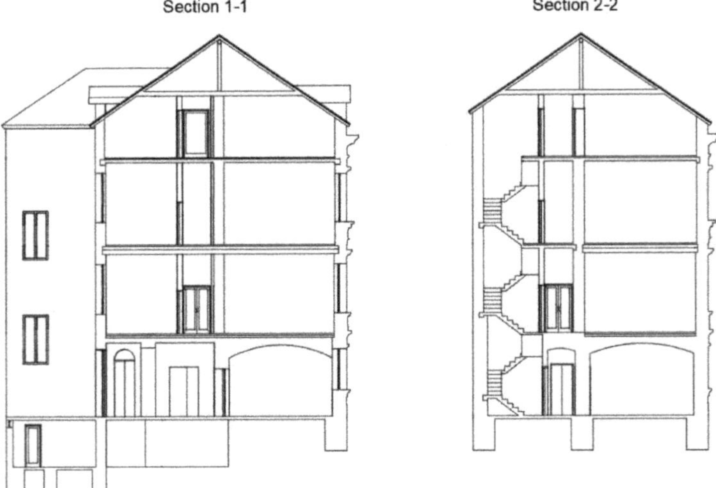

Figure 9. Transversal building sections.

3. Methodology

3.1. Assessment Procedure

The first step in a complete post-earthquake building assessment is a rapid, preliminary assessment of the usability [25,26] of all buildings damaged in the earthquake. Additionally, it is of great importance to preserve the three-dimensional data of the facades of culturally-protected goods in the form of point clouds obtained by laser scanning or drone imaging. The mentioned data can also be used to assess existing structures for the creation of a 3D numerical model. Similar technology was used in the following papers [6,27,28]. In cases where it is needed, detailed assessment and available Non-Destructive Testing (NDT) assessment methods [29] are favorably used. A rapid preliminary assessment is conducted as early as possible after the earthquake, bearing in mind the safety of civil engineers in the field. In Croatia, this type of assessment consisted of a quick visual inspection of individual elements of the load-bearing structure, stating the appropriate degree of damage and deciding on the classification of the building into one of six possible categories (Figure 10): U1 Usable without limitations (Green label), U2 Usable with recommendations

(Green label), PN1 Temporary unusable—detailed inspection needed (Yellow label), PN2 Temporary unusable—emergency interventions needed (Yellow label), N1 Unusable due to external impacts (Red label) and N2 Unusable due to damage (Red label).

Figure 10. Six categories of usability divided into three original labels (in Croatian) [30].

3.2. Rapid Preliminary Assessment Results

A rapid assessment of the building in question was conducted on 23 March 2020. After a quick visual inspection of individual elements of the load-bearing structure, a decision was made to classify the building as temporarily unusable (Yellow label) with a recommendation for a detailed assessment (PN1). Basic conclusions of the preliminary assessment are:

- There is visible damage in the form of cracks on the wall coverings, arches (Figure 11a), vaults and ceilings (Figure 11b) on all floors;
- Separation and local decay of plaster;
- Minor local damage is visible on structural elements (walls, columns, arches);
- In the eastern part of the building, diagonal cracks are visible on the load-bearing walls.

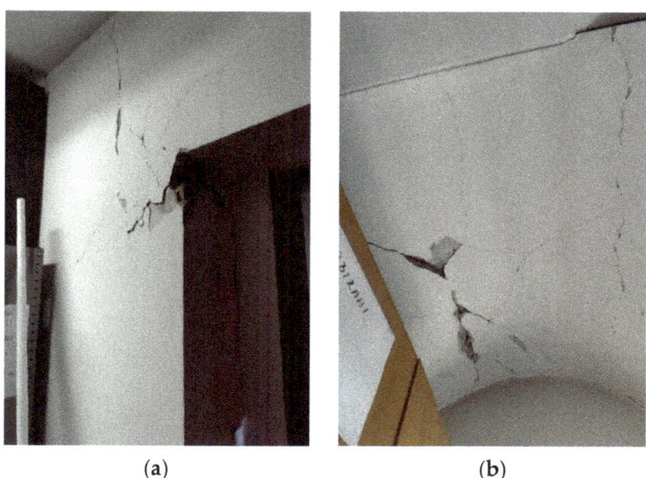

(a)   (b)

Figure 11. Cracks on the 1st floor: wall, lintel (**a**) and ceiling (**b**) (photo credit: I. Hafner).

The second floor and attic suffered minor damage, while the most severe damage is found on the eastern (Figure 12a,b) and central staircase wings (Figure 13a,b). Recommendations were given that the building can be used with a restriction in the zones where there is a danger of plaster falling. Additionally, the eastern and western staircase can be used with a restriction in the number of people, while the central staircase is not to be used until a detailed assessment is conducted.

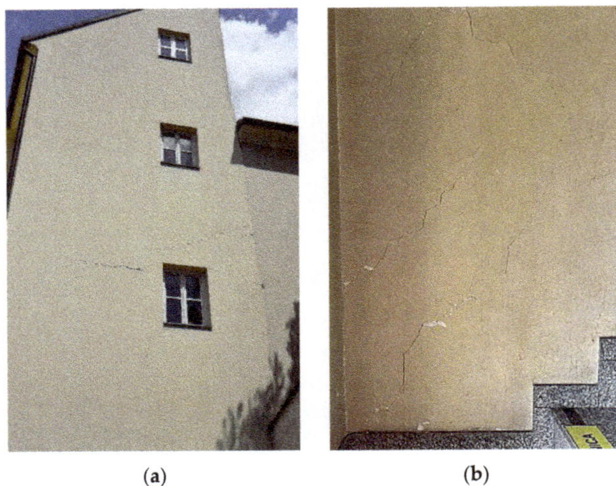

(a)   (b)

Figure 12. Cracks on the eastern staircase: exterior (**a**) and interior (**b**) (photo credit: I. Hafner).

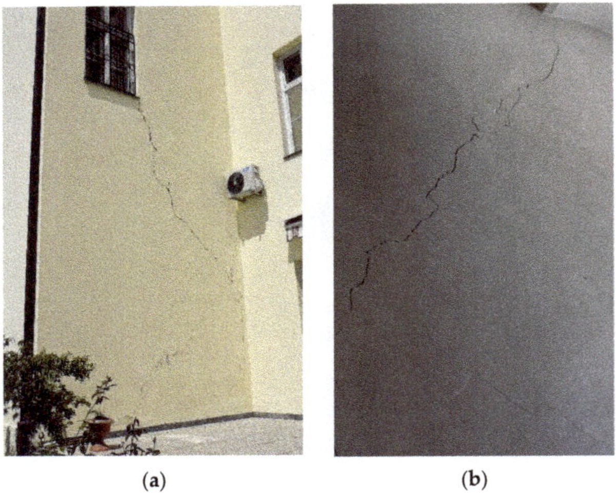

(a)   (b)

Figure 13. Diagonal cracks on the central staircase: exterior (**a**) and interior (**b**) (photo credit: I. Hafner).

3.3. Detailed Assessment Results

According to the current standards for the design of structures—a series of Eurocodes, HRN EN 1990–1998 and the relevant national annexes, the building that is the subject of this study is in the range of peak ground acceleration of 0.255 g; that is, the expected earthquake intensity is IX according to EMS-98 scale for a return period of 475 years. No geotechnical tests have been performed for the site in question for this article, but based on empirical data, a category B foundation soil (deposits of very compacted sand, gravel or hard clay, at least several tens of meters deep) or category C (deep deposits of compacted or medium—compacted sand, gravel or hard clay with a thickness of several tens of meters to several hundred meters) can be assumed. Moreover, based on the latest findings obtained from the research of the Croatian Geological Institute in cooperation with the University of Zagreb, a seismic microzonation map was prepared according to Eurocode 8 standards for the Zagreb area [31]. According to the mentioned seismic microzonation (2017–2019), the soil in the immediate vicinity of the assessed building belongs to the category of soil type C. Soil type C causes a certain amplification of the soil shaking, which must be taken into account when assessing the condition of the structure.

All damage, structural and non-structural, is photographed and described. They are plotted in the floor plans of the building (Figures 14–16). The building was inspected from the air by an unmanned aerial vehicle, and no damage was observed to the main load-bearing structure or the building's roof structure. Decorative crosses, statues and reliefs were also inspected. For the purposes of digital preservation, the 3D model of the building was made on the basis of photogrammetric images.

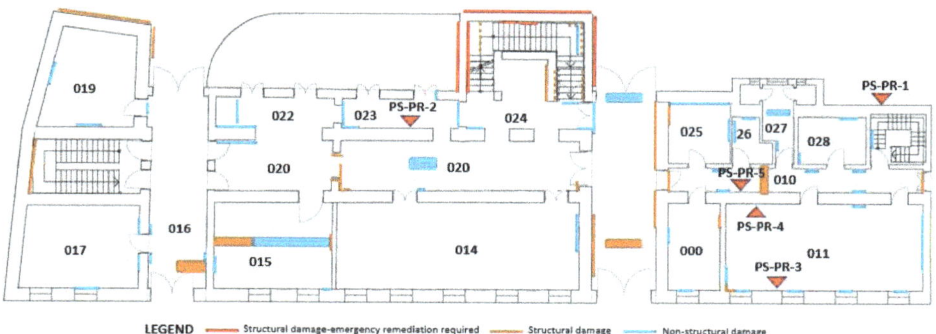

Figure 14. Ground floor of the building—damage scheme and shear strength testing positions.

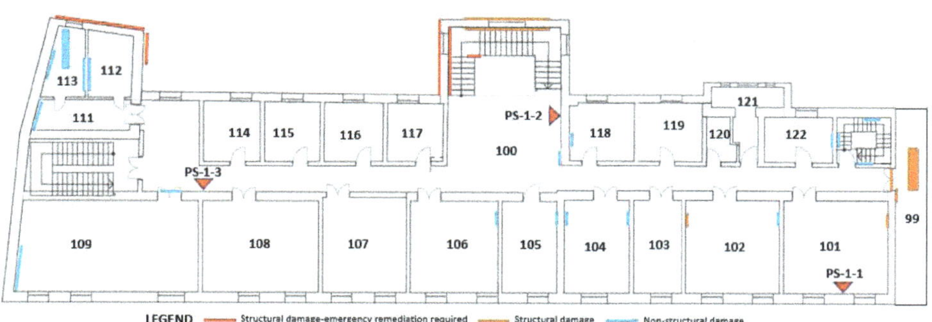

Figure 15. 1st floor of the building—damage scheme and shear strength testing positions.

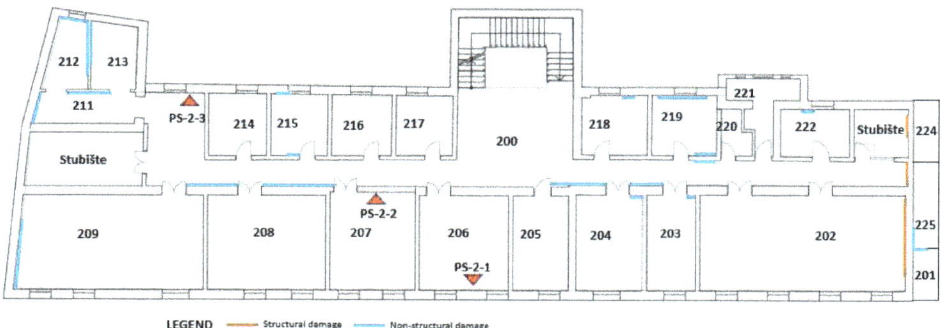

Figure 16. 2nd floor of the building—damage scheme and shear strength testing positions.

A detailed inspection of the building revealed the following damage: at the ground floor, the damage is visible in the form of cracks on the wall coverings, arches, vaults and ceilings, as well as separation and local decay of plaster. Cracks in the barrel vaults are mostly parallel to the supporting joints, probably due to lateral movements during the earthquake. They are the result of the occurrence of tensile stresses perpendicular to the supporting joint. Such cracks can cause hinge formation and consequent loss of stability if they propagate deep enough. Fortunately, the cracks in the assessed building are narrow and mostly found in the plaster. Minor local damage to structural elements (walls, columns, arches) is also visible. In the central core of the building where the main staircase is located and in the eastern part of the building, diagonal cracks are visible on the load-bearing walls, which can also be seen on the north side of the building. Damage is visible on all floors in the form of cracks and falling plaster on the walls. Minor local damage to the walls on the 1st floor is also visible, and cracks at the joints of partition walls and ceilings are locally visible. In the central wing of the building where the main staircase is located and in the eastern wing of the building, diagonal cracks are visible on the load-bearing walls, which can also be seen on the north side of the building.

The 2nd floor and attic suffered minor damage. Particular attention should be paid to the central part of the building, occupying the wing with the staircase. The formation of cracks on the central wing transverse walls is clearly visible, indicating a possible failure mechanism out-of-plane of the entire wing. A wedge was made, and the cracks were interconnected and propagated inside the building (they also appear in the stair beams). There was no displacement of the walls out-of-plane, but the preconditions for its failure were met. The central wing needs to be strengthened as soon as possible as part of the entire building's renovation.

The east wing was also damaged at the ground floor and 1st floor level. The cracks that appeared propagated were through the entire wall of the south facade of the wing. It is unfavorable that the cracks are joined and continue to the transversely-joined walls and lintels. The cause of such cracks can be the slight contribution of a torsional response of the building as a whole, where the boundary elements are the most loaded ones, and their failure occurs. Additionally, that part with the building is connected to the neighboring building. Although this can generally have a positive effect on the whole building, in the case of walls on the east wing, such a boundary condition can cause additional forces. If the walls are not well connected to the diaphragms by a tensile compression connection, this can cause them to fail. Since there has been no displacement of the wall out-of-plane, it is not in danger of collapsing, but it should be strengthened soon, and further damage propagation should be prevented. Minor damage in the form of cracks on the walls can be seen on the west and east staircases.

The building can be used in its entirety except for the main staircase. Depending on the possibilities in the future, a static and seismic analysis of the existing condition of the building should be made, and regardless of whether the entire building will be reinforced, the main staircase and other walls with cracks along the entire width of the walls must be repaired and reinforced. Before that, it is necessary to do all the research work to determine the characteristics of the masonry and other necessary data for the structural analysis.

3.4. In Situ Masonry Shear Strength Tests

In order to assess the condition of the structure after the earthquake and corresponding analysis of the existing condition of the load-bearing structure, in situ tests were carried out. Determination of shear strength (mortar in the composition of load-bearing masonry) of solid brick masonry [32,33] was performed "in situ" using a small hydraulic press "Holmatro" with a load capacity of 200 kN. The mortar was moved horizontally in the vicinity of one brick in order to determine the shear strength. At the same time, the structure of the existing wall was minimally damaged. A total of eleven positions on the load-bearing walls were selected for testing the shear strength of the masonry (Figures 14–16): five positions on the ground floor (PS-PR-1 to PS-PR-5), three positions on the 1st floor (PS-1-1 to PS-1-3) and three positions on the 2nd floor (PS-2-1 to PS-2-3). The shear strength test of the masonry was carried out in nine places. After removing the plaster, it was found that due to the method of masonry (no bricks were found in the longitudinal direction—the inner part of the wall is built of bricks "on edge"), conditions to perform the test were not met for positions PS-PR-3 and PS-PR-4.

Results of the Shear Strength Tests

The shear strength of the masonry was obtained based on the registered horizontal force H_{umax} acting on one brick at the time of reaching the shear strength in that brick and the corresponding mortar area on both sides of the shear is transmitted (Ag + Ad). The test method can be seen in Figure 17 and photographs (Figures 18 and 19).

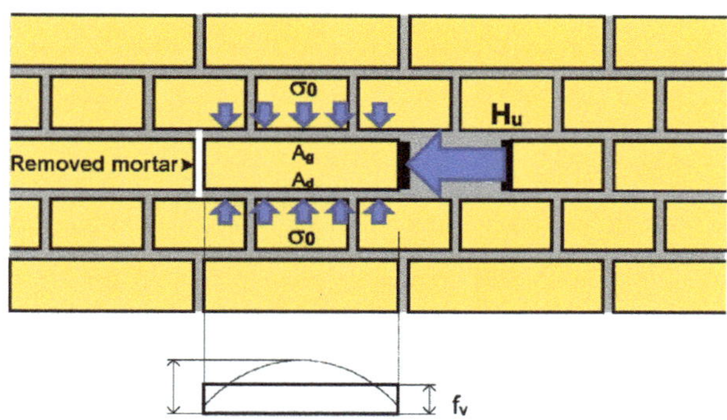

Figure 17. Masonry shear strength test method.

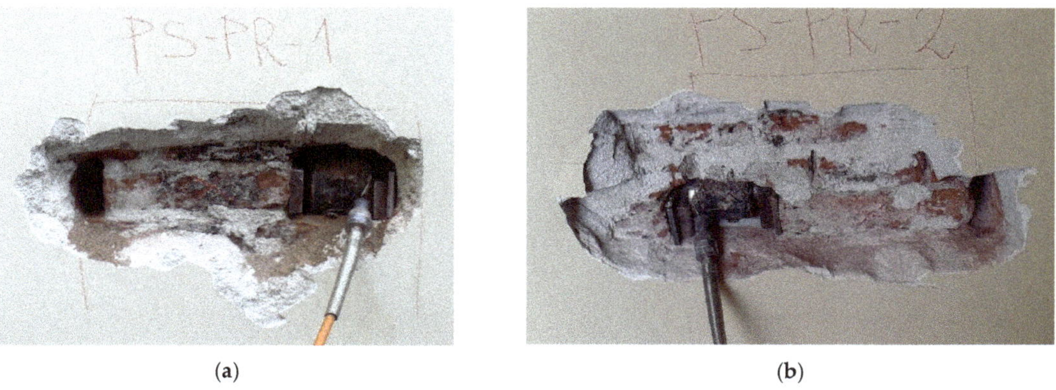

(a)   (b)

Figure 18. Masonry shear strength tests at measuring positions PS-PR-1 (a) and PS-PR-2 (b) on the ground floor (photo credit: L. Lulić).

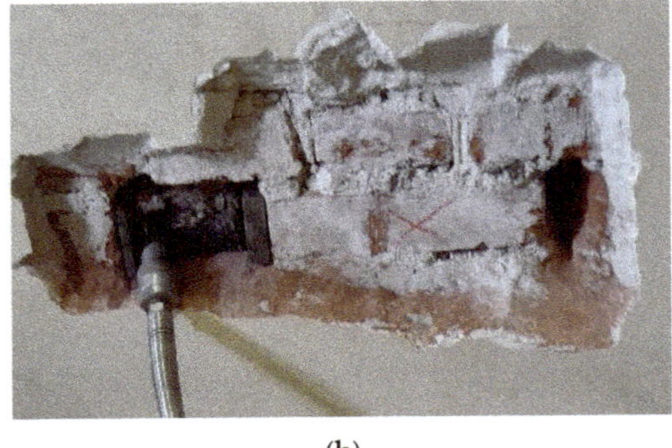

(a)　　　　　　　　　　　　　　　　　　(b)

Figure 19. Masonry shear strength tests at measuring positions PS-1-1 (**a**) and PS-1-3 (**b**) on the 1st floor (photo credit: L. Lulić).

The masonry shear strength test results in the load-bearing walls of the ground floor, 1st floor and 2nd floor are shown in the following Table 1. The values from testing site positions PS-PR-5, PS-1-1, PS-1-2 and PS-2-3 were disregarded because of significant deviations from other results (Figure 20).

Table 1. Masonry shear strength test results in load-bearing walls.

Floor	Testing Site	h (cm)	A_h (cm^2)	$H_{u, max}$ (kN)	Shear Strength f_v (MPa)
Ground floor	PS-PR-1	45	784	55.5	0.708
	PS-PR-2	60	812	58.3	0.717
	PS-PR-3	75	-	-	-
	PS-PR-4	60	-	-	-
	PS-PR-5	60	504	16.3	0.323
1st floor	PS-1-1	50	448	17.6	0.393
	PS-1-2	50	728	121.9	1.675
	PS-1-3	70	526	24.4	0.464
2nd floor	PS-2-1	45	783	44.7	0.571
	PS-2-2	55	812	40.6	0.500
	PS-2-3	55	840	94.8	1.129

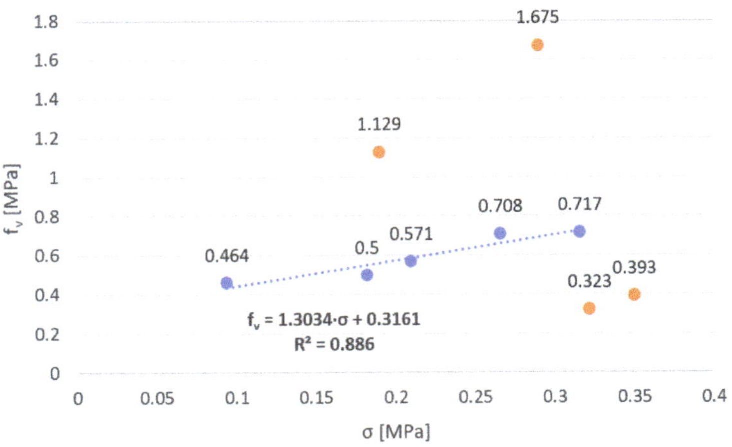

Figure 20. f_v (shear strength)—σ_0 (vertical stress) diagram.

It is the shear strength f_v with the contribution of σ_0—vertical stress. During the test, each measuring position is precisely located to calculate the vertical load (G_0), that is, vertical stress (σ_0) from the numerical model. Therefore, for each test site, in addition to the floor plan position, the height of the measuring position from the upper edge of the ceiling structure (h) is recorded. When analyzing the shear strength of mortar, the vertical constant load is taken into account, that is, vertical stresses at a particular test position. Shear strength according to Mohr–Coulomb law is calculated by Equation (1).

$$f_{vm} = \mu \cdot \sigma_0 + f_{vm0} \tag{1}$$

It can be seen from the diagram that the shear strength without vertical pressure is 0.316 MPa and that coefficient of friction is 1.303. Due to the high value in comparison to EC standard recommendation, the friction coefficient was taken as $\mu = 0.40$ (according to [34]). The obtained shear strength of masonry without vertical pressure, i.e., cohesion, is higher than the one provided by the regulations for the case when there are no tests ($f_{vm0} = 0.32$ MPa > 0.10 MPa). The results show that the quality of masonry is good in contrast to similar buildings from that period.

3.5. Numerical Modeling

The 3D numerical model of the assessed building is obtained in 3Muri software. The macro-element approach is adopted due to computational efficiency and high precision [35]. Its versatility in modeling (implementing elements of various materials, realistic floor stiffnesses, strengthening and many more) makes it highly valuable in a region where a vast majority of building stock is made of masonry. Similar case studies in 3Muri software were used as a base for our research [36–38].

Macro-element approach implies equivalent-frame method which uses non-linear beam elements. Macro-elements (or non-linear beam elements) are divided into three categories which are piers, spandrels and rigid nodes. In piers and spandrels, all deformation is concentrated, and they are connected with rigid nodes. Figure 21 shows an equivalent frame model made of mentioned macro-elements.

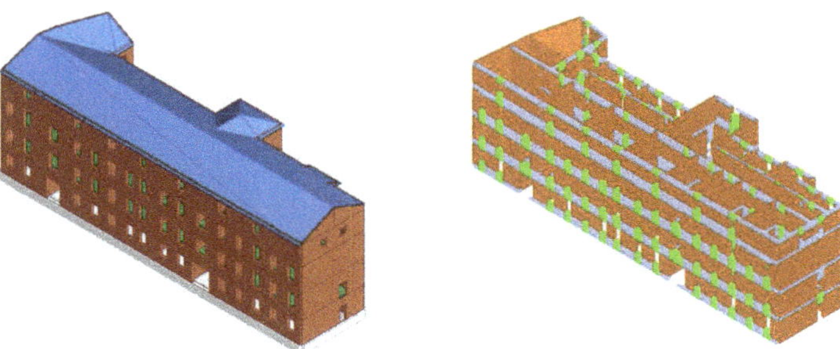

Figure 21. 3D model and 3D equivalent frame in 3Muri.

Non-linear static pushover analysis [39,40] allows us to check the overstrength ratio used in linear analysis and it gives us more detailed insight into critical elements, possible failure mechanisms, and global behavior of the building as a whole. Pushover analysis is performed with constant gravity loads and monotonically-increasing horizontal loads. Two different distributions of the horizontal loads along the structure's height are used for the pushover analysis. The first distribution has a uniform pattern where the horizontal load is proportional to the mass of the building. The second distribution has a modal pattern where the horizontal load is distributed along with the building's height proportionally to the first vibration mode shape of the building determined through elastic analysis (Figures 22 and 23). These horizontal loads are applied at the location of the masses in the model, i.e., at each floor level in the center of masses. Moreover, accidental eccentricity is taken into account to cover uncertainty in the calculation of the center of masses of the building. The 5% of the building's length perpendicular to seismic load direction is taken into account on each side for both x (longitudinal) and y (transverse) directions.

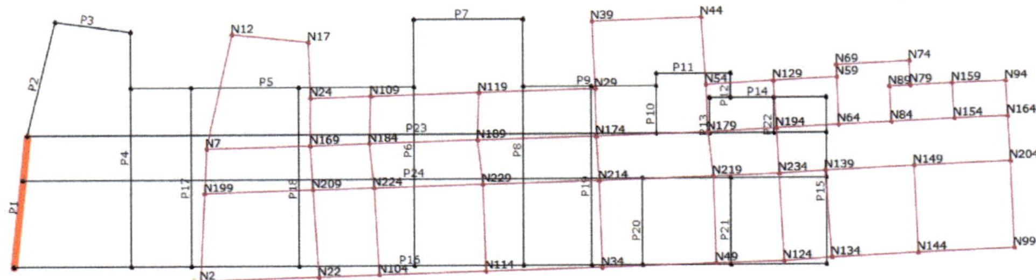

Figure 22. Mode shape used for pushover in y-direction, T = 0.37 s.

Figure 23. Mode shape used for pushover in x-direction, T = 0.24 s.

In each incremental step, internal forces are redistributed according to the element equilibrium, and stiffness is degraded in plastic range. Additionally, ductility is controlled with maximum drift which is 0.004 for shear failure and 0.008 for bending failure [34,41]. Generally, masonry walls have three main failure modes as described in [41–44]. Turnsek–Cacovic constitutive law is used as diagonal cracking is usually the dominant failure mode for existing unreinforced masonry structures [45,46]. In the work of [47], diagonal failure strength is correlated with shear strength used for Turnsek–Cacovic constitutive law. Shear strength obtained by in situ testing is compared with shear strength approximation from the visual MQI method which is explained in more detail in [48] and further developed in [49,50]. Compared shear strengths are very close for this case study which implies good precision of the MQI method. Cracked stiffness of vertical elements is used in a model as recommended in [34], so that cracking that occurs during lifetime because of expected earthquakes of a smaller magnitude is taken into account. Shear and flexural stiffness are taken as half value of initial stiffness.

In a 3D model, floors are modeled as horizontally rigid diaphragms, which is precise enough due to the real in-plane stiffness of the horizontal floor structures. Axial in-plane stiffness of rigid diaphragms in software is infinite and the mass of the real slab is taken into account. Many similar old masonry buildings have an unfavorable distribution of seismic forces due to traditional flexible timber floors [22]. In seismic analysis, the roof is excluded from the load-bearing structure because it does not significantly affect the response of the structure and does not contribute to the global resistance of the structure. Although it was left out of the structural part, its contribution in the form of load on the structure itself was not neglected.

The mean values of material characteristics used in the numerical model (Table 2) are a combination of the literature review [34,51] and on-site testing. Regarding experimental in situ tests and detailed inspection of the structure knowledge level 2 (normal knowledge) can be defined. Based on the achieved knowledge level, confidence factor was taken as 1.2.

Table 2. Masonry material characteristics.

Material Characteristic	Value
Modulus of elasticity	3000 N/mm^2
Shear modulus	1200 N/mm^2
Specific weight	18 kN/m^3
Mean compressive strength of masonry	6.63 N/mm^2
Shear strength	0.14 N/mm^2
Characteristic compressive strength of masonry	5.53 N/mm^2
Confidence factor	1.2
Partial safety factor for material	1
Shear drift	0.0053
Bending drift	0.0107
Final creep coefficient	0.5

According to the work in [52], the building is classified as regular in height but irregular in floor plan, requiring 3D modeling. The building is classified as a torsional stiff system.

First, static analysis is performed according to [53]. Next, the seismic analysis is done. The educational building belongs to importance class III because its seismic resistance is of great importance given the consequences associated with a collapse. Hence, importance factor is γ_I = 1.2. Three PGA values are used for two limit states. According to the new law "Law on the Reconstruction of Earthquake-Damaged Buildings in the City of Zagreb, Krapina-Zagorje County and Zagreb County (NN 102/2020)" [54], ultimate limit state return period can be different depending on the level of strengthening for old masonry buildings damaged in the recent earthquakes. Limit state of significant damage with a return period of 475 and limit state of damage limitation with a return period of 95 years were checked [55]. In the new law [54], the return period of 225 years which corresponds to a probability of exceedance of 20% in 50 years is introduced for a limit state of significant damage.

Elastic response spectrums for acceleration are calculated for all three return periods taking into account parameters for soil type C, which is found on the location of the building. Altogether, 24 pushover analyses are performed; for x- and y-direction in both orientation, with two load distributions, without and with −/+ 5% of accidental eccentricity.

The result of the performed seismic analysis is a capacity curve that shows the ratio of the shear force in the base of the structure and the displacement of the control node. The control node was selected in the immediate vicinity of the center of mass and is located on the top floor of the building. Obtained capacity curves for all 24 analyses can be seen in Figure 24. Bilinearized pushover curves for the x- and y-direction are shown in Figures 25 and 26. Total base shear in kN is plotted on the y-axis and the displacement of the control nodes in mm is plotted on the x-axis.

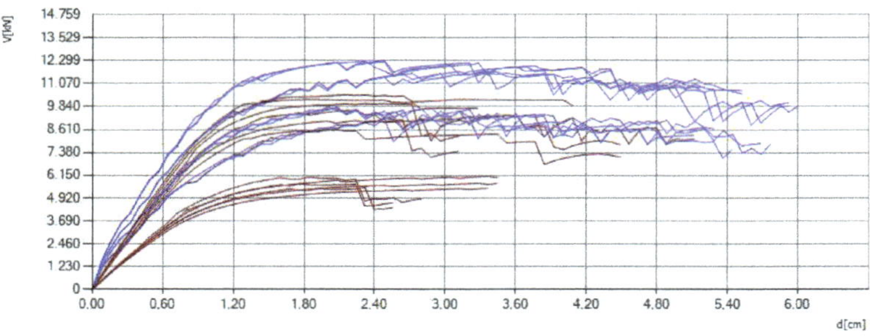

Figure 24. Pushover curves for the x- (blue) and y- (red) direction.

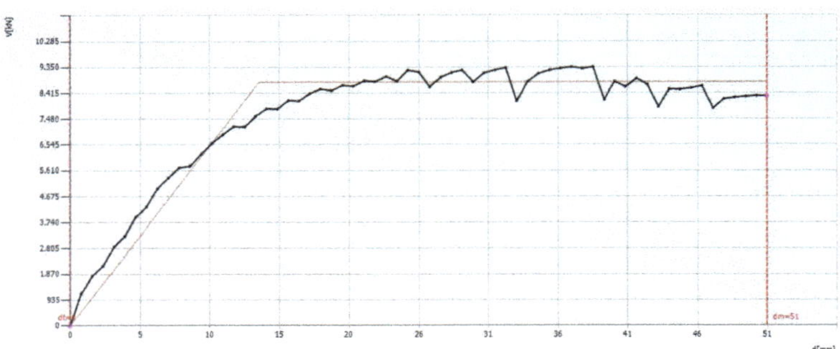

Figure 25. The most relevant pushover curve for the x-direction.

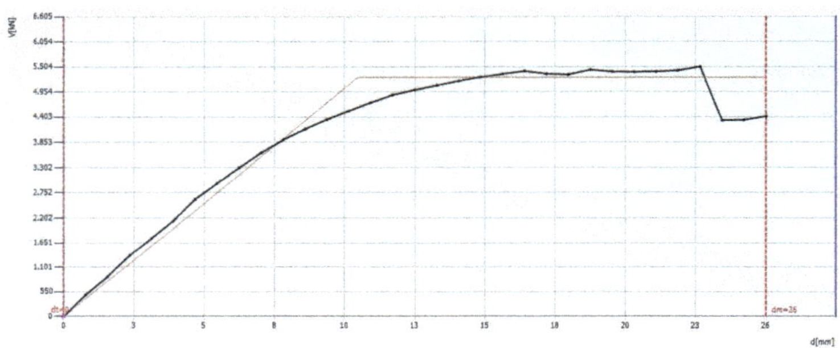

Figure 26. The most relevant pushover curve for the y-direction.

Figures 27 and 28 show the damaged state at the last step of the pushover curves for the x- and y-direction where yellow color represents shear damage, and the red color represents bending damage.

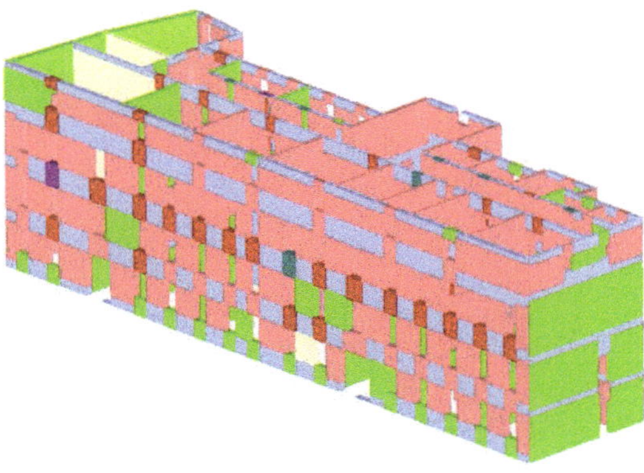

Figure 27. Damage at maximum displacement capacity for pushover in the x-direction.

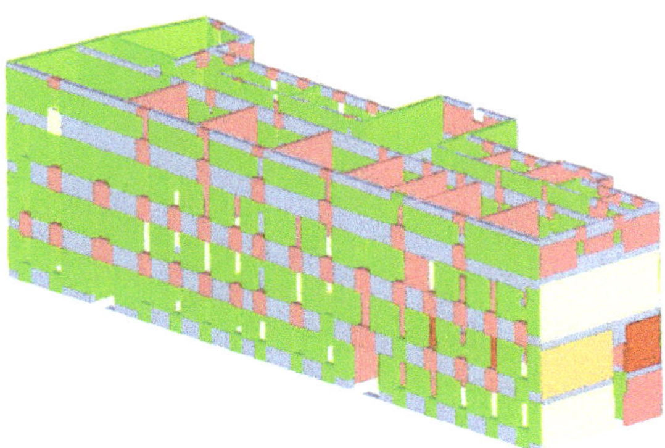

Figure 28. Damage at maximum displacement capacity for pushover in the y-direction.

After the response of the structure, the capacity of the structure is obtained and checks are carried out according to the basic requirements relating to the state of structural damage, defined by limit states. Parameters for equivalent SDOF systems from Figures 27 and 28 are shown in Table 3. These parameters are obtained during bilinearization based on the equivalent energy principal and are used for target displacement determination.

Table 3. SDOF parameters for pushover in x- and y-direction.

Parameter	Value (x-Direction)	Value (y-Direction)
T^* (s)	0.411	0.433
m^* (kg)	2,725,590	2,339,506
w (kN)	45,733	45,733
M (kg)	4,661,901	4,661,901
m^*/M (%)	58.47	50.18
Γ	1.31	1.41
F^*y (kN)	6719	3736
d^*y (cm)	1.06	0.76
d^*m (cm)	3.91	1.81

Results are also given in the form of a parameter α (Table 4), where α is the ratio between the limit capacity acceleration of the building and reference peak ground acceleration on type A ground. Parameter α is given for all limit states. A problem arises with old masonry buildings that often cannot be strengthened to such an extent that today's building codes regarding seismic resistance are satisfied. Hence, a new legal document [54] that followed recent earthquakes allows for different levels of earthquake resistance after reconstruction.

Table 4. α values.

Return Period	α (x-Direction)	α (y-Direction)
475	0.633	0.291
225	0.894	0.411
95	0.560	0.363

For the return period of 475 years, none of the 24 analyses satisfies the limit state of significant damage, while for the return period of 225 years, only 8 of the 24 analyses satisfies the same limit state. For a return period of 95 years, none of the 24 analyses satisfies the limit state of limited damage.

The target displacement is determined in accordance with Eurocode 8 (Appendix B) [52]. Since the building's natural period of vibration is less than the period of T_C, the target displacement is determined by the procedure for short periods shown in Figure 29. Buildings with short natural periods of vibration do not comply with the equal displacement rule [56] as buildings with medium-long and long natural periods of vibration.

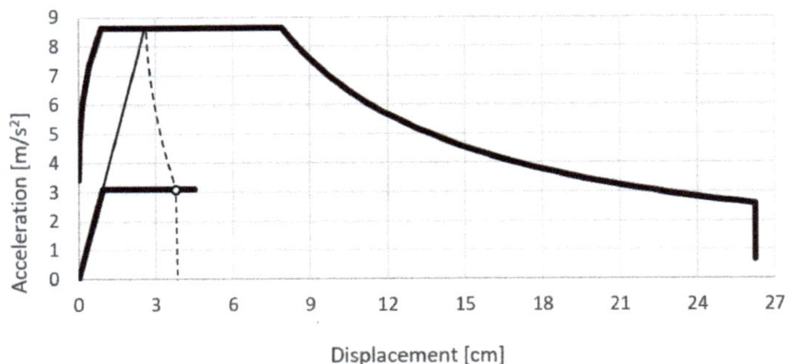

Figure 29. Target displacement determination for short periods.

Since 3Muri does not consider the bending failure of walls out-of-plane during the global non-linear static analysis, it is necessary to subsequently perform an analysis of the bending of the walls out-of-plane. This takes into account the seismic action perpendicular to the walls. The analysis is performed for the boundary condition near collapse. A return period of 475 years is used. Walls are considered to have passed the out-of-plane bending check if their M_{Rd}/M_{Ed} ratio is greater than or equal to 1.0. In Figure 30, walls that did not satisfy the check are colored red.

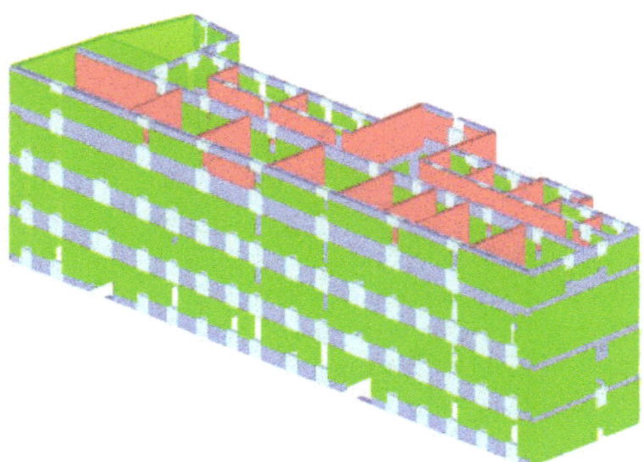

Figure 30. Bending out-of-plane results.

The 3Muri program does not consider out-of-plane loss of stability of local mechanisms in the global analysis either. It assumes that proper connection is established between walls and between walls and diaphragms. That way, out-of-plane local mechanisms are prevented so that the global in-plane response of the building can be evaluated [43]. Therefore, the resistance to local losses of stability is checked in a special program module; certain parts of a single wall are checked and the interaction of parts of several individual walls that can together form different local mechanisms. Figure 31 shows an example of a local mechanism where the kinematic block is colored red.

Local mechanisms are arbitrarily defined according to the structure's shape, common failure mechanisms and earthquake damage. The emergence of local mechanisms is often due to the poor interconnection of walls and walls with floor structures. Linear kinematic analysis is used. Defining a local mechanism consists of three steps. To begin with, it is necessary to define a kinematic block that is a part of a wall that is considered absolutely rigid, and that is subject to movement or tilting relative to another block or the rest of the wall. Then, the boundary conditions are defined and finally, the load needs to be set. Some of the possible local failure mechanisms of the observed structure and its resistance to the shown forms of failure are presented below (Figure 32 and Table 5). Parameter α represents the ratio between the spectral acceleration of the activation of the mechanisms and spectral acceleration of the seismic demand.

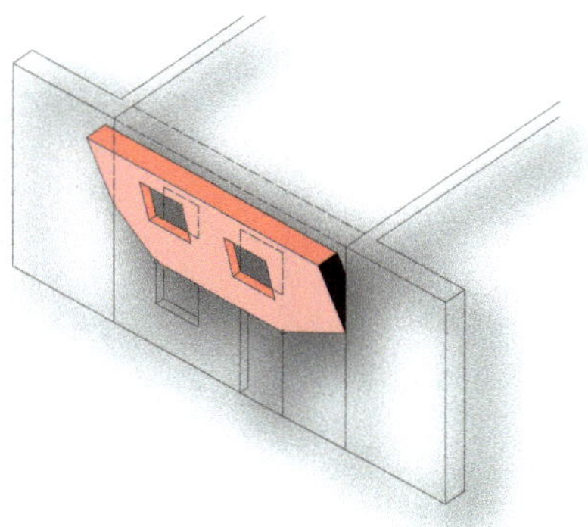

Figure 31. Example of a local mechanism.

Figure 32. Local mechanism LM1 (east wing), LM2 and LM3 (central wing).

Table 5. Results for local mechanisms' analysis.

Local Mechanism	α
LM1	4.93
LM2	2.10
LM3	0.53

After the performed analyses, the real damage was compared with the damage distribution previewed in 3Muri (Figures 33–36). Description of building's capacity through displacement rather than forces allows us to better understand and accurately predict a building's response in the form of a damage initiation and propagation throughout all phases of analysis all the way until the formation of failure mechanism and collapse. That is possible, thanks to incremental non-linear static pushover analysis that shows us damage patterns for every macro-element in every step of the analysis. Compared results are very similar to real damage, which implies good accuracy of the used analysis and software.

Figure 33. First comparison (photo credit: T. Kišiček).

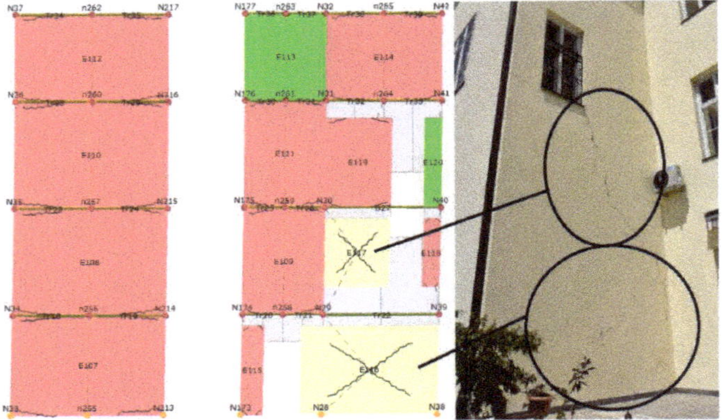

Figure 34. Second comparison (photo credit: T. Kišiček).

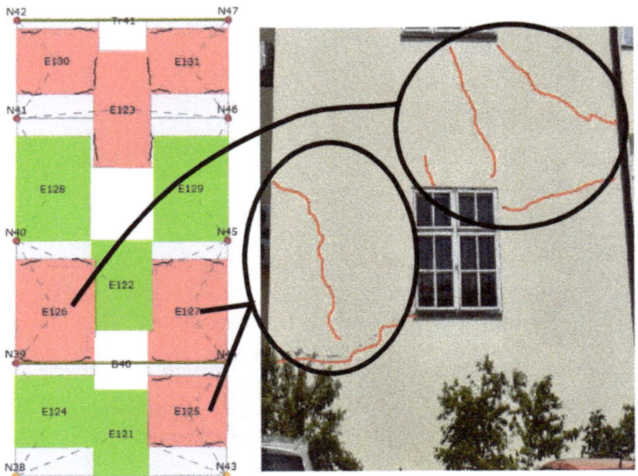

Figure 35. Third comparison (photo credit: T. Kišiček).

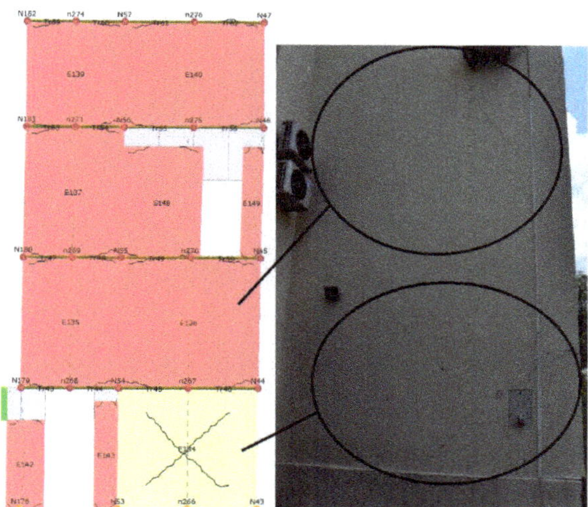

Figure 36. Fourth comparison (photo credit: T. Kišiček).

Another important piece of information is that the studied building is a part of a masonry building aggregate, which is not rare in Zagreb and other European cities. The behavior of buildings in the aggregate is extremely complex, where many parameters affect their response during an earthquake. Moreover, buildings at the ends of aggregates are often more severely damaged than those inside aggregates during earthquakes [57,58]. Modeling of the entire aggregate or, at least, adjacent buildings is recommended. Adjacent buildings can be modeled as mutually independent, mutually interconnected by the same walls, or mutually interconnected by various connections that accurately simulate real coherence. Those connections transfer axial and/or shear load between adjacent buildings in aggregate and they can be modeled as linear or non-linear elements [58–60]. Due to the lack of data on neighboring buildings, modeling of only the observed building is often resorted to. This is on the safe side in the case of a unit inside the aggregate when deformations are observed, but there is a risk of misinterpretation of the failure mechanism [61].

The observed building was modeled as an isolated unit due to insufficient data on neighboring buildings. In order to roughly assess the effect of neighboring buildings on the observed building, the vulnerability index proposed in the articles [62,63] is used. The index is based on five categories depending on the height of the surrounding buildings, the position of the building in the unit, the mismatch of floor heights, differences in material characteristics between buildings and differences in the areas of openings on the facade walls. This method was developed as an upgrade to the vulnerability index based on the ten parameters described in [64]. According to the mentioned five parameters, the buildings in the aggregate have a minimal impact on the observed building, and it can be assumed that the modeling of the building is as isolated as on the safe side.

4. Discussion and Conclusions

Based on similar case studies from the recent earthquakes in Italy [37,38,65–69], the assessment of the existing educational building is made in order to determine its seismic resistance. The building was damaged in the recent earthquakes in Croatia. The building was not constructed according to the principles of seismic design, but the reconstruction in 1997 partially improved the condition of the structure by adding transverse walls and replacing old wooden beams with reinforced concrete floors. Such rigid diaphragms enable good connection of all walls and thus, better behavior of the building in an earthquake. That is why increasing the stiffness of the traditional timber floors in old masonry buildings is usually one of the first measures in seismic retrofitting. On the other hand, new research suggests that replacing traditional wooden floors with rigid diaphragms, i.e., RC floors, can induce some unwanted consequences such as cracks on the edges of the two materials or, in the worst scenario, disintegration and collapse of the masonry walls. However, for earthquakes with expected smaller magnitudes as the ones in Zagreb, this measure is convenient and serves to reinforce the existing structure to horizontal actions.

The assessment of the existing unconfined masonry structure was performed using 3Muri with an equivalent frame method. Rather than linear analysis, non-linear static (pushover) seismic analysis is used due to its various benefits. Limit state checks were carried out for the return periods of 95, 225 and 475 years. The results are in line with the expected behavior with respect to the existing wall distribution in the structure. The structure is less rigid and has greater displacements in the y-direction. Additionally, the capacity of the structure is smaller for the y-direction. There is also a small eccentricity between the center of rigidity and the center of mass that causes a slight but negative impact of torsion on the global behavior of the structure. The critical elements are the walls of the central stairwell and the transverse walls on the west side of the building Additionally, bending failure out-of-plane was checked. Local failure mechanisms were also analyzed using linear kinematic analysis. A comparison of the actual damage with the damage obtained as a result of the conducted non-linear static seismic analysis in the 3Muri program was also conducted.

After assessing the building's behavior in an earthquake, it is obvious that strengthening is needed to raise the seismic resistance level of the building. Damage caused by an earthquake must be repaired to prevent the progression of damage and the possible threat to the global resistance and stability of the building.

As Croatia was hit by two severe earthquakes last year, the knowledge of 'build back better' is fully appreciated. That means that sustainable materials and innovative concepts [70–73] should be used and energy efficiency ensured [74,75]. Various methods are used for strengthening of masonry buildings [76–78] and materials such as FRP and TRM will probably be adopted due to their compatibility and reversibility.

This study provides a detailed insight into the behavior of the building during the earthquake, as well as the extent and distribution of damage and critical elements for earthquakes of different intensities.

Author Contributions: Conceptualization: M.S., L.L. and T.K.; methodology: M.S. and T.K.; software: L.L.; validation: M.S. and T.K.; investigation: I.H., L.L. and K.O.; writing—original draft preparation: M.S., K.O., L.L. and I.H.; writing—review and editing: L.L., M.S. and T.K.; visualization: L.L., K.O., I.H. and M.S.; supervision: M.S. and T.K.; project administration: M.S. All authors have read and agreed to the published version of the manuscript.

Funding: This research was funded by the Croatian Science Foundation, grant number UIP-2019-04-3749 (ARES project—Assessment and rehabilitation of existing structures—development of contemporary methods for masonry and timber structures), project leader: Mislav Stepinac.

Institutional Review Board Statement: Not applicable.

Informed Consent Statement: Not applicable.

Data Availability Statement: The data presented in this study are available on request from the corresponding author. The data are not publicly available due to privacy reasons.

Conflicts of Interest: The authors declare no conflict of interest.

References and Note

1.   Stepinac, M.; Lourenço, P.B.; Atalić, J.; Kišiček, T.; Uroš, M.; Baniček, M.; Šavor Novak, M. Damage classification of residential buildings in historical downtown after the ML5.5 earthquake in Zagreb, Croatia in 2020. *Int. J. Disaster Risk Reduct.* **2021**, *55*, 102140. [CrossRef]

2.   Federal Republic of Yugoslavia (FRY) Official Journal. Ordinance on Technical Standards for the Construction of High-Rise Buildings in Seismic Areas. No. 31/81, 49/82, 29/83, 20/88, 52/90, 1 February 2021.

3. Federal Republic of Yugoslavia (FRY) Official Journal. Ordinance on Temporary Technical Regulations for Construction in Seismic Areas. No. 39/64, 1 February 2021.
4. ARES Project 1st Workshop. Available online: www.grad.hr/ares (accessed on 15 February 2021).
5. Atalić, J.; Šavor Novak, M.; Uroš, M. Seismic risk for Croatia: Overview of research activities and present assessments with guidelines for the future. *Građevinar.* **2019**, *71*, 923–947.
6. Stepinac, M.; Gašparović, M. A Review of Emerging Technologies for an Assessment of Safety and Seismic Vulnerability and Damage Detection of Existing Masonry Structures. *Appl. Sci.* **2020**, *10*, 5060. [CrossRef]
7. Ortega, J.; Vasconcelos, G.; Rodrigues, H.; Correia, M.; Ferreira, T.M.; Vicente, R. Use of post-earthquake damage data to calibrate, validate and compare two seismic vulnerability assessment methods for vernacular architecture. *Int. J. Disaster Risk Reduct.* **2019**, *39*, 101242. [CrossRef]
8. Rodríguez, A.S.; Rodríguez, B.R.; Rodríguez, M.S.; Sánchez, P.A. 9-Laser Scanring and Its Applications to Damage Detection and Monitoring in Masonry Structures. *Woodhead Publishing Series in Civil and Structural Engineering, Long-term Performance and Durability of Masonry Structures*; Ghiassi, B., Lourenço, P.B., Eds.; Woodhead Publishing, 2019; pp. 265–285. Available online: https://www.sciencedirect.com/science/article/pii/B9780081021101000091 (accessed on 15 February 2021).
9. Yavari, S.; Chang, S.E.; Elwood, K.J. Modeling post-earthquake functionality of regional health care facilities. *Earthq. Spectra.* **2010**, *26*, 869–892. [CrossRef]
10. Marshall, J.D.; Jaiswal, K.; Gould, N.; Turner, F.; Lizundia, B.; Barnes, J.C. Post-earthquake building safety inspection: Lessons from the canterbury, New Zealand, earthquakes. *Earthq. Spectra.* **2013**, *29*, 1091–1107. [CrossRef]
11. Didier, M.; Baumberger, S.; Tobler, R.; Esposito, S.; Ghosh, S.; Stojadinovic, B. Improving post-earthquake building safety evaluation using the 2015 Gorkha, Nepal, Earthquake rapid visual damage assessment data. *Earthq. Spectra.* **2017**, *33*, 415–438. [CrossRef]
12. Zhang, Y.; Burton, H.V.; Sun, H.; Shokrabadi, M. A machine learning framework for assessing post-earthquake structural safety. *Struct. Saf.* **2018**, *72*, 1–16. [CrossRef]
13. Kim, T.; Song, J.; Kwon, O.S. Pre- and post-earthquake regional loss assessment using deep learning. *Earthq. Eng. Struct. Dyn.* **2020**, *49*, 657–678. [CrossRef]
14. Naito, S.; Tomozawa, H.; Mori, Y.; Nagata, T.; Monma, N.; Nakamura, H.; Fujiwara, H.; Shoji, G. Building-damage detection method based on machine learning utilizing aerial photographs of the Kumamoto earthquake. *Earthq. Spectra.* **2020**, *36*, 1166–1187. [CrossRef]
15. Bialas, J.; Oommen, T.; Rebbapragada, U.; Levin, E. Object-based classification of earthquake damage from high-resolution optical imagery using machine learning. *J. Appl. Remote Sens.* **2016**, *10*, 036025. [CrossRef]
16. Republic of Croatia Ministry of Construction and Physical Planning. Proposal of the Long-Term Strategy for Mobilising Investment in the Renovation of the National Building Stock of the Republic of Croatia. 2014. Available online: https://mgipu.gov.hr/UserDocsImages//dokumenti/Engleska//HR-Art4BuildingStrategy_en.pdf (accessed on 15 February 2021).
17. Papadimitriou, P.; Kapetanidis, V.; Karakonstantis, A.; Spingos, I.; Kassaras, I.; Sakkas, V.; Kouskouna, V.; Karatzetzou, A.; Pavlou, K.; Kaviris, G.; et al. First Results on the Mw=6.9 Samos Earthquake of 30 October 2020. Bull. *Geol. Soc. Greece.* **2020**, *56*, 251–279. [CrossRef]
18. Günaydin, M.; Atmaca, B.; Demir, S.; Altunişik, A.C.; Hüsem, M.; Adanur, S.; Ateş, Ş.; Angin, Z. Seismic damage assessment of masonry buildings in Elazığ and Malatya following the 2020 Elazığ-Sivrice earthquake, Turkey. *Bull. Earthq. Eng.* **2021**, *19*, 2421–2456. [CrossRef]
19. Casapulla, C.; Argiento, L.U.; Maione, A. Seismic safety assessment of a masonry building according to Italian Guidelines on Cultural Heritage: Simplified mechanical-based approach and pushover analysis. *Bull. Earthq. Eng.* **2018**, *16*, 2809–2837. [CrossRef]
20. Grillanda, N.; Valente, M.; Milani, G.; Chiozzi, A.; Tralli, A. Advanced numerical strategies for seismic assessment of historical masonry aggregates. *Eng. Struct.* **2020**, *212*, 110441. [CrossRef]
21. Valente, M.; Milani, G. Damage assessment and collapse investigation of three historical masonry palaces under seismic actions. *Eng. Fail. Anal.* **2019**, *98*, 10–37. [CrossRef]
22. Ortega, J.; Vasconcelos, G.; Rodrigues, H.; Correia, M. Assessment of the influence of horizontal diaphragms on the seismic performance of vernacular buildings. *Bull. Earthq. Eng.* **2018**, *16*, 3871–3904. [CrossRef]
23. Endo, Y.; Pelà, L.; Roca, P. Review of Different Pushover Analysis Methods Applied to Masonry Buildings and Comparison with Nonlinear Dynamic Analysis. *J. Earthq. Eng.* **2017**, *21*, 1234–1255. [CrossRef]
24. Herak, M.; Allegretti, I.; Herak, D.; Ivančić, I.; Kuk, V.; Marić, K.; Markušić, S.; Sović, I. Seismic Hazard Map of Croatia for a Return Periods of 95, 225 and 475 years. Available online: http://seizkarta.gfz.hr/karta.php (accessed on 15 February 2021).
25. Uroš, M.; Šavor Novak, M.; Atalić, J.; Sigmund, Z.; Baniček, M.; Demšić, M.; Hak, S. Post-earthquake damage assessment of buildings–procedure for conducting building inspections. *Gradjevinar* **2020**, *72*, 1089–1115. [CrossRef]
26. Stepinac, M.; Rajčić, V.; Barbalić, J. Inspection and condition assessment of existing timber structures. *Gradjevinar* **2017**, *69*, 861–873. [CrossRef]
27. Dall'Asta, A.; Leoni, G.; Meschini, A.; Petrucci, E.; Zona, A. Integrated approach for seismic vulnerability analysis of historic massive defensive structures. *J. Cult. Herit.* **2019**, *35*, 86–98. [CrossRef]

28. Betti, M.; Bonora, V.; Galano, L.; Pellis, E.; Tucci, G.; Vignoli, A. An Integrated Geometric and Material Survey for the Conservation of Heritage Masonry Structures. *Heritage* **2021**, *4*, 35. [CrossRef]
29. Stepinac, M.; Kisicek, T.; Renić, T.; Hafner, I.; Bedon, C. Methods for the assessment of critical properties in existing masonry structures under seismic loads-the ARES project. *Appl. Sci.* **2020**, *10*, 1576. [CrossRef]
30. The Database of Usability Classification, Croatian Centre of Earthquake Engineering (HCPI—Hrvatski Centar za Potresno Inženjerstvo), Faculty of Civil Engineering, University of Zagreb and The City of Zagreb. June 2020. Available online: https://www.hcpi.hr/ (accessed on 15 February 2021).
31. Seismic and Geological Micro-Zoning of a Part of the City of Zagreb, Book 1—Seismic and Geological Micro-Zoning, Croatian Geological Institute, Zagreb. 2019. Available online: https://www.hgi-cgs.hr/studija-seizmicka-i-geoloska-mikronizacija-dijela-grada-zagreba/ (accessed on 15 February 2021).
32. Krolo, J.; Damjanović, D.; Duvnjak, I.; Smrkić, M.F.; Bartolac, M.; Košćak, J. Methods for determining mechanical properties of walls. *Gradjevinar* **2021**, *73*, 127–140. [CrossRef]
33. Methods of Test for Masonry—Part 3: Determination of Initial Shear Strength. Available online: https://standards.iteh.ai/catalog/standards/cen/bbead1e6-bd33-4a25-bc75-c023da989f65/en-1052-3-2002 (accessed on 15 February 2021).
34. Eurocode 8: Design of Structures for Earthquake Resistance—Part 3: Assessment and Retrofitting of Buildings. Available online: https://www.phd.eng.br/wp-content/uploads/2014/07/en.1998.3.2005.pdf (accessed on 15 February 2021).
35. Mouyiannou, A.; Rota, M.; Penna, A.; Magenes, M. Identification of Suitable Limit States from Nonlinear Dynamic Analyses of Masonry Structures. *J. Earthq. Eng.* **2014**, *18*, 231–263. [CrossRef]
36. Lamego, P.; Lourenço, P.B.; Sousa, M.L.; Marques, R. Seismic vulnerability and risk analysis of the old building stock at urban scale: Application to a neighbourhood in Lisbon. *Bull. Earthq. Eng.* **2017**, *15*, 2901–2937. [CrossRef]
37. Formisano, A.; Marzo, A. Simplified and refined methods for seismic vulnerability assessment and retrofitting of an Italian cultural heritage masonry building. *Comput. Struct.* **2017**, *180*, 13–26. [CrossRef]
38. Malcata, M.; Ponte, M.; Tiberti, S.; Bento, R.; Milani, G. Failure analysis of a Portuguese cultural heritage masterpiece: Bonet building in Sintra. *Eng. Fail. Anal.* **2020**, *115*, 104636. [CrossRef]
39. Fajfar, P.; Fischinger, M. N2—a method for non-linear seismic analysis of regular buildings. In Proceedings of the 9th World conference in earthquake engineering, Tokyo-Kyoto, Japan, 2–9 August 1998; pp. 111–116.
40. Cerovečki, A.; Kraus, I.; Morić, D. N2 building design method. *Gradjevinar* **2018**, *70*, 509–518. [CrossRef]
41. 3muri User Manual 12.2.1. Available online: https://www.3muri.com/en/brochures-and-manuals/ (accessed on 11 January 2021).
42. Kišiček, T.; Stepinac, M.; Renić, T.; Hafner, I.; Lulić, L. Strengthening of masonry walls with FRP or TRM. *Gradjevinar* **2020**, *72*, 937–953. [CrossRef]
43. Lagomarsino, S.; Penna, A.; Galasco, A.; Cattari, S. TREMURI program: An equivalent frame model for the nonlinear seismic analysis of masonry buildings. *Eng. Struct.* **2013**, *56*, 1787–1799. [CrossRef]
44. Penna, A.; Lagomarsino, S.; Galasco, A. A nonlinear macroelement model for the seismic analysis of masonry buildings. *Earthq. Eng. Struct. Dyn.* **2014**, *43*, 159–179. [CrossRef]
45. Turnšek, V.; Čačovič, F. Some Experimental Results on the Strength of Brick Masonry Walls. Available online: http://www.hms.civil.uminho.pt/ibmac/1970/149.pdf (accessed on 8 February 2021).
46. Tomaževič, M. Shear resistance of masonry walls and Eurocode 6: Shear versus tensile strength of masonry. *Mater. Struct. Mater. Constr.* **2009**, *42*, 889–907. [CrossRef]
47. Borri, A.; Castori, G.; Corradi, M.; Speranzini, E. Shear behavior of unreinforced and reinforced masonry panels subjected to in situ diagonal compression tests. *Constr. Build. Mater.* **2011**, *25*, 4403–4414. [CrossRef]
48. Borri, A.; Corradi, M.; Castori, G.; De Maria, A. A method for the analysis and classification of historic masonry. *Bull. Earthq. Eng.* **2015**, *13*, 2647–2665. [CrossRef]
49. Borri, A.; Corradi, A.; De Maria, A.; Sisti, R. Calibration of a visual method for the analysis of the mechanical properties of historic masonry. *Procedia Struct. Integr.* **2018**, *11*, 418–427. [CrossRef]
50. Borri, A.; Corradi, M.; De Maria, A. The Failure of Masonry Walls by Disaggregation and the Masonry Quality Index. *Heritage* **2020**, *3*, 65. [CrossRef]
51. Ghiassi, B.; Vermelfoort, A.T.; Lourenço, P.B. *Masonry Mechanical Properties*; Ghiassi, B., Milani, G., Eds.; Woodhead Publishing, 2019. Chapter 7. pp. 239–261. Available online: https://www.sciencedirect.com/science/article/pii/B9780081024393000075 (accessed on 8 February 2021). [CrossRef]
52. Eurocode 8: Design of Structures for Earthquake Resistance—Part 1: General Rules, Seismic Actions and Rules for Buildings. Available online: https://www.phd.eng.br/wp-content/uploads/2015/02/en.1998.1.2004.pdf (accessed on 8 February 2021).
53. Eurocode 6: Design of Masonry Structures—Part 1-1: General Rules for Reinforced and Unreinforced Masonry Structures. Available online: https://www.phd.eng.br/wp-content/uploads/2015/02/en.1996.1.1.2005.pdf (accessed on 8 February 2021).
54. Law on the Reconstruction of Earthquake-Damaged Buildings in the City of Zagreb, Krapina-Zagorje County and Zagreb County (NN 102/2020). Available online: https://www.zakon.hr/z/2656/Zakon-o-obnovi-zgrada-o%C5%A1te%C4%87enih-potresom-na-podru%C4%8Dju-Grada-Zagreba,-Krapinsko-zagorske-%C5%BEupanije,-Zagreba%C4%8Dke-%C5%BEupanije,-Sisa%C4%8Dko-moslava%C4%8Dke-%C5%BEupanije-i-Karlova%C4%8Dke-%C5%BEupanije (accessed on 9 February 2021).
55. Eurocode 8: Design of Structures for Earthquake Resistance—Part 3: Assessment and Retrofitting of Buildings—National Annex.

56. Sürmeli, M.; Yüksel, E. An adaptive modal pushover analysis procedure (VMPA-A) for buildings subjected to bi-directional ground motions. *Bull. Earthq. Eng.* **2018**, *16*, 5257–5277. [CrossRef]

57. Fagundes, C.; Bento, R.; Cattari, S. On the seismic response of buildings in aggregate: Analysis of a typical masonry building from *Azores*. *Structures* **2017**, *10*, 184–196. [CrossRef]

58. Senaldi, I.; Magenes, G.; Penna, A. Numerical Investigations on the Seismic Response of Masonry Building Aggregates. *Adv. Mater. Res.* **2010**, *133–134*, 715–720. [CrossRef]

59. Formisano, A.; Massimilla, A. A Novel Procedure for Simplified Nonlinear Numerical Modeling of Structural Units in Masonry Aggregates. *Int. J. Archit. Herit.* **2018**, *12*, 1162–1170. [CrossRef]

60. Stavroulaki, M.E. Dynamic Behavior of Aggregated Buildings With Different Floor Systems and Their Finite Element Modeling. *Front. Built Environ.* **2019**, *5*, 138. [CrossRef]

61. Tomić, I.; Penna, A.; DeJong, M.; Butenweg, C.; Correia, A.; Candeias, P.; Senaldi, I.; Guerrini, G.; Malomo, D.; Beyer, K. Seismic Testing of Adjacent Interacting Masonry Structures (AIMS). In Proceedings of the 17th World Conference on Earthquake Engineering, Sendai, Japan, 27 September–2 October 2021.

62. Formisano, A.; Landolfo, R.; Mazzolani, F.M.; Florio, G. A Quick Methodology for Seismic Vulnerability Assessment of Historical Masonry Aggregates. 2010. Available online: https://www.researchgate.net/publication/266617700_A_quick_methodology_for_seismic_vulnerability_assessment_of_historical_masonry_aggregates (accessed on 8 February 2021).

63. Formisano, A.; Florio, G.; Landolfo, R.; Mazzolani, F.M. Numerical calibration of an easy method for seismic behaviour assessment on large scale of masonry building aggregates. *Adv. Eng. Softw.* **2015**, *80*, 116–138 [CrossRef]

64. Benedetti, D.; Benzoni, G.; Parisi, M.A. Seismic vulnerability and risk evaluation for old urban nuclei. *Earthq. Eng. Struct. Dyn.* **1988**, *16*, 183–201. [CrossRef]

65. Formisano, A. Theoretical and Numerical Seismic Analysis of Masonry Building Aggregates: Case Studies in San Pio Delle Camere (L'Aquila, Italy). *J. Earthq. Eng.* **2017**, *21*, 227–245. [CrossRef]

66. da Porto, F.; Munari, M.; Prota, A.; Modena, C. Analysis and repair of clustered buildings: Case study of a block in the historic city centre of L'Aquila (Central Italy). *Constr. Build. Mater.* **2013**, *38*, 1221–1237. [CrossRef]

67. Lucibello, G.; Brandonisio, G.; Mele, E.; De Luca, A. Seismic damage and performance of Palazzo Centi after L'Aquila earthquake: A paradigmatic case study of effectiveness of mechanical steel ties. *Eng. Fail. Anal.* **2013**, *34*, 407–430. [CrossRef]

68. Chieffo, N.; Formisano, A. Comparative Seismic Assessment Methods for Masonry Building Aggregates: A. Case Study. *Front. Built Environ.* **2019**, *5*, 123. [CrossRef]

69. Boschi, S.; Borghini, A.; Pintucchi, B.; Bento, R.; Milani, G. Seismic vulnerability of historic masonry buildings: A case study in the center of Lucca. *Procedia Struct. Integr.* **2018**, *11*, 169–176. [CrossRef]

70. Fortunato, G.; Funari, M.F.; Lonetti, P. Survey and seismic vulnerability assessment of the Baptistery of San Giovanni in Tumba (Italy). *J. Cult. Herit.* **2017**, *26*, 64–78. [CrossRef]

71. Stepinac, M.; Šušteršič, I.; Gavrić, I.; Rajcic, V. Seismic design of timber buildings: Highlighted challenges and future trends. *Appl. Sci.* **2020**, *10*, 1380. [CrossRef]

72. Funari, M.F.; Mehrotra, A.; Lourenço, P.B. A Tool for the Rapid Seismic Assessment of Historic Masonry Structures Based on Limit Analysis Optimisation and Rocking Dynamics. *Appl. Sci.* **2021**, *11*, 942. [CrossRef]

73. Funari, M.F.; Spadea, S.; Lonetti, P.; Fabbrocino, F.; Luciano, R. Visual programming for structural assessment of out-of-plane mechanisms in historic masonry structures. *J. Build. Eng.* **2020**, *31*, 101425. [CrossRef]

74. Milovanović, B.; Bagarić, M. How to achieve Nearly zero-energy buildings standard. *Gradjevinar* **2020**, *72*, 703–720. [CrossRef]

75. Valluzzi, M.R.; Salvalaggio, M.; Croatto, G.; Dorigatti, G.; Turrini, U. Nested Buildings: An Innovative Strategy for the Integrated Seismic and Energy Retrofit of Existing Masonry Buildings with CLT Panels. *Sustainability* **2021**, *13*, 1188. [CrossRef]

76. Ortega, J.; Vasconcelos, G.; Rodrigues, H.; Correia, M.; Lourenço, P.B. Traditional earthquake resistant techniques for vernacular architecture and local seismic cultures: A literature review. *J. Cult. Herit.* **2017**, *27*, 181–196. [CrossRef]

77. Skejić, D.; Lukačević, I.; Ćurković, I.; Čudina, I. Application of steel in refurbishment of earthquake-prone buildings. *Gradjevinar* **2020**, *72*, 955–966. [CrossRef]

78. Kouris, L.A.S.; Triantafillou, T.C. State-of-the-art on strengthening of masonry structures with textile reinforced mortar (TRM). *Constr. Build. Mater.* **2018**, *188*, 1221–1233. [CrossRef]

sustainability

Article

Assessment of Mechanical Properties and Structural Morphology of Alkali-Activated Mortars with Industrial Waste Materials

Iman Faridmehr [1], Chiara Bedon [2,*], Ghasan Fahim Huseien [3], Mehdi Nikoo [4] and Mohammad Hajmohammadian Baghban [5]

1   Institute of Architecture and Construction, South Ural State University, Lenin Prospect 76, 454080 Chelyabinsk, Russia; faridmekhri@susu.ru
2   Department of Engineering and Architecture, University of Trieste, Via Alfonso Valerio 6/1, 34127 Trieste, Italy
3   Department of Building, School of Design and Environment, National University of Singapore, Singapore 117566, Singapore; eng.gassan@yahoo.com
4   Young Researchers and Elite Club, Ahvaz Branch, Islamic Azad University, Ahvaz 61349-37333, Iran; m.nikoo@iauahvaz.ac.ir
5   Department of Manufacturing and Civil Engineering, Norwegian University of Science and Technology (NTNU), 2815 Gjøvik, Norway; mohammad.baghban@ntnu.no
*   Correspondence: chiara.bedon@dia.units.it; Tel.: +39-040-558-3837

Citation: Faridmehr, I.; Bedon, C.; Huseien, G.F.; Nikoo, M.; Baghban, M.H. Assessment of Mechanical Properties and Structural Morphology of Alkali-Activated Mortars with Industrial Waste Materials. *Sustainability* 2021, 13, 2062. https://doi.org/10.3390/su13042062

Academic Editor: José Ignacio Alvarez
Received: 18 January 2021
Accepted: 10 February 2021
Published: 14 February 2021

Publisher's Note: MDPI stays neutral with regard to jurisdictional claims in published maps and institutional affil-iations.

Copyright: © 2021 by the authors. Licensee MDPI, Basel, Switzerland. This article is an open access article distributed under the terms and conditions of the Creative Commons Attribution (CC BY) license (https://creativecommons.org/licenses/by/4.0/).

Abstract: Alkali-activated products composed of industrial waste materials have shown promising environmentally friendly features with appropriate strength and durability. This study explores the mechanical properties and structural morphology of ternary blended alkali-activated mortars composed of industrial waste materials, including fly ash (FA), palm oil fly ash (POFA), waste ceramic powder (WCP), and granulated blast-furnace slag (GBFS). The effect on the mechanical properties of the Al_2O_3, SiO_2, and CaO content of each binder is investigated in 42 engineered alkali-activated mixes (AAMs). The AAMs structural morphology is first explored with the aid of X-ray diffraction, scanning electron microscopy, and Fourier-transform infrared spectroscopy measurements. Furthermore, three different algorithms are used to predict the AAMs mechanical properties. Both an optimized artificial neural network (ANN) combined with a metaheuristic Krill Herd algorithm (KHA-ANN) and an ANN-combined genetic algorithm (GA-ANN) are developed and compared with a multiple linear regression (MLR) model. The structural morphology tests confirm that the high GBFS volume in AAMs results in a high volume of hydration products and significantly improves the final mechanical properties. However, increasing POFA and WCP percentage in AAMs manifests in the rise of unreacted silicate and reduces C-S-H products that negatively affect the observed mechanical properties. Meanwhile, the mechanical features in AAMs with high-volume FA are significantly dependent on the GBFS percentage in the binder mass. It is also shown that the proposed KHA-ANN model offers satisfactory results of mechanical property predictions for AAMs, with higher accuracy than the GA-ANN or MLR methods. The final weight and bias values given by the model suggest that the KHA-ANN method can be efficiently used to design AAMs with targeted mechanical features and desired amounts of waste consumption.

Keywords: fly ash; granulated blast-furnace slag: palm oil fly ash; ordinary Portland cement; recycled ceramics; green mortar; alkali-activated mix design; compressive strength; embodied energy; CO_2 emission

1. Introduction

There is an environmental concern worldwide regarding the production of ordinary Portland cement (OPC), as it is widely use in the construction sector. It is commonly accepted that OPC manufacturing causes serious pollution issues, including a considerable amount of CO_2 emissions. The majority of annually produced concrete (10 billion metric tons of concrete) contains OPC [1]. Every ton of OPC generates approximately one ton of CO_2, and around 7% of all the global CO_2 emissions are ascribed to OPC production and its corresponding raw material extraction process [2]. It was also estimated that OPC manufacturing would quadruple over the next three decades, which is expected to lead to extensive environmental impacts [3]. To address this concern, using industrial by-products (green concrete philosophy), instead of raw material extraction in conventional concrete, is recognized as a practical solution for a sustainable and cleaner concrete production. Meanwhile, the disposal of industrial waste materials is associated with undesirable ecological impacts. In contrast, their recycling largely contributes to sustainable design, saving natural resources and preventing waste dumping into landfills. The application of industrial wastes with environmentally friendly (i.e., low energy consumption and low CO_2 emission) and inexpensive properties for partial or full replacement of OPC in concrete have attracted the attention of many researchers [4–7]. Allalouex et al. [8] explored the effects of calcined halloysite nano-clay (CHNC) on the physico-mechanical properties and microstructure of high volume slag (HVS) cement mortar. The study in [9] investigated the feasibility of novel industrial waste-co-fired blended ash (CBA) in the development of alkali-activated masonry mortar and reinforced alkali-activated mortar. The design and preparation of Mater-Bi/halloysite nanocomposite materials that could be employed as bioplastics alternative to the petroleum derived products was investigated by Lisuzzo et al. [10]. A novel green protocol for the consolidation and protection of waterlogged archeological woods with wax microparticles has been designed in [11].

This research considers four different industrial waste materials, granulated blast-furnace slag (GBFS), palm oil fly ash (POFA), fly ash (FA), and waste ceramic powder (WCP) in ternary blended alkali-activated mortars. FA, burnt coal by-product, is among the most attractive industrial wastes for producing AAMs because of several unique properties, such as high levels of SiO_2 and Al_2O_3, low cost and embodied energy, and availability at a large scale in many countries. Ogawa et al. [12] investigated the contribution of fly ash to the compressive strength development of mortars cured at different temperatures on the basis of the cementing efficiency factor, they concluded that this factor is significantly affected by the curing temperature. An experimental study on the stress-strain characteristics of alkali-activated slag (AAS) and alkali-activated class C fly ash (FAC) mortars subjected to axial compression was presented in [13]. The authors concluded that the brittleness index estimated from the stress-strain characteristics increases with an increase of activator concentration. In addition, Alkali-activated slag mortar displays highly brittle behaviour marked by no softening behaviour followed by sudden and total failure.

POFA is a by-product mostly produced by various agriculture industries (by burning agricultural waste) in Southeast Asia countries. There is no market value for POFA, and its application is typically limited to landfilling in lagoons and ponds associated with significant environmental pollution. POFA is classified as a pozzolanic substance that is rich in silica content. Accordingly, such an abundant agricultural waste can be used as a partial substitute for OPC, or as a concrete binder to improve strength and durability. The use of an optimum level of POFA, ground granulated blast furnace slag (GGBS), and low calcium FA with manufactured sand (M-sand) to produce geopolymer mortar was investigated by Islam et al. [14]. They concluded that the increase in the POFA content beyond 30% reduces the compressive strength. In another study, the effects of exposing POFA/FA-based geopolymer mortar to elevated temperatures at an early stage in terms of microstructural and compressive strength was investigated in [15]. They concluded that that replacement of the 0–100% of POFA in FA-based geopolymer mortar expedites the start of micro-pore formation, due to exposure to high temperatures and shifts the strength peak from 300 °C to 500 °C.

Furthermore, WCP has high durability to harsh environmental (sulfuric acid or sulfate) conditions. Generally, this material waste has mainly been recycled as a filler for applications such as gardening and tartan floors. To follow a sustainable design approach, several studies proposed the application of ceramic waste in concrete and mortar [16]. The main drawback of using ceramic waste (fine or coarse aggregates) as a replacement of limestone aggregates is its extensive water absorption. Therefore, the durability of such a mix may turn out to be its major deficiency. Harsh environmental agents such as carbonation, chloride, and deleterious salts may easily penetrate the concrete and negatively influence its mechanical properties. Samadi et al. [17] investigated the long-term performance, mechanical properties, and durability of a mortar comprising ceramic waste as supplementary cementitious material and ceramic particles as fine aggregates. In this study, the structure morphology and thermal traits of the designed mixes were characterized using scanning electron microscopy (SEM), thermogravimetric analysis (TGA), differential thermal analysis (DTA), X-Ray Diffraction (XRD), and Fourier-transform infrared spectroscopy (FTIR) measurements. Mechanical and microstructural properties of mortars incorporating ceramic waste powder exposed to the hydrochloric acid solution was investigated in [18]. GBFS (a by-product of iron and steel production) has high SiO_2, and CaO content levels in GBFS provide mechanical properties similar to OPC. It can be used as an OPC replacement producing calcium silicate hydrates (C-S-H), which are strength-enhancing compounds that improve the concrete's strength, durability, and appearance. The effect of granulated blast furnace slag on the self-healing capability of mortar incorporating crystalline admixture was investigated by Li et al. [19]. The results of this study indicated that the mortar with crystalline admixture and 10 wt % GBFS has the highest self-healing capability, and the healing product is mainly composed of calcium carbonate.

Several researchers proposed the consumption of industrial waste materials in AAM designs as a sustainable substitute for OPC [20–22]. However, limited research has been done about the mechanical properties of ternary blended alkali-activated mortars composed of industrial waste materials. In this paper, the mechanical properties of zero cement binder with ternary blended AAMs composed of industrial waste materials are investigated using experimental tests at different ages (1, 3, 7, and 28 days of curing). The effects of the SiO_2, CaO, and Al_2O_3 contents of each binder mass on mechanical properties are investigated using 42 engineered alkali-activated mixes (AAMs). The structural morphology of AAMs is also investigated by the aid of X-ray diffraction (XRD), scanning electron microscopy (SEM), and Fourier-transform infrared spectroscopy (FTIR) measurements. Furthermore, an optimized artificial neural network (ANN) combined with a metaheuristic Krill Herd algorithm (KHA-ANN) and an ANN-combined genetic algorithm (GA-ANN) are developed and compared with a multiple linear regression (MLR) model. From the three different models, the comparative analysis of collected predictions of mechanical properties for the studied AAMs shows the potential of the KHA-ANN model.

2. Materials and Test Methods

2.1. Material Properties of Industrial Wastes

With the aid of the X-ray fluorescence spectroscopy (XRF) test, the chemical compositions of the studied waste materials were determined, as shown in Table 1. It was revealed that the main component in WCP, FA, and POFA was SiO_2 (72.6%, 57.2%, and 64.2%, respectively), while in the GBFS it was CaO (51.8%). SiO_2, Al_2O_3, and CaO are important oxides throughout the hydration process and production phases of the C-(A)-S-H gels.

However, the low contents of Al_2O_3 and CaO in WCP require adding materials containing high amounts of Al_2O_3 (FA) and CaO (GBFS) to produce high-performance alkali-activated binders. According to ASTM C518-15 [23], FA and WCP are classified as class F pozzolans, due to the existence (higher than 70%) of $SiO_2 + Al_2O_3 + Fe_2O_3$.

2.2. Design of AAMs

Ternary blended AAMs were examined to determine the influence of calcium oxide on the geopolymerization process. Using trial mixes, the optimum values of sodium silicate to sodium hydroxide ratio, sodium hydroxide molarity, binder to aggregate ratio, and alkaline solution to binder ratio were selected as 0.75, 4 M, 1, and 0.4, respectively, where these values were fixed for all AAMs. Analytical grade sodium silicate solution (Na_2SiO_3), comprised of H_2O (55.80 wt %), SiO_2 (29.5 wt %), and Na_2O (14.70 wt %), was used as an alkali activator to prepare the proposed AAMs. The sodium hydroxide (NH) pellet was dissolved in water to make the alkaline solution with 4 M concentration. This solution was first cooled for around 24 h and then added to sodium silicate (NS) solution in order to make an alkaline activator solution with a modulus ratio (SiO_2 to Na_2O) of 1.02. This ratio of NS to NH was fixed to 0.75 for all the alkaline mixtures.

Table 1. Physical and chemical features of waste materials.

Material	GBFS	FA	POFA	WCP
Physical characteristics				
Specific gravity	2.9	2.2	1.96	2.6
particle size (μm)	12.8	10	8.2	35
Chemical composition (% mass)				
SiO_2	30.8	57.20	64.20	72.6
Al_2O_3	10.9	28.81	4.25	12.6
Fe_2O_3	0.64	3.67	3.13	0.56
CaO	51.8	5.16	10.20	0.02
MgO	4.57	1.48	5.90	0.99
K_2O	0.36	0.94	8.64	0.03
Na_2O	0.45	0.08	0.10	13.5
SO_3	0.06	0.10	0.09	0.01
Loss on ignition (LOI)	0.22	0.12	1.73	0.13

Four ternary blended AAMs were investigated, where at each level, the GBFS percentage as a source of CaO remained constant to a minimum of 20% in the replacement process and a maximum of 70%, as presented in Table 2. Using the SEM test, the effects of each industrial waste replacement on the contents of SiO_2, CaO, and Al_2O_3 and the AAMs geopolymerization process can be seen.

Table 2. Ternary blended alkali-activated mixes (AAMs) and contents of constituents in each mix design.

| No. | High-Volume FA Mix Design | | | | | |
| | Binder (% Mass) | | | Ratio of Chemical Composition | | |
	FA	GBFS	POFA	SiO_2:Al_2O_3	CaO:SiO_2	CaO:Al_2O_3
1	70	30	0	2.10	0.39	0.82
2	70	20	10	2.31	0.28	0.66
3	60	40	0	2.15	0.51	1.10
4	60	30	10	2.38	0.39	0.94
5	60	20	20	2.62	0.29	0.76
6	50	50	0	2.22	0.65	1.43
7	50	40	10	2.47	0.51	1.26
8	50	30	20	2.74	0.40	1.08
9	50	20	30	3.02	0.29	0.89

Table 2. *Cont.*

	High-Volume POFA Mix Design					
No.	POFA	GBFS	FA			
1	70	30	0	8.63	0.42	3.61
2		20	10	7.04	0.32	2.23
3	60	40	0	7.32	0.53	3.86
4		30	10	6.13	0.41	2.54
5		20	20	5.33	0.31	1.66
6	50	50	0	6.25	0.65	4.08
7		40	10	5.34	0.52	2.80
8		30	20	4.72	0.41	1.94
9		20	30	4.27	0.31	1.31
	High-Volume GBFS Mix Design					
No.	GBFS	FA	POFA			
1	70	30	0	2.38	0.97	2.32
2		20	10	2.85	0.97	2.77
3		10	20	3.53	0.96	3.41
4		0	30	4.57	0.96	4.41
5	60	40	0	2.29	0.80	1.83
6		30	10	2.69	0.80	2.15
7		20	20	3.25	0.79	2.59
8		10	30	4.06	0.79	3.23
9		0	40	5.35	0.79	4.25
10	50	50	0	2.22	0.65	1.43
11		40	10	2.57	0.65	1.66
12		30	20	3.04	0.65	1.97
13		20	30	3.68	0.65	2.39
14		10	40	4.65	0.65	3.03
15		0	50	6.25	0.65	4.07
	High-Volume WCP Mix Design					
No.	WCP	FA	GBFS			
1	70	0	30	5.09	0.26	1.31
2		10	20	4.62	0.17	0.79
3	60	0	40	4.79	0.37	1.77
4		10	30	4.35	0.27	1.19
5		20	20	4.01	0.18	0.74
6	50	0	50	4.48	0.50	2.24
7		10	40	4.08	0.39	1.59
8		20	30	3.77	0.29	1.09
9		30	20	3.53	0.20	0.70

The essential features of Table 2 can be summarized as follows:

- For AAMs with a high volume of FA, CaO content is observed to increase with increasing GBFS content and decreasing FA and POFA contents. However, Al_2O_3 content decreases with increasing GBFS and POFA contents and decreasing FA content.
- For AAMs with a high volume of POFA, the reduction in the POFA content leads to progressive reduction of the silicate content. Furthermore, the Al_2O_3 content increases when the ratio of GBFS to FA increases, whereas the CaO content significantly decreases.
- For AAMs with a high volume of GBFS, an increase in the GBFS content leads to an increase in CaO content. However, by decreasing GBFS content and increasing POFA content, SiO_2 content increases.
- For AAMs with a high volume of WCP, finally, an increase in the content of WCP leads to an enhancement of SiO_2 content. Additionally, the replacement of WCP by increasing the amount of GBFS results in an increased CaO content.

2.3. Experimental Test Procedure

After preparing the AAMs, the experimental program discussed herein involved the casting process, where the resulting mortar was poured into the molds using the two-layers pouring method. To eliminate air pockets within the mixture, each layer was subjected to vibration for 15 s. After finishing the casting process, the AAMs were cured for 24 h in an ambient atmosphere (with a temperature $24 \pm 1.5\ °C$ and a relative humidity of 75%). Figure 1 illustrates the typical procedure to obtain AAMs with varying ratios of industrial wastes.

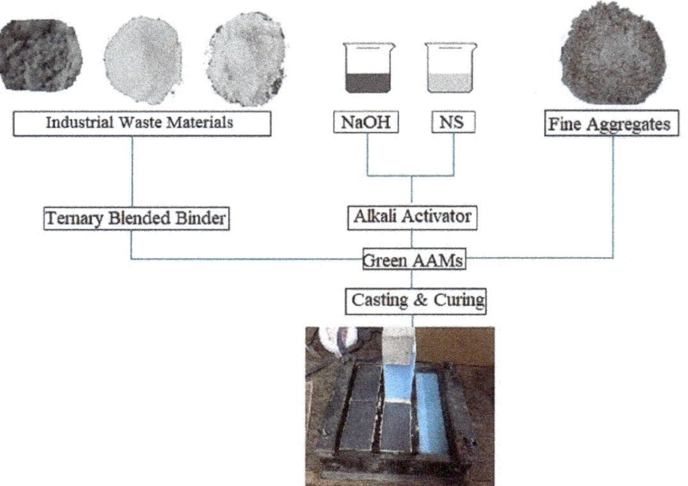

Figure 1. Procedure of producing green alkali-activated mixes (AAMs) mortars.

A standard test rig was used for the experimental derivation of compressive strength (CS), tensile strength (TS), and flexural strength (FS). All the mechanical properties were recorded at the age of 1, 3, 7, 28 days, following ASTM C109-109M [24]. Cube molds of the dimension (50 × 50 × 50) mm were prepared for hardened tests of compressive strength. Prisms of the dimension (40 × 40 × 160) mm were used to prepare the samples for flexural strength test, while for tensile strength tests, cylinders were prepared with a diameter of 75 mm and depth of 150 mm. For the compressive test, a universal testing machine was used, and a constant loading rate of 2.5 kN/s was applied to all tested specimens. The equivalent compressive strength values were hence recorded automatically on the basis of the specimens' size. A similar approach was applied to calculate the TS of specimens in compliance with ASTM C496/C496M [25], in which the splitting TS can be predicted as:

$$TS = 2P/\tau DL \tag{1}$$

In Equation (1), TS is the splitting tensile capacity (MPa), P is the maximum axial load resistance (N), D is the diameter of the cylinder (mm), and L is the length of the cylinder (mm).

The flexural strength—also called modulus of rupture, bend strength, or fracture strength—is another essential measure of the mechanical performance for brittle materials. This is defined as the material's ability to resist deformation under load. The flexural strength test was carried out using the ASTM C78/C78M [26] procedure, where adequately cured prismatic specimens were tested. Similar to CS and TS, the universal testing machine was employed, in which the FS value is expressed as:

$$FS = 3FL/2bd^2 \tag{2}$$

where F is the applied load at the fracture point (N), L is length of the support span (mm), b is the width (mm), and d is the thickness (mm).

All the CS, TS, and FS tests were performed on three specimens for each design mix, after each curing age, and their average prediction was considered. Several test methods such as X-ray diffraction (XRD) and scanning electron microscopy (SEM) were also adopted to access the microstructural properties of alkali-activated binder incorporating industrial wastes. The XRD test is a rapid and straightforward technique for the non-destructive characterization of crystalline materials. It provides information on phases, preferred crystal orientations, and structural parameters. In this research study, the alkali-activated paste powders were scanned in the 2-theta range of 5 to 60 degrees at the step size of 0.02 degrees. To analyze XRD data, the MDI Jade software (version 6.5) and Match software (version 3.10.2.173) were used to verify the glassy nature of the specimens. SEM with high magnifications was finally used to examine the surface morphology of the tested specimens. Operating conditions consisted in a beam energy of 20 keV, beam current of 726 pA, and count time of 10 s with 3500 counts per second, reported as relative atomic concentrations.

At the first stage, the alkali-activated samples were collected from the specimens subjected to the CS test setup, and then each sample was sowed on to the double cellophane sheets followed by attaching to the coin. In the second stage, each sample was placed in the brass stub holder and dried for 5 minutes using IR radiations before using a Blazer sputter coater to cover with gold. The SEM was in fact performed by coating the mortar samples with a thin layer of gold (1.5 nm to 3 0 nm its thickness) prior to the analysis. The resultant patterns were thus monitored using 20 kV with 1000× magnification, at a working distance in the range of 1 to 50 mm, depending on the operational conditions. Significant morphology images were captured immediately after, by selecting reasonably high image magnifications.

3. Test Results

3.1. Mechanical Properties

Table 3 shows the binder constitution and 28-day CS, FS, and TS for all 42 studied AAMs. Table 3 shows that the highest mechanical properties were achieved by AAMs with a high volume of GBFS, while the AAMs with a high volume of POFA resulted in the lowest mechanical properties. The mechanical properties in AAMs with a high volume of WCP also were not satisfactory.

However, increasing the GBFS percentage in the binder mass improved mechanical behaviour in this category. The mechanical features in AAMs with a high-volume FA were significantly dependent on the percentage of GEFS in the binder mass, where substituting the GBFS by POFA significantly decreased the CS. Overall, the average CS of studied

AAMs was 61.3 MPa, which is very satisfactory with much lower embodied energy and CO_2 emission compared to traditional cement-based mortar.

Table 3. Binder constitutions of AAMs, with the corresponding mechanical properties.

AAM Design	Binder (% Mass)				Mechanical Properties		
	FA	GBFS	WCP	POFA	28-Day CS (MPa)	28-Day TS (MPa)	28-Day FS (MPa)
High Volume FA							
1	0.700	0.300	0.000	0.000	78.180	4.560	8.270
2	0.700	0.200	0.000	0.100	65.890	4.340	7.150
3	0.600	0.400	0.000	0.000	80.510	5.180	10.780
4	0.600	0.300	0.000	0.100	81.700	4.770	10.680
5	0.600	0.200	0.000	0.200	52.600	3.080	6.000
6	0.500	0.500	0.000	0.000	80.460	5.060	10.740
7	0.500	0.400	0.000	0.100	76.900	4.860	9.490
8	0.500	0.300	0.000	0.200	70.400	4.640	8.120
9	0.500	0.200	0.000	0.300	46.240	3.460	6.430
High Volume POFA							
10	0.000	0.300	0.000	0.700	34.530	2.240	4.140
11	0.100	0.200	0.000	0.700	23.040	1.370	2.760
12	0.000	0.400	0.000	0.600	45.960	3.310	5.880
13	0.100	0.300	0.000	0.600	37.800	2.620	4.590
14	0.200	0.200	0.000	0.600	28.800	2.180	3.580
15	0.000	0.500	0.000	0.500	55.640	4.760	7.570
16	0.100	0.400	0.000	0.500	47.100	3.660	6.010
17	0.200	0.300	0.000	0.500	40.600	2.980	4.880
18	0.300	0.200	0.000	0.500	36.800	2.570	4.370
High Volume GBFS							
19	0.300	0.700	0.000	0.000	85.090	5.800	10.860
20	0.200	0.700	0.000	0.100	97.750	7.160	12.180
21	0.100	0.700	0.000	0.200	86.400	5.670	10.770
22	0.000	0.700	0.000	0.300	70.530	5.440	10.260
23	0.400	0.600	0.000	0.000	80.680	5.560	10.710
24	0.300	0.600	0.000	0.100	72.440	5.140	9.540
25	0.200	0.600	0.000	0.200	71.930	5.040	9.560
26	0.100	0.600	0.000	0.300	70.840	4.890	9.360
27	0.000	0.600	0.000	0.400	70.220	4.770	9.140
28	0.500	0.500	0.000	0.000	80.460	5.060	10.740
29	0.400	0.500	0.000	0.100	80.430	5.110	10.540
30	0.300	0.500	0.000	0.200	67.220	4.810	8.530
31	0.200	0.500	0.000	0.300	65.140	4.420	7.980
32	0.100	0.500	0.000	0.400	56.340	4.360	7.540
33	0.000	0.500	0.000	0.500	55.640	4.280	7.570

Table 3. *Cont.*

	High Volume WCP						
34	0.000	0.300	0.700	0.000	34.020	2.680	4.620
35	0.100	0.200	0.700	0.000	22.400	1.750	2.970
36	0.000	0.400	0.600	0.000	68.440	5.670	9.260
37	0.100	0.300	0.600	0.000	52.080	5.160	7.670
38	0.200	0.200	0.600	0.000	46.760	3.940	6.580
39	0.000	0.500	0.500	0.000	74.120	5.300	10.120
40	0.100	0.400	0.500	0.000	66.190	4.990	9.070
41	0.200	0.300	0.500	0.000	60.170	4.710	8.220
42	0.300	0.200	0.500	0.000	56.470	4.510	7.710
Average					61.3	4.33	7.927
ST.DEV.					18.70	1.23	2.47

3.2. Structural Morphology

The microstructure and composition of industrial waste hydration products are crucial for defining the properties of hardened mortar. The main hydration products are calcium silicate hydrate gels (C-S-H) and calcium hydroxide (CH), aluminate/aluminoferrite phases (AFt and AFm type phases) [27]. Among these hydration products, the most important one is C-S-H gel because it is the principal binding phase in cement-based systems, and it is responsible for strength, density, and so forth. By using morphology tests, microstructure and chemical composition define the final properties of the hardened mortar.

3.2.1. Scanning Electron Microscopy (SEM)

The high-resolution, SEM images are useful within materials science to test the quality of materials, ensuring that they are an appropriate option for the required purpose and can be used to predict and prevent material failure.

The SEM images showing the microstructure of AAMs with a high volume of FA contents are shown in Figure 2a, b and c with 50%, 60%, and 70% FA, respectively. Figure 2a reveals that the microstructure of the sample containing a high amount of FA was less dense compared to other matrices, due to a lack of C-S-H gel formation. Once the content of FA was decreased and replaced by GBFS, the microstructure of AAMs was enhanced and provided a denser surface, as shown in Figure 2b and c. The SEM images also acknowledged that in AAMs with 70% FA, the gel structure and unreacted/partially reacted particles were extensively distributed over the surface, whereas by decreasing the FA content to 60% and 50%, a dense gel dominated the surface of AAMs.

Figure 3 shows the effects of POFA replacement by GBFS on AAMs with 60% FA. The SEM image clearly indicates that the design mixes with zero POFA (60% FA and 40% GBFS) have a dense structure (Figure 3a). On the other hand, once the POFA percentage, replaced by GBFS, was increased from 0% to 20%, the dense gel phase was decreased, and the AAMs enclosed unreacted/partially reacted phase (Figure 3b). Increasing the content of POFA affected the calcium content and hence reduced the C-S-H product formation with poor microstructure. This particular issue also negatively affects the mechanical properties, where the CS decreased from 80.5 to 52.6 MPa by replacing POFA with GBFS from 0% to 20%, as shown in Table 3.

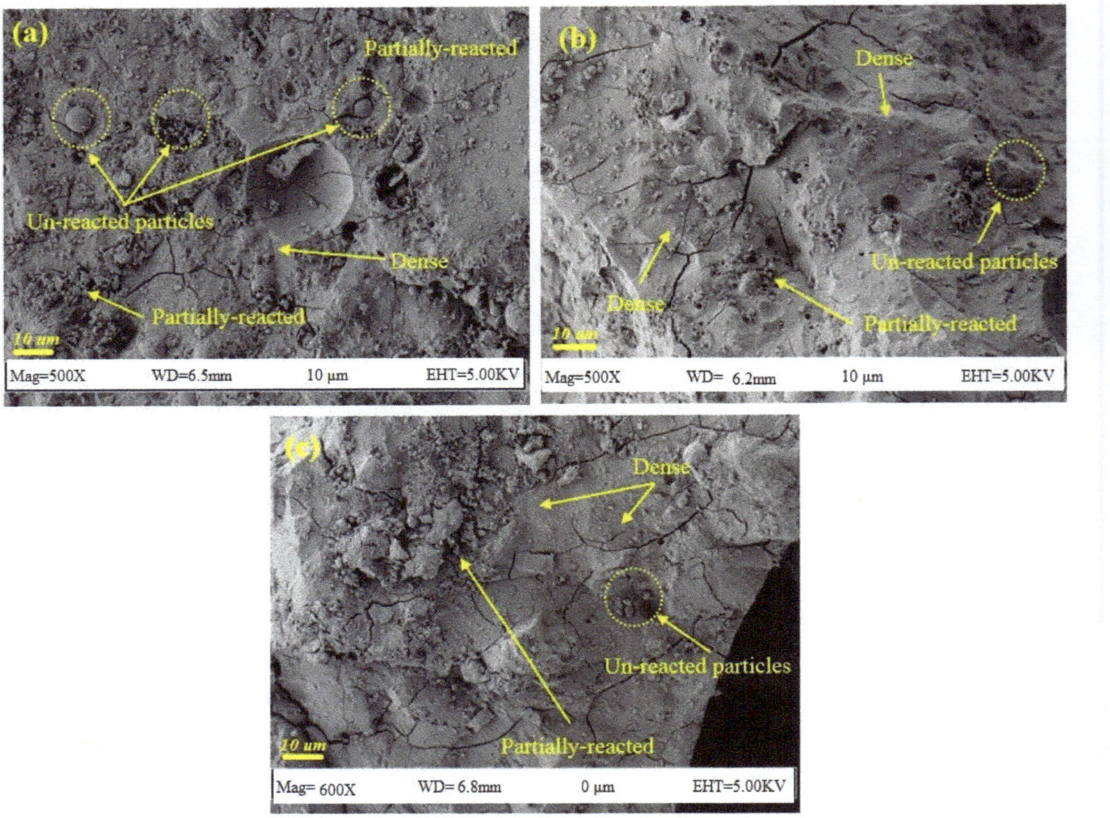

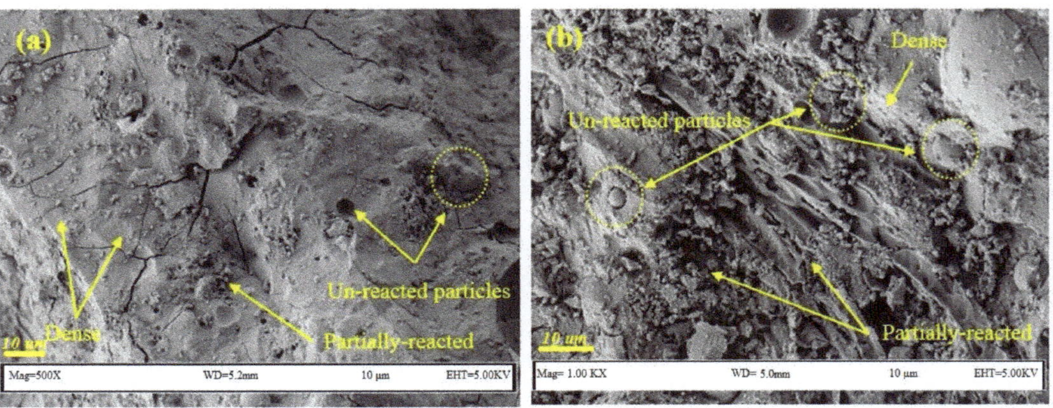

Figure 2. Scanning electron microscopy (SEM) images of AAMs containing a high volume of fly ash (FA), with evidence of (**a**) 70% FA, (**b**) 60 % FA, (**c**) 50% FA.

Figure 3. SEM images of palm oil fly ash (POFA) replacement by granulated blast-furnace slag (GBFS) in AAMs with 60% FA, with evidence of (**a**) 0% POFA, (**b**) 20% POFA.

Figure 4 shows SEM images of AAMs with a high volume of WCP. The result indicates the creation of a larger amount of crystalline $Ca(OH)_2$ in hexagonal plate-like structures (unreacted particles/partially reacted matrix) in AAMs with 60% and 70% of

WCP (Figure 4a,b). This issue can be explained by the hydration process of an excess amount of CaO in the GBFS and FA that produced an elevated amount of $Ca(OH)_2$ crystals. When GBFS was mixed with FA, it shaped a hydrated binding cement paste (HCP) of calcium silicate hydrate (C-S-H). The formation of a massive amount of C-S-H crystals in AAMs with 50% of WCP made it denser, and this was attributed to the pozzolanic reaction of SiO_2 with $Ca(OH)_2$ during the hydration process (Figure 4c). Subsequently, such dense microstructures contributed to enhancing the CS of mix designs.

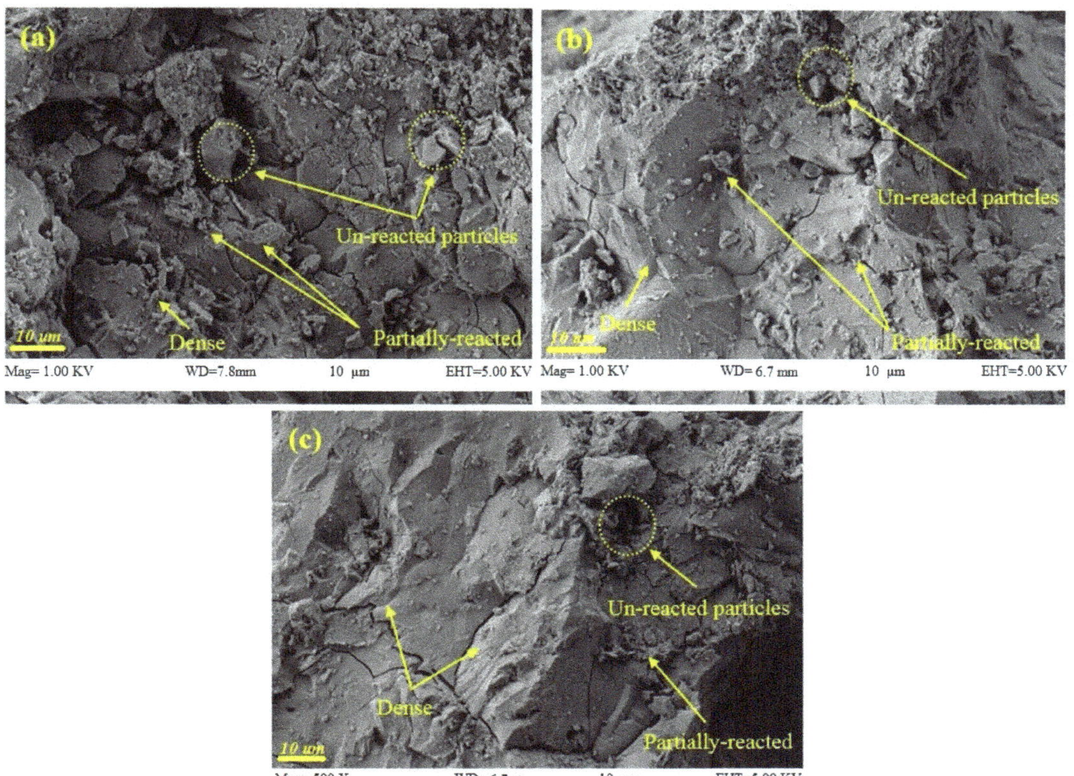

Figure 4. SEM images of AAM designs with a high volume of waste ceramic powder (WCP), with evidence of (**a**) 70% WCP, (**b**) 60% WCP, and (**c**) 50 % WCP.

3.2.2. X-ray Diffraction (XRD)

To identify the crystal phase of geopolymerization products and evaluate the effects of silicate, Al, and Ca on the derived C-S-H gel, with different substitution ratios of industrial wastes in the AAMs at the age of 28 days, XRD patterns were recorded between 0 and 60 degrees.

Figure 5 shows the XRD patterns of AAMs containing a high volume of FA, where the peak intensity corresponds to crystalline quartz (SiO_2) and mullite ($3Al_2O_32SiO_2$ or $2Al_2O_3 SiO_2$) phases in AAMs with 60% and 70% FA. The crystalline phases' peak intensity was increased with an increasing amount of FA from 50% to 60% to 70%, and the C-S-H gel peak was replaced by the quartz peak (SiO_2), where more quartz appeared to be non-reactive with 70% of FA content. The products formed due to the reaction between the glassy fraction of FA and GBFS that consisted of very poor crystalline phases. Kumar et al. [28] have reported that the FA consists of reactive and refractory glass, where the reactive glass only participates in the geopolymerization. The broad and diffused background peak with a maxima of around 10 degrees emerged from the short-range order of the hydrotalcite ($Mg_6Al_2CO_3OH_{16}$ $4H_2O$) in all AAMs with different content of FA. The peaks around 24 and 33.8 degrees were also allocated to nepheline ($Na_3KAl_4Si_4O16$).

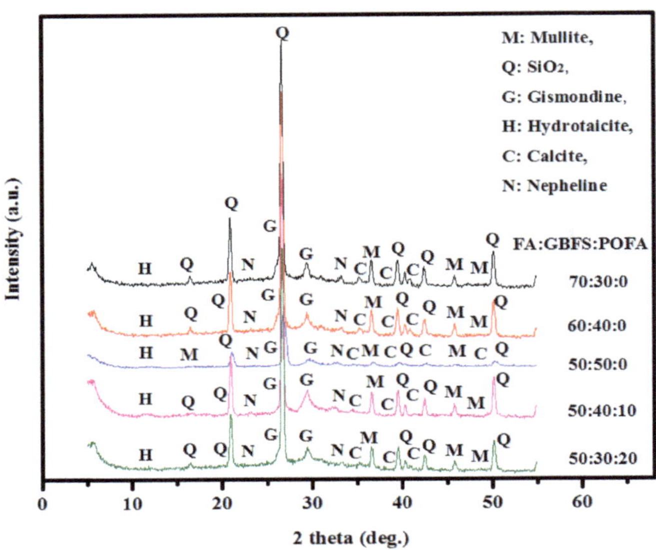

Figure 5. X-ray diffraction (XRD) pattern of AAMs containing a high volume of FA and POFA replaced by GBFS containing 50% FA.

Figure 5 also shows the XRD patterns of AAMs with 50% FA and a different GBFS replacement ratio by POFA. Figure 5 indicates that peak intensity corresponds to the quartz (SiO_2) as the POFA replacement was increased in AAMs (with maxima around 28 degrees) while the peak intensity is dedicated to C-S-H in the design mix with 0% POFA. Furthermore, the C-S-H gel peak intensity at 31 degrees was enhanced when the POFA content was increased to 10% and 20%, improving the CS of AAMs. There is a reduction in the C-S-H gel peak intensity in a design mix with a 30% POFA. In this particular design mix, the quartz peak at 28 degrees showed low intensity with higher width. The low product of C-S-H gel reduced the CS from 70.4 to 46.2 MPa as the POFA replaced by GBFS increased from 20% to 30%, as shown in Table 3.

3.2.3. Fourier-Transform Infrared Spectroscopy (FTIR)

In various AAMs matrices, the formation of reaction products and geopolymerization degree were identified using FTIR measurements. The FTIR test is a chemical analysis method that searches for Si-O and Al-O reaction zones in AAMs. In such a matrix, the development of C-S-H gel occurred with the dissolution of minerals, due to the alkaline activators' addition to the base materials. This resulted in the release of Al via the hydroxylation, which in turn caused the attachment of OH^- ions present in the alkali to form the Al–O–Al bond by rupturing the weak bonds. Then, a negatively charged Al in IV fold coordination was liberated. Finally, a charge balance was achieved by Ca, which reacted preferentially over Na [29].

GBFS contains higher CaO than FA and POFA, leading to a high potential for Ca solubility in the mixture. The quantity of soluble Ca depends on the volume of GBFS present in the mixture, which directly affects the C-S-H gel. The unit oligomer of (Si–O–Al) Ca could be built up in chains, sheets, or a three-dimensional framework through the polycondensation process to result in the product hardening [30]. Furthermore, Davidovits described the Ca-based AAMs as the condensation resulting from the hydroxylation of the gehlenite and akermanite phase to form cyclo Ca-ortho-sialate-disiloxo (C_3AS_3) and Ca-disiloxonate hydrate C-S-H, respectively. Figure 6 shows the FTIR spectra and FTIR fingerprint zone of AAMs with a high volume of FA.

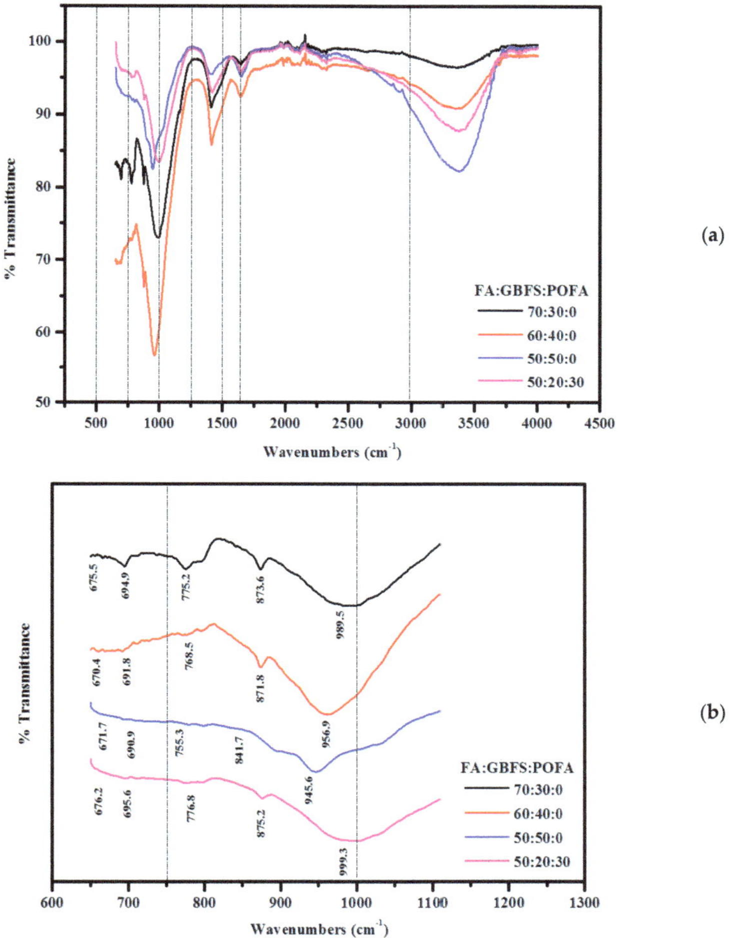

Figure 6. (**a**) Fourier-transform infrared spectroscopy (FTIR) spectra and (**b**) FTIR fingerprint zone of AAMs with a high volume of FA.

Following the information provided in Figure 6, Table 4 demonstrates the FTIR peak positions and band assignments.

Table 4. Fourier-transform infrared spectroscopy (FTIR) peak positions and band assignments.

AAM	Ratio		f_c' (MPa)	Band Position (cm^{-1}) and Assignments				
FA:GBFS:POFA	Si/Al	Ca/Si		Al-O	Si-O	AlO$_4$	CSH	C(N)ASH
70:30:0	2.10	0.39	78.2	675.5	694.9	775.2	873.6	989.5
60:40:0	2.38	0.39	80.5	670.4	691.8	768.5	871.8	956.9
50:50:0	2.22	0.65	80.6	671.7	690.9	755.3	871.4	945.6
50:20:30	3.02	0.29	46.2	676.2	695.6	776.8	875.2	999.3

The critical outcomes from Table 4 can be discussed as follows:

- By increasing the FA content from 50% to 70%, the band of the C(N)ASH gel product increases from 945.6 to 989.5 cm^{-1}.
- By increasing the FA content from 50% to 70%, the C-S-H and Si-O band frequency increases, leading to a less homogenous structure of AAMs for 60% and 70% FA content, and subsequently smaller silicate re-organization compares to the 50% FA mix design.
- The band of 945.6 cm^{-1} presented in the AAMs containing 50% FA, 50% GBFS, and 0% POFA achieves the highest CS of 80.6 MPa.
- The broad band at 1647 cm^{-1} and a weak peak at 3353 cm^{-1} are allocated to the stretching vibrations of O–H bonds and H–O–H bending vibrations of interlayer adsorbed H$_2$O molecules, respectively [31]. On the basis of these values, increasing the FA content is found to cause structural changes in the examined mixes, which can be attributed to the reduction of C-S-H formation with C(N)-A-S-H type gels and to the increase of the amount of Si and Al. These changes slow down the rate of geopolymerization and negatively affect the mechanical strength of AAMs.
- GBFS released soluble Ca that displaced the Si atoms from Si–O bonds, leading to a reduction in the vibrational frequency. Therefore, it can be concluded that the frequency of vibration decreases with the increase in the molecular molar mass of the attached atoms.
- The addition of FA or POFA caused the increment in the SiO$_2$/Al$_2$O$_3$ ratio and the vibrational frequency of Si–O–Si (Al).
- The band corresponding to AlO$_4$ shifted from 755.3 cm^{-1} to 768.5 to 775.2 cm^{-1} in samples prepared with 50%, 60%, and 70% FA binder, respectively.

4. Prediction of Mechanical Properties Using ANN

4.1. Methods

4.1.1. Feed-Forward ANN and Krill Herd Algorithm

The multilayer feed-forward network provides a reliable feature for the ANN structure and, therefore, was used in this research. The multilayer feed-forward network comprises three individual layers: the input layer, where the data are defined to the model; the hidden layer/s, where the input data are processed; and finally, the output layer, where the results of the feed-forward ANN are produced. Each layer contains a group of nodes referred to as neurons that are connected to the proceeding layer. The neurons in hidden and output layers consist of three components; weights, biases, and an activation function that can be continuous, linear, or nonlinear. Standard activation functions include nonlinear sigmoid functions (logsig, tansig) and linear functions (poslin, purelin) [32]. Once the architecture of a feed-forward ANN (number of layers, number of neurons in each layer, activation function for each layer) is selected, the weight and bias levels should be adjusted using training algorithms. One of the most reliable ANN training algorithms is the backpropagation (BP) algorithm, which distributes the network error to arrive at the best fit or minimum error [33,34].

The Krill Herd algorithm is a smart group algorithm for optimization in engineering disciplines [35]. This algorithm has been used to achieve more efficiency in civil engineering. It refers to a novel metaheuristic algorithm to determine the weight optimization of each ANN model. In this algorithm, krill individuals search for food in different places and are presented as different decision variables. The objective is to calculate the distance between krill individuals and the availability of extra food, related to the cost. Therefore, the time-dependent position of a krill individual is measured by functional processes, which include the process of foraging, search displacement, and random physical diffusion [36]. In the process of foraging motion, the individual krill velocity is always affected by another krill displacement in the multidimensional search space, where the velocity changes dramatically and dynamically on the basis of the internal influence parameters, including the influence of the target group and the repulsive effect.

Equations (3) to (8) can efficiently formulate the displacement description of a krill individual [36]:

$$\theta_i^{new} = \epsilon_i \theta_i^{max} + \mu_n \theta_i^{old} \tag{3}$$

$$\epsilon_i = \epsilon_i^{local} + \epsilon_i^{target} \tag{4}$$

$$\epsilon_i^{local} = \sum_{i=0}^{Ns-1} f_{ij} x_{ij} \tag{5}$$

$$f_{ij} = \frac{f_i - f_j}{f_w - f_b} \tag{6}$$

$$x_{ij} = \frac{x_i - x_j}{|f_w - f_b| rand(0.1)} \tag{7}$$

$$\epsilon_i^{target} = 2 \left(rand(0.1) + \frac{i}{i_{max}} \right) f_i^{best} x_i^{best} \tag{8}$$

where θ_i^{max} represents the highest motion created and θ_i^{old} is the motion created; μ_n represents the algebraic magnitude of the motion created, while the target effects are represented by ϵ_i^{local} and ϵ_i^{target}.

Moreover, f_w and f_b are the worst and the best population positions, respectively; f_i and f_j are the ith and j^{th} krill individual proportions. The current number and the highest number are provided by i_{max}. To identify the neighboring members of each krill individual, a sensor distance parameter (SD_i) is used [36]:

$$SD_i = \frac{1}{5n_p} \sum_{i=0}^{n_p-1} |x_i - x_j| \tag{9}$$

where n_p represents the number of krill individuals in the population, while x_i and x_j are the positions of ith and j^{th} krill, respectively [36].

4.1.2. Input and Output Parameters

Considering the information about the contents of constituents in each AAM, as provided in Table 2, and their mechanical properties at the age of 1, 3, 7, 28 days (as summarized in Table 3), Table 5 defines the input and output parameters along with their statistical properties that were used for training the ANN model.

Table 5. Statistical parameters of input and output parameters.

Statistical Index	Average	ST.DEV.	Max	Min	Type
FA	0.25	0.21	0.7	0	Input
GBFS	0.41	0.16	0.7	0.2	Input
WCP	0.12	0.24	0.7	0	Input
POFA	0.22	0.23	0.7	0	Input
Si:Al (ratio)	4.03	1.58	8.63	2.1	Input
Ca:Si (ratio)	0.52	0.23	0.97	0.17	Input
Ca:Al (ratio)	2.05	1.09	4.41	0.66	Input
Age (day)	9.75	10.79	28	1	Input
Compressive Strength (MPa)	40.61	20.55	97.75	7.61	Output
Tensile Strength (MPa)	2.94	1.52	7.16	0.34	Output
Flexural Strength (MPa)	5.1	2.77	12.18	0.64	Output

Since the behaviour and number of input data should be statistically evaluated against the output data, Figure 7 also shows the histograms of output data (CS, TS, and FS mechanical properties).

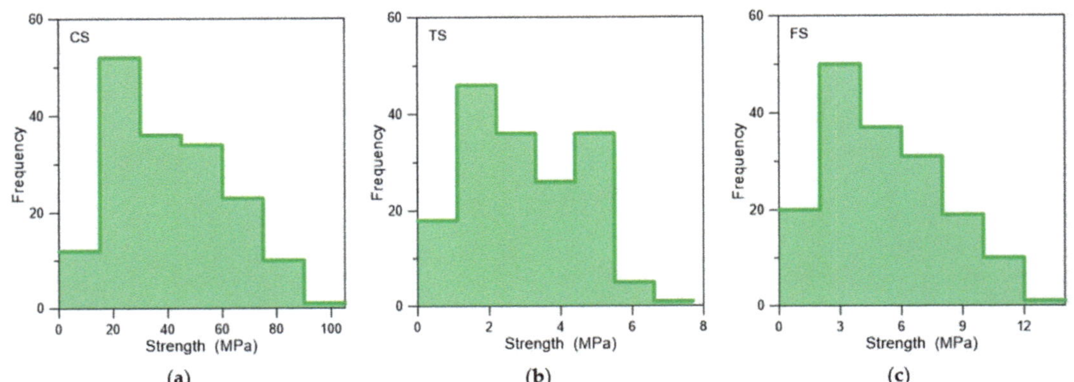

Figure 7. Histogram of output data for (**a**) compressive strength (CS), (**b**) tensile strength (TS), and (**c**) flexural strength (FS) parameters.

4.1.3. Developing the Topology and Structure of Feed-Forward ANN

The ANN used in this study is a new feed-forward model. Up to 70% of input data (out of 168 samples) was used for training, and the remaining 30% part was considered for testing the network. According to the characteristics of the available input data and the number of outputs, a two-layer ANN was proposed in the initial attempt, and its adequacy was evaluated by using several measures. Table 6 shows the structure and topology of this feed-forward ANN.

Table 6. Structure and topology of the proposed feed-forward artificial neural network (ANN).

Features of ANN						
Number of Input	Number of Output	Neural Network	Hidden Layer	Node	Learning Role	Transfer Function
8	3	newff	2	9-8	Levenberg–Marquardt	tansig

To optimize the weights of the proposed ANN model, the Krill Herd algorithm was used as a new metaheuristic algorithm in building materials technology. Table 7 shows the features of KHA, once applied in the proposed feed-forward ANN.

Table 7. Properties of the Krill Herd algorithm.

Features of Krill Herd Algorithm				
Number of Krills	Minimum Number of Krill Herd	Maximum Iteration	Maximum Induced Speed	Dmax
10	2	200	0.01	0.005

Table 8 shows the statistical results of training and testing of the proposed ANN model combined with the Krill Herd algorithm (KHA-ANN).

To evaluate the adequacy of the proposed KHA-ANN model, more in detail, the comparative input of Table 8 is based on the following statistical indicators:

$$MAE = \frac{1}{n}\sum_{i=1}^{n}|P_i - O_i| \quad \text{Mean Absolute Error} \tag{10}$$

$$MSE = \frac{1}{n}\sum_{i=1}^{n}(P_i - O_i)^2 \quad \text{Mean Squared Error} \tag{11}$$

$$RMSE = \left[\frac{1}{n}\sum_{i=1}^{n}(P_i - O_i)^2\right]^{\frac{1}{2}} \quad \text{Root Mean Squared Error} \tag{12}$$

$$AAE = \frac{\left|\sum_{i=1}^{n}\frac{(O_i - P_i)}{O_i}\right|}{n} \times 100 \quad \text{Average Absolute Error} \tag{13}$$

$$EF = 1 - \frac{\sum_{i=1}^{n}(P_i - O_i)^2}{\sum_{i=1}^{n}(\overline{O}_i - O_i)^2} \quad \text{Model Efficiency} \tag{14}$$

$$VAF = \left[1 - \frac{var(O_i - P_i)}{var(O_i)}\right] \times 100 \quad \text{Variance Account Factor} \tag{15}$$

where O_i is an experimental observation and P_i is a prediction.

Table 8. Statistical results of training and testing the Krill Herd algorithm (KHA-ANN) model.

Statistical Index	Compressive Strength	Flexural Strength	Tensile Strength
Training			
MAE	2.21	0.31	0.20
MSE	10.13	0.20	0.08
RMSE	3.18	0.45	0.29
AAE %	0.07	0.08	0.10
EF	0.97	0.97	0.96
VAF %	0.98	0.97	0.96
Testing			
MAE	2.29	0.37	0.24
MSE	11.72	0.30	0.14
RMSE	3.42	0.55	0.37
AAE %	0.06	0.07	0.09
EF	0.97	0.97	0.94
VAF %	0.97	0.96	0.94

The results of Table 8 indicate that the proposed KHA-ANN model has an acceptable error in the above-mentioned statistical indicators. Error criteria for training and testing data are calculated using data values in the main range of variables and not in the normal range. The preliminary results indicated that the average value of R^2 for the compressive, flexural, and tensile strength at the training and testing stage was equal to 0.975, 0.972, and 0.965, respectively, confirming the high accuracy of the proposed KHA-ANN model with 8-9-8-3 structure. Figure 8 shows the structure of the proposed KHA-ANN model.

Table 9 provides the final weights for both hidden layers estimated by the KHA-ANN model. Using the values of the weights between the different ANN layers, the three output parameters (compressive, flexural, and tensile strength) can be thus determined and predicted.

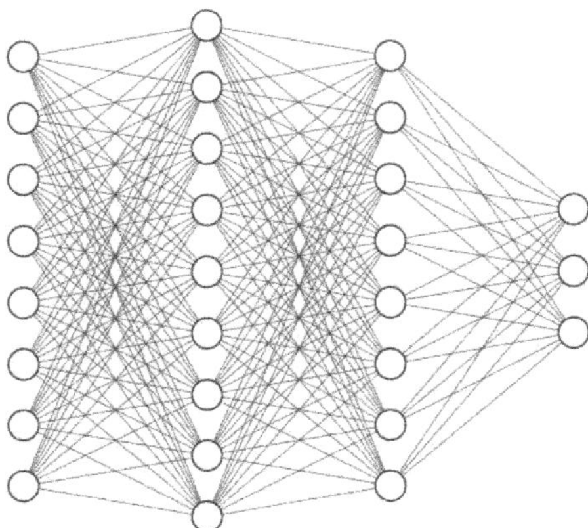

Figure 8. Feed-forward artificial neural network (ANN) with 8-9-8-3 structure.

Table 9. Final weight and bias values for the KHA-ANN model.

IW									b_1
−0.2316	−0.1623	−0.0825	−0.2644	−0.1643	−0.0004	−0.3556	−0.3701		0.1468
0.0154	−0.1933	0.2731	0.3086	−0.3697	−0.3144	0.4251	−0.0591		0.1912
0.3718	−0.0961	−0.0270	−0.3663	0.0026	−0.2619	0.3216	0.0096		0.3349
0.1158	−0.2180	−0.0022	−0.2767	−0.0707	0.2570	0.4840	0.1595		0.3154
0.3578	0.1603	0.0146	−0.3111	−0.2367	0.0690	0.0009	0.0303		0.2471
0.4683	0.1633	−0.0891	0.0365	0.0634	−0.3246	0.1379	0.3169		−0.2807
−0.1659	0.3423	−0.0292	−0.3112	−0.5617	−0.1967	−0.3909	0.0561		−0.1984
−0.1554	−0.1560	−0.2129	0.2389	−0.0872	−0.1272	0.1465	−0.0184		−0.1573
0.0656	0.3110	0.0569	0.2420	−0.2830	−0.1525	0.0169	−0.2753		0.1651
LW1									b_2
0.0936	0.3101	0.4358	0.1354	0.1660	0.2729	−0.2360	0.2942	0.3647	0.4161
0.0232	0.0799	0.1747	−0.0157	−0.0259	−0.0059	−0.0148	0.2845	0.2798	0.2902
0.0508	−0.1987	−0.1151	0.2572	0.1240	0.1459	0.1177	0.1984	−0.2452	0.0451
0.0317	−0.2141	−0.0854	0.0013	−0.0459	0.0460	0.0470	0.0261	−0.0151	0.1342
0.1973	−0.0587	−0.0486	0.3834	−0.4181	0.0272	−0.1700	0.1561	0.0631	−0.0106
0.1484	0.2095	0.1868	−0.0200	0.4514	0.3818	−0.3466	0.1404	0.0909	0.3694
0.1452	0.1178	0.1755	0.1312	0.0498	−0.1287	0.0839	0.5858	0.1076	−0.2783
−0.2980	−0.1555	0.1522	0.2578	−0.3475	−0.0006	−0.2633	−0.0045	0.1366	−0.2593
LW2									b_3
−0.6219	−0.4363	−0.0018	−0.7521	0.7479	0.1300	−0.5880	−0.7886		0.2419
−0.9147	0.0772	0.0716	−0.0193	−0.4594	0.2806	0.8959	−0.7159		0.1474
0.2704	0.3903	−0.1096	0.7060	−0.5831	−0.1659	−0.8359	−0.6671		−0.8958

IW: weight values for the input layer; LW1: weight values for the first hidden layer; LW2: weight values for the second hidden layer; b_1: bias values for the first hidden layer; b_2: bias values for the second hidden layer; b_3: bias values for the output layer

4.2. KHA-ANN Model Validation

To validate the proposed KHA-ANN model used in this study, a multiple linear regression (MLR) model and a genetic algorithm combined with an ANN (GA-ANN) are considered in this study. In an MLR model, two or more independent variables have a major effect on the dependent variable, as shown in the following equation,

$$y = f(x_1, x_2, \ldots)y = a_0 + a_1 x_1 + a_2 x_2 + \ldots \tag{16}$$

where y is a dependent variable; x_1, x_2, $\ldots$, are independent variables; a_1, a_2, $\ldots$ are equation coefficients. In this paper, different models of MLR are examined for input and output variables. The most suitable coefficients for the MLR model for each output parameter are given as:

$$CS = -4.5 + 19.9\,FA + 182.2\,GBFS - 3.4\,WCP - 0.74\,Si:Al - 73.3\,Ca:Si - 2.31\,Ca:Al + 1.2338\,Age \tag{17}$$

$$FS = -0.75 + 1.99\,FA + 22.95\,GBFS + 0.10\,WCP - 0.129\,Si:Al - 8.75\,Ca:Si - 0.294\,Ca:Al + 0.16890\,Age \tag{18}$$

$$TS = -0.03 + 0.70\,FA + 12.74\,GBFS + 0.131\,WCP - 0.100\,Si:Al - 4.66\,Ca:Si - 0.211\,Ca:Al + 0.08493\,Age \tag{19}$$

For the second evaluation, a genetic algorithm combined with an ANN (GA-ANN) is concerned, and its characteristics are summarized in Table 10.

Table 10. Characteristics of the genetic algorithm combined with an ANN (GA-ANN).

Parameter	Value	Parameter	Value
Max generations	100	Crossover (%)	50
Recombination (%)	15	Crossover method	Single point
Lower bound	−1	Selection mode	1
Upper bound	+1	Population size	150

Table 11 shows the statistical results for input parameters provided by all three investigated models. The results of KHA-ANN, GA-ANN, and MLR models are also shown in Figure 9 for all the three output parameter of CS, TS, FS. The collected results indicate that the KHA-ANN model provides more accurate results, compared to the GA-ANN and MLR models.

Table 11. Statistical results for input parameters.

Model Name	Statistical Index	Compressive Strength	Tensile Strength	Flexural Strength
KHA-ANN	MAE	2.24	0.21	0.33
	MSE	10.60	0.10	0.23
	RMSE	3.26	0.32	0.48
	AAE %	0.07	0.09	0.08
	EF	0.97	0.96	0.97
	VAF %	0.97	0.96	0.97
GA-ANN	MAE	2.79	0.19	0.29
	MSE	75.74	0.15	0.44
	RMSE	8.70	0.39	0.67
	AAE %	0.07	0.07	0.06
	EF	0.82	0.93	0.94
	VAF %	0.82	0.93	0.94
MLR	MAE	6.87	0.55	0.92
	MSE	74.21	0.47	1.37
	RMSE	8.61	0.68	1.17
	AAE %	0.21	0.25	0.24
	EF	0.82	0.80	0.82
	VAF %	0.82	0.80	0.82

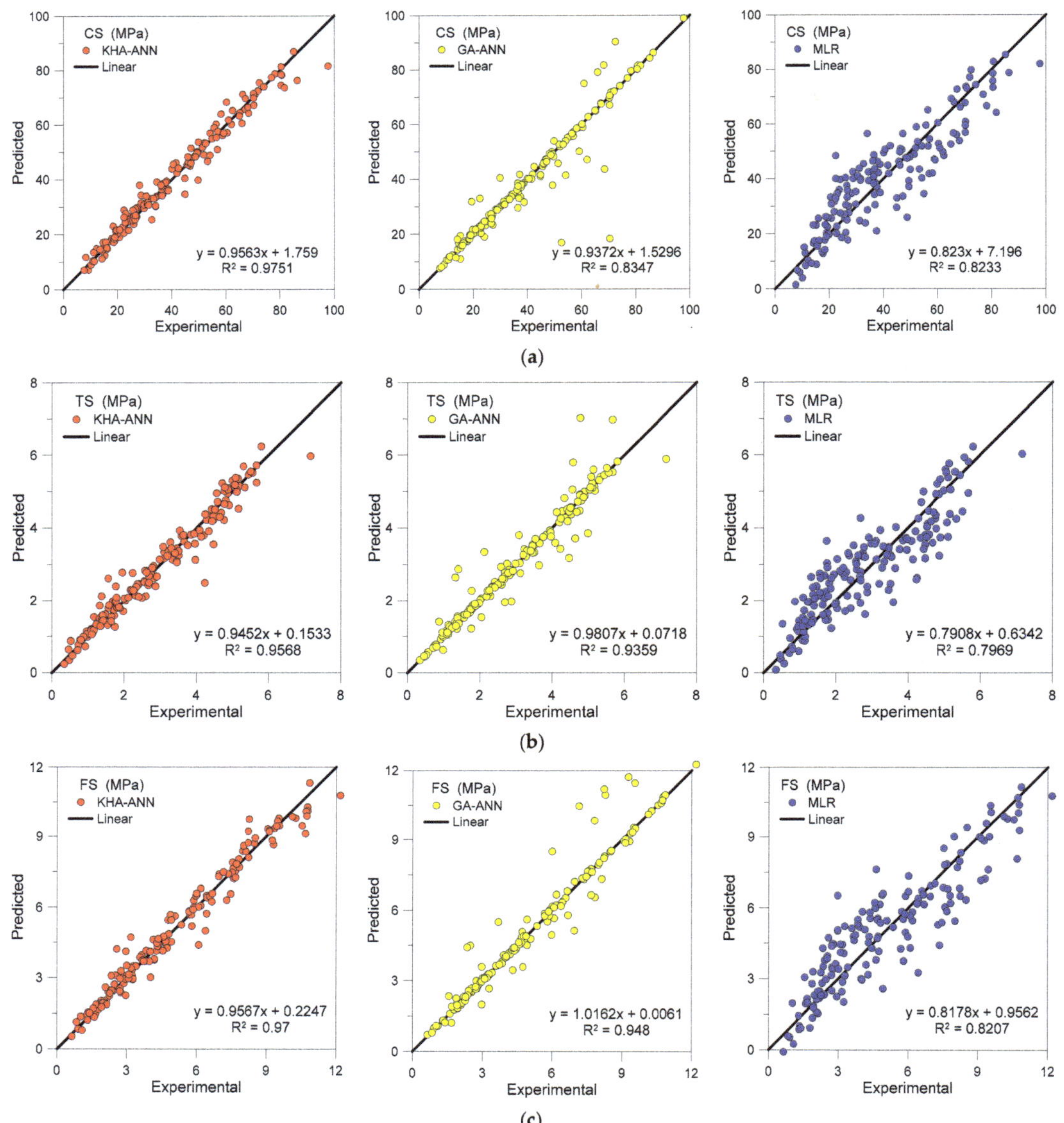

Figure 9. Experimentally observed versus predicted values of (**a**) compressive strength (CS), (**b**) tensile strength (TS) and (**c**) flexural strength (FS), using all the available data.

Given that all three outputs have nonlinear behaviour and are not a function of any fixed trend, the MLR model is found not able to provide reliable results. Its R^2 value and EF index for all the CS, TS, and FS predictions is in fact characterized by the lowest value, compared to KHA-ANN and GA-ANN models. Meanwhile, the estimated R^2 value for the GA-ANN model is also not satisfactory. However, the EF index predicted by this model can be seen as more reliable, compared to the MLR model. Nevertheless, the KHA-ANN model provides the most reliable results for both R^2 and EF indexes, and thus indicating its high potential and accuracy for the purpose of this study.

Such a confirmation is also acknowledged in Figure 10, where the KHA-ANN model estimates are shown to be more accurate for the ratio of observational to computational values (for all the input parameters), compared to the other two approaches, GA-ANN and MLR.

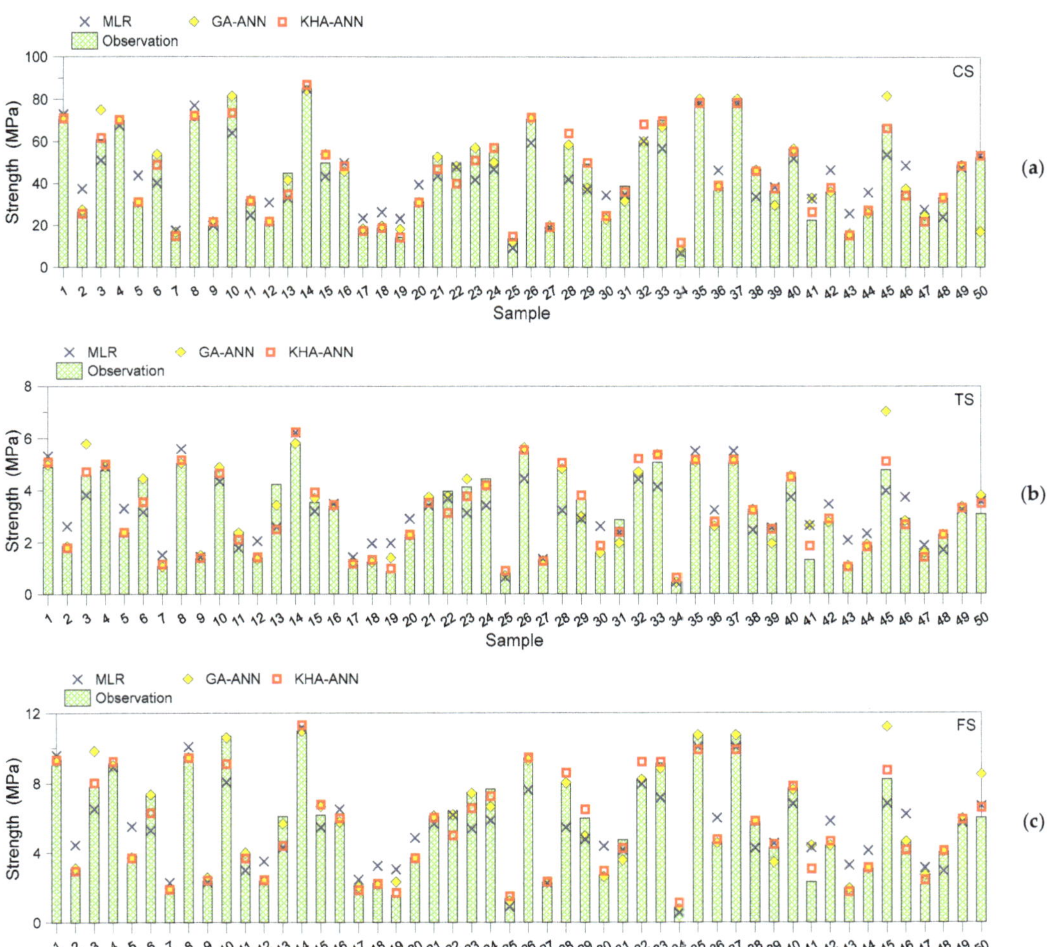

Figure 10. Observational to computational values for the estimation of (**a**) CS, (**b**) TS, and (**c**) FS parameters (50 out of 168 samples).

5. Discussion

The overall analysis of mechanical properties and structural morphology of all alkali-activated mix categories can be summarized in the following outcomes:

- The results indicate that AAMs with a high volume of GBFS provide the highest mechanical properties. The SEM results also confirm that the alkali activation in this category can produce hydration products in the form of C-S-H, which is the primary reaction product for OPC strength development.
- In the design mixes with high volume FA, by replacing FA with GBFS, a dense gel dominated the AAMs structure, and this results in the further improvement of the observed mechanical properties. The fact is that GBFS has a significant contribution

to the geopolymerization process in AAMs, thus leading the paste to reach higher mechanical performances.

- An increase in the POFA content was found to reduce the mechanical properties of AAMs. The morphology tests also confirmed that increasing the POFA percentage in AAMs can be associated to the rise in the unreacted silicate and reduced the C-S-H product, which significantly affected the CS and microstructure of AAMs. The results show that the ratio of $SiO_2:Al_2O_3$ above 3.5 negatively affects the overall AAMs mechanical properties and microstructure.

- The mechanical parameters of AAMs containing high volume WCP was observed as significantly lower than the other mixes. The ratio of $SiO_2:Al_2O_3$ was relatively high in this category, which negatively affected the forming C-S-H gels in geopolymerization process. The sodium oxide content (Na2O) was also observed at a high ratio (13.5%) in the WCP chemical composite, compared to 0.08% with FA and 0.45% with GBFS. The negative effect of a high content of silica and low calcium content contributed to the rather poor mechanical performance of these design mixes.

Different amounts of waste were investigated to optimize the mechanical properties and consumption of waste materials in each ternary blended AAMs. Therefore, depending on the availability of the particular waste material and required mechanical properties, a proper design mix can be selected. While GBFS and FA are commonly used in the concrete industry, WCP and POFA have lower interest. This research developed new environmentally-friendly AAMs with WCP, as the main binder, combined by GBFS and FA for various construction applications. The results confirmed that using 50% to 70% of WCP in AAMs provided a considerable (36–70 MPa) CS for many building and construction purposes. By exciding this value, the calcium oxide content was reduced and negatively affected the formation of C-S-H gel. The mix prepared with 50% WCP, 40% GBFS, and 10% FA (mix number 40) may represent optimum AAMs in this category, once the consumption of WCP and relatively high CS are needed.

Furthermore, to take advantage of the plentiful amount of POFA as a waste by-product provided by palm oil industries in several South-East Asian nations, this research proposed ternary blended AAMs containing 50% to 70% POFA as the binder mass and different amounts of GBFS and FA. FA is an abundant and cheap waste by-product, and its application in the production of geopolymer and alkali-activated mortar/concrete could lead to sustainable development. On the other hand, the inclusion of GBFS produces secondary hydration that resulted in higher C-S-H formation. This research confirms that the GBFS/FA ratio affects the structural morphology and mechanical properties in AAMs containing high volume POFA. GBFS-rich AAMs had higher mechanical properties and a more dense structure. Depending on the application and the availability of the waste material, each of the AAMs can be selected. For instance, if a high CS is required, the mix number 15 can be considered (around 55 MPa). On the other hand, if using FA has a high priority, mix number 18 would be recommended, and mix number 10 consumes a high amount of POFA with reasonable CS for many building and construction purposes.

From the developed algorithms to support the mix design, finally, it is also concluded that the KHA-ANN model, on the basis of its final weight and bias values, can be efficiently used to design AAMs with targeted mechanical properties, in which the desired amounts of waste consumption can be optimized on the basis of available local waste materials.

6. Concluding Remarks

This study assessed the mechanical properties of ternary blended alkali-activated mortars composed of industrial waste materials, using experimental tests and structural morphology evaluation. To examine the effect of each binder mass percentage on the compressive, tensile and flexural mechanical strength values, 42 engineered AAMs were investigated. By using the available experimental test database, three different models were presented to estimate the mechanical properties of AAMs depending on binder mass constituents. The following provides the main findings of this research:

1. Test results indicated that the highest mechanical properties were achieved by AAMs with a high volume of GBFS, while the AAMs with a high volume of POFA resulted in the lowest mechanical properties. Furthermore, the average CS, FS, and TS of studied AAMs were predicted in 61.3, 7.92, and 4.33 MPa, respectively, which, compared to traditional cement-based mortar, is highly satisfactory from a mechanical point of view, but also characterized by reduced embodied energy and CO_2 emission.

2. SEM images revealed that the microstructure of the AAMs containing a high amount of FA was less dense compared to other matrices, due to lack of C-S-H gel formation, where, by the replacement of FA with GBFS, the microstructure of AAMs was enhanced and provided a denser surface. Besides the SEM images have shown that increasing the content of POFA has affected the calcium (CaO) content, as well as the creation of a larger amount of crystalline $Ca(OH)_2$ in hexagonal plate-like structures (unreacted particles/partially reacted matrix) in AAMs with 60% and 70% of WCP, both producing poor microstructure and mechanical properties.

3. The XRD pattern of AAMs containing a high volume of FA indicated that the peak intensity corresponds to crystalline quartz (SiO_2) and mullite ($3Al_2O_32SiO_2$ or $2Al_2O_3 SiO_2$) phases, in which by increasing the FA content, the C-S-H gel peak was replaced by a quartz peak where more quartz appeared to be non-reactive. Furthermore, the XRD tests have shown that by replacing GBFS with POFA in AAMs containing 50% FA, the peak intensity corresponded to the quartz, whereas the peak intensity was dedicated to C-S-H gel in the AAMs with 0% POFA (50% FA +50% GBFS).

4. FTIR spectra and the FTIR fingerprint zone reveal that by increasing the FA content from 50% to 70% in AAMs, the band of C(N)ASH gel product increased from 945.6 to 989.5 cm^{-1} along with increasing the C-S-H and Si–O band frequency, leading to the less homogenous structure and smaller silicate re-organization, and subsequently negatively affected the mechanical properties of AAMs. Furthermore, the results indicated that GBFS released soluble Ca that displaced the Si atoms from Si–O bonds, leading to a reduction in the vibrational frequency.

The ANN combined with the metaheuristic Krill Herd algorithm provided satisfactorily results to estimate the mechanical properties of AAMs compared to the ANN combined with the genetic algorithm and multiple linear regression models where statistical indexes such as R^2 value, EF, and VAF had higher values, indicating a lower error of this model. Furthermore, the Krill Herd algorithm optimization can also be used as a powerful tool in optimizing ANN weights. By using the optimized weight and bias of KHA-ANN, it is possible to design AAMs with targeted mechanical properties and simultaneously manage the consumption of waste materials depending on their availability.

Author Contributions: This research article results from a joint collaboration of the involved authors. G.F.H. developed the experimental tests; M.N. developed the ANN model; I.F. compared the ANN results; C.B. and M.H.B. verified the manuscript structure and supervised the overall research study. All authors have read and agreed to the published version of the manuscript.

Funding: This research work was supported by funding (I.F.) from the Federal State Autonomous Educational Institution of Higher Education South Ural State University (National Research University).

Institutional Review Board Statement: The study did not require ethical approval.

Informed Consent Statement: Not applicable.

Data Availability Statement: Supporting data will be shared upon request.

Conflicts of Interest: The authors declare no conflict of interest.

References

1. Siddique, R.; Singh, G.; Singh, M. Recycle option for metallurgical by-product (Spent Foundry Sand) in green concrete for sustainable construction. *J. Clean. Prod.* **2018**, *172*, 1111–1120. [CrossRef]
2. Turner, L.K.; Collins, F.G. Carbon dioxide equivalent (CO_2-e) emissions: A comparison between geopolymer and OPC cement concrete. *Constr. Build. Mater.* **2013**, *43*, 125–130. [CrossRef]

3. Scrivener, K.L.; John, V.M.; Gartner, E.M.; UN Environment. Eco-efficient cements: Potential economically viable solutions for a low-CO$_2$ cement-based materials industry. *Cem. Concr. Res.* **2018**, *114*, 2–26. [CrossRef]

4. Katare, V.D.; Madurwar, M.V.; Raut, S. Agro-Industrial Waste as a Cementitious Binder for Sustainable Concrete: An Overview. In *Sustainable Waste Management: Policies and Case Studies*; Springer International Publishing: Berlin/Heidelberg, Germany, 2019; pp. 683–702.

5. Al-Kutti, W.; Nasir, M.; Johari, M.A.M.; Islam, A.S.; Manda, A.A.; Blaisi, N.I. An overview and experimental study on hybrid binders containing date palm ash, fly ash, OPC and activator composites. *Constr. Build. Mater.* **2018**, *159*, 567–577. [CrossRef]

6. Almalkawi, A.T.; Balchandra, A.; Soroushian, P. Potential of Using Industrial Wastes for Production of Geopolymer Binder as Green Construction Materials. *Constr. Build. Mater.* **2019**, *220*, 516–524. [CrossRef]

7. Vedrtnam, A.; Bedon, C.; Barluenga, G. Study on the Compressive Behaviour of Sustainable Cement-Based Composites Under One-Hour of Direct Flame Exposure. *Sustainability* **2020**, *12*, 10548. [CrossRef]

8. Allalou, S.; Kheribet, R.; Benmounah, A. Effects of calcined halloysite nano-clay on the mechanical properties and micro-structure of low-clinker cement mortar. *Case Stud. Constr. Mater.* **2019**, *10*, e00213.

9. Shelote, K.M.; Gavali, H.R.; Bras, A.; Ralegaonkar, R.V. Utilization of Co-Fired Blended Ash and Chopped Basalt Fiber in the Development of Sustainable Mortar. *Sustainability* **2021**, *13*, 1247. [CrossRef]

10. Lisuzzo, L.; Cavallaro, G.; Milioto, S.; Lazzara, G. Effects of halloysite content on the thermo-mechanical performances of composite bioplastics. *Appl. Clay Sci.* **2020**, *185*, 105416. [CrossRef]

11. Lisuzzo, L.; Hueckel, T.; Cavallaro, G.; Sacanna, S.; Lazzara, G. Pickering Emulsions Based on Wax and Halloysite Nanotubes: An Ecofriendly Protocol for the Treatment of Archeological Woods. *ACS Appl. Mater. Interfaces* **2021**, *13*, 1651–1661. [CrossRef]

12. Ogawa, Y.; Uji, K.; Ueno, A.; Kawai, K. Contribution of fly ash to the strength development of mortars cured at different temperatures. *Constr. Build. Mater.* **2021**, *276*, 122191. [CrossRef]

13. Kumarappa, D.B.; Peethamparan, S. Stress-strain characteristics and brittleness index of alkali-activated slag and class C fly ash mortars. *J. Build. Eng.* **2020**, *32*, 101595. [CrossRef]

14. Islam, A.; Alengaram, U.J.; Jumaat, M.Z.; Bashar, I.I. The development of compressive strength of ground granulated blast furnace slag-palm oil fuel ash-fly ash based geopolymer mortar. *Mater. Des.* **2014**, *56*, 833–841. [CrossRef]

15. Ranjbar, N.; Mehrali, M.; Alengaram, U.J.; Metselaar, H.S.C.; Jumaat, M.Z. Compressive strength and microstructural analysis of fly ash/palm oil fuel ash based geopolymer mortar under elevated temperatures. *Constr. Build. Mater.* **2014**, *65*, 114–121. [CrossRef]

16. Mohammadhosseini, H.; Lim, N.H.A.S.; Tahir, M.M.; Alyousef, R.; Alabduljabbar, H.; Samadi, M. Enhanced performance of green mortar comprising high volume of ceramic waste in aggressive environments. *Constr. Build. Mater.* **2019**, *212*, 607–617. [CrossRef]

17. Samadi, M.; Huseien, G.F.; Mohammadhosseini, H.; Lee, H.S.; Lim, N.H.A.S.; Tahir, M.M.; Alyousef, R. Waste ceramic as low cost and eco-friendly materials in the production of sustainable mortars. *J. Clean. Prod.* **2020**, *266*, 121825. [CrossRef]

18. Mohit, M.; Ranjbar, A.; Sharifi, Y. Mechanical and microstructural properties of mortars incorporating ceramic waste powder exposed to the hydrochloric acid solution. *Constr. Build. Mater.* **2021**, *271*, 121565. [CrossRef]

19. Li, G.; Liu, S.; Niu, M.; Liu, Q.; Yang, X.; Deng, M. Effect of granulated blast furnace slag on the self-healing capability of mortar incorporating crystalline admixture. *Constr. Build. Mater.* **2020**, *239*, 117818. [CrossRef]

20. Awoyera, P.; Adesina, A.; SivaKrishna, A.; Gobinath, R.; Kumar, K.R.; Srinivas, A. Alkali activated binders: Challenges and opportunities. *Mater. Today: Proc.* **2020**, *27*, 40–43. [CrossRef]

21. Adesina, A. Properties of Alkali Activated Slag Concrete Incorporating Waste Materials as Aggregate: A Review. *Mater. Sci. Forum* **2019**, *967*, 214–220. [CrossRef]

22. Goldaran, R.; Lotfollahi-Yaghin, M.A.; Aminfar, M.H.; Turer, A. Investigation of attenuation and acoustic wave propagation path caused by corrosion for reliability assessment of prestressed pipe monitoring using Acoustic Emission technique. *Modares Mech. Eng.* **2017**, *17*, 306–314.

23. ASTM International. *C618-15 Standard Specification for Coal Fly Ash and Raw or Calcined Natural Pozzolan for use in Concrete*; ASTM International: West Conshohocken, PA, USA, 2015.

24. ASTM International. *Cement, Standard Test Method for Compressive Strength of Hydraulic Cement Mortars (Using 2-in or [50-mm] Cube Specimens)*; ASTM International: West Conshohocken, PA, USA, 2013.

25. ASTM International. *C496/C496M-11, Standard Test Method for Splitting Tensile Strength of Cylindrical Concrete Specimens*; ASTM International: West Conshohocken, PA, USA, 2004; pp. 469–490.

26. ASTM International. *C78/C78M Standard Test Method for Flexural Strength of Concrete (Using Simple Beam with Third-Point Loading)*; ASTM International: West Conshohocken, PA, USA, 2010.

27. Richardson, I. The calcium silicate hydrates. *Cem. Concr. Res.* **2008**, *38*, 137–158. [CrossRef]

28. Kumar, S.; Kumar, R.; Mehrotra, S.P. Influence of granulated blast furnace slag on the reaction, structure and properties of fly ash based geopolymer. *J. Mater. Sci.* **2010**, *45*, 607–615. [CrossRef]

29. García-Lodeiro, I.; Fernandez-Jimenez, A.; Palomo, A.; Macphee, D.E. Effect of calcium additions on N–A–S–H cementitious gels. *J. Am. Ceram. Soc.* **2010**, *93*, 1934–1940. [CrossRef]

30. Yusuf, M.O.; Johari, M.A.M.; Ahmad, Z.A.; Maslehuddin, M. Evolution of alkaline activated ground blast furnace slag–ultrafine palm oil fuel ash based concrete. *Mater. Des.* **2014**, *55*, 387–393. [CrossRef]

31. Nath, S.; Maitra, S.; Mukherjee, S.; Kumar, S. Microstructural and morphological evolution of fly ash based geopolymers. *Constr. Build. Mater.* **2016**, *111*, 758–765. [CrossRef]
32. Nikoo, M.; Hadzima-Nyarko, M.; Nyarko, E.K.; Nikoo, M. Determining the natural frequency of cantilever beams using ANN and heuristic search. *Appl. Artif. Intell.* **2018**, *32*, 309–334. [CrossRef]
33. Haykin, S. *Neural Networks: A Comprehensive Foundation*; Prentice Hall: Upper Saddle River, NJ, USA, 1999.
34. Bishop, C.M. *Pattern Recognition and Machine Learning*; Springer: New York, NY. USA, 2006.
35. Asteris, P.G.; Nozhati, S.; Nikoo, M.; Cavaleri, L.; Nikoo, M. Krill herd algorithm-based neural network in structural seismic reliability evaluation. *Mech. Adv. Mater. Struct.* **2018**, *26*, 1146–1153. [CrossRef]
36. Bolaji, A.L.; Al-Betar, M.A.; Awadallah, M.A.; Khader, A.T.; Abualigah, L.M. A comprehensive review: Krill Herd algorithm (KH) and its applications. *Appl. Soft Comput.* **2016**, *49*, 437–446. [CrossRef]

 *sustainability*

Article

Study on the Compressive Behaviour of Sustainable Cement-Based Composites under One-Hour of Direct Flame Exposure

Ajitanshu Vedrtnam [1,2,*], Chiara Bedon [3] and Gonzalo Barluenga [2]

[1] Department of Mechanical Engineering, Invertis University, Bareilly 243123, India
[2] Departamento de Arquitectura, Universidad de Alcalá, 28801 Madrid, Spain; gonzalo.barluenga@uah.es
[3] Department of Engineering and Architecture, University of Trieste, 34127 Trieste, Italy; chiara.bedon@dia.units.it
* Correspondence: ajitanshu.m@invertis.org

Received: 6 October 2020; Accepted: 12 December 2020; Published: 16 December 2020

Abstract: Fire is a significant threat to human life and civil infrastructures. Builders and architects are hankering for safer and sustainable alternatives of concrete that do not compromise with their design intent or fire safety requirements. The aim of the present work is to improve the residual compressive performance of concrete in post-fire exposure by incorporating by-products from urban residues. Based on sustainability and circular economy motivations, the attention is focused on rubber tire fly ash, aged brick powder, and plastic (PET) bottle residuals used as partial sand replacement. The selected by-products from urban residues are used for the preparation of Cement-Based Composites (CBCs) in two different proportions (10% and 15%). Thermal CBC behaviour is thus investigated under realistic fire scenarios (i.e., Direct Flame (DF) for 1 h), by following the International Organization for Standardization (ISO) 834 standard provisions, but necessarily resulting in nonuniform thermal exposure for the cubic specimens. The actual thermal exposure is further explored with a Finite Element (FE) model, giving evidence of thermal boundaries effects. The post-fire residual compressive strength of heated concrete and CBC samples is hence experimentally derived, and compared to unheated specimens in ambient conditions. From the experimental study, the enhanced post-fire performance of CBCs with PET bottle residual is generally found superior to other CBCs or concrete. The structure–property relation is also established, with the support of Scanning Electron Microscopy (SEM) micrographs. Based on existing empirical models of literature for the prediction of the compressive or residual compressive strength of standard concrete, newly developed empirical relations for both concrete and CBCs are assessed.

Keywords: concrete; cement-based composites (CBCs); compressive strength; fire exposure; thermal boundaries; finite element (FE) numerical modelling; empirical formulations

---

1. Introduction

A series of consistent unfortunate fire disasters has highlighted the need for safer structural materials [1]. Concrete is safer under fire due to noncombustibility and low thermal conductivity in comparison to other widely used construction materials, such as glasses [2,3], steel [4], composites [5], and wood [6]. However, the physical, chemical and mechanical properties of concrete deteriorate due to fire exposure [7,8]. The structural and occupant safety in fire depends on the behaviour of concrete; thus, the performance of concrete under elevated temperature is extensively investigated [9–12]. The performance of concrete under elevated temperature deteriorates due to thermal incompatibility of constituents, aggregate and cement paste interface debonding, aggregate deformation, calcium silicate hydrate (CSH) gel disruption, cement paste chemical transformation and internal pressure resulted from entrapped steam [13]. The higher concrete grade suffers higher strength loss [14]. The heating rate and time of exposure affects concrete performance in fire [15]. The concrete suffers negligible up to 200 °C as only free water evaporates (exceptions are also reported [16]), at 300 °C bounded water starts evaporating; thus, CSH starts irreversibly dehydrating. At 400 °C, the strength of concrete further deteriorates, however the residual compressive strength does not suffer much [17]. At 530 °C, $Ca(OH)_2$ converts into CaO, which results in shrinkage of cement paste 33% by volume [18]. The compressive strength of concrete can hence reduce down to 90% at 750 °C [19]. A multitude of parameters, however, like the nature and setup of the experiments, or the mix design parameters, govern these findings [15].

The performance of concrete depends on fillers and adhesives at normal and elevated temperature [20]. A number of supplementary cementitious materials (SCMs) can be used in for improving the fire performance of concrete [21]. The inclusion of suitable filler may prevent thermally induced explosive spalling if their melting temperature is lower than the temperature at which explosive spalling occurs [15]. The melted fillers could be absorbed by the matrix resulting in a conduit left for steam. A further permeable network could be created that would allow outward migration of steam for sinking the pore pressure.

The first objective of the present study is to assess and support the use of waste and harmful materials as fillers for improving the post-fire residual performance of standard concrete. Some studies of literature [22–27] are dedicated to the assessment of specific properties and effects of concrete-based solutions inclusive of PET residuals, with some efforts for the expected post-fire performance [28]. However, the cited documents are often related to standard methods and procedures, that often do not reproduce a real boundary condition for the tested materials.

To this aim, rubber fly ash and Polyethylene Terephthalate (PET) plastic bottle residual and aged brick powder are used in this paper for the prepared CBCs. Moreover, special attention is spent for the realistic thermal exposure of cubic samples in fire conditions. In general, the standard ISO 834 thermal history (or a fixed rate of heating) is imposed for different time durations in dedicated furnaces to expose the investigated specimens to elevated temperature [15,29]. In a few cases, the concrete specimens were exposed to real flame [30].

As such, this research study focuses on the performance of CBCs to 1 h of Direct Flame (DF) thermal exposure. A brief description of materials and methods is presented in Section 2. The final objective of this work is in fact to compare the residual compressive strength of concrete and CBCs specimens, after subjecting the samples to nonuniform, realistic fire exposure. To this aim, it is well known that Finite Element (FE) numerical models can be used to realistically predict the thermo or mechanical (and even combined) behavior of concrete (or other general materials in fire conditions). Reliable modelling strategies for concrete in fire are reported in [31,32]. In this paper, with the support of the ABAQUS/Standard computer software [33], the attention is focused on the thermal boundaries and expected predictions for the tested cubic specimens, showing the potential and limits of the FE method (Section 3). An additional support for the interpretation of test results is derived from Scanning Electron Microscopy (SEM) images. Finally, based on the experimental observations, a set of empirical formulations are proposed in Section 4, for a reliable and efficient prediction of the compressive mechanical parameters of the tested materials under elevated temperatures.

2. Experiments, Materials and Methods

2.1. Preparation of the Specimens

A total of 126 cubical samples (three for every reading) were prepared for fire and compression testing, by casting them in steel molds (150 mm × 150 mm × 150 mm their size) with the assistance of a vibrator. The samples were removed from the steel molds after 24 h and cured in water for 28 days at room temperature (26–32 °C).

Some standard American Society for Testing and Materials, (ASTM) test procedures were thus followed for evaluating the physical properties of concrete. The amount of ordinary Portland cement (OPC) type I, natural sand, aggregate (19 mm) and water were 335 kg/m^3, 871 kg/m^3, 1083 kg/m^3, and 225 kg/m^3 respectively for the concrete cubes. The water to cement ratio was 0.67. The OPC purchased from the local vendor complies with ASTM C150 [34]. The natural sand and aggregate conform to ASTM C33-16 [35]. The specific gravity, absorption, and loose density (kg/m^3) of natural sand and aggregate were 2.39 and 2.61, 1% and 0.39%, 2002.4 and 1818.1, respectively. The CBC samples included of rubber fly ash or PET plastic bottle residual or aged brick powder as a replacement for the natural sand in two different proportions, i.e., 10% or 15%. The particle size of the SCMs were less than 1.18 mm for aged brick powder and plastic residual, and 90 μm for rubber fly ash obtained by the sieve analysis.

Table 1 presents, in accordance with [36], some basic mechanical properties for the material components in use.

Table 1. Basic mechanical properties of concrete and CBC components in use.

Material	Young Modulus (GPa)	Density (kg/m^3)	Poisson's Ratio
Cement	21.9	2050.3	0.279
Fine Aggregate	26.59	1472	0.24
Coarse Aggregate	23.49	1434	0.13

In this regard, it is worth mentioning that the PET residual was obtained from previous experimental steps (Figure 1) in which bottles were burned for achieving added-value products (grease, useful chemicals) [37]. The COH-RC method includes heating plastics into a close chamber having oxygen but no oxygen supply during the process at different temperatures and directing the exhaust gases into another container filled with water for condensation [37]. The residual in the reactor of Figure 1a was thus converted into the powder before it could be used for the tested CBCs (Figure 1b). The rubber tyre fly ash was then obtained after burning it, while the waste brick powder was used directly as urban residual.

(a)

(b)

Figure 1. Production process for the PET bottle residual following constrained oxygen heating and residual condensation Method [37]: (**a**) waste plastic treatment reactor and (**b**) PET (EC number: 607-507-1) bottle residual before crushing.

2.2. Fire and Compression Testing

All the specimens were heated in the furnace under 1 h of DF exposure (Figure 2), following—even for limited exposure time only—the general standardized provisions from the ISO 834 document [38]. This resulted in the time–temperature furnace variation that is shown in Figure 3, as well as in a nonuniform temperature distribution on the surface and in the volume of each tested cube. According to Figure 2c, it can be perceived that the lateral faces of each cube are also subject to severe thermal exposure.

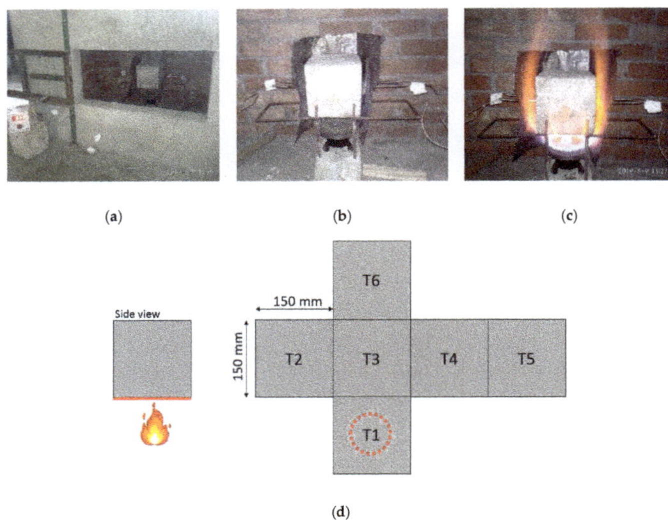

Figure 2. Selected photographs of fire testing for the CBC specimens: (**a**) furnace view; (**b**) setup and (**c**) heating stage, with (**d**) thermal loading scheme.

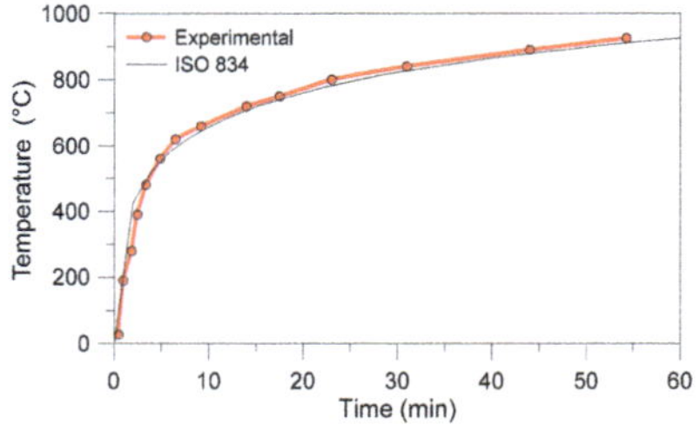

Figure 3. Experimentally measured time–temperature history in the furnace (1 h in DF) and corresponding standard ISO 834 fire curve.

The surface temperature of the specimens was measured using an infrared thermocouple, a thermal imaging camera (Testo-868, Universal 16 channel, RS4R5, accuracy: ±2 °C) and some additional K-type thermocouples (chromel/Alumel, range −200 to 1300 °C, accuracy: ±2.2 °C). The temperature increase was recorded on all the faces of the tested cubes. For clarity of presentation, the nonuniform surface temperature was then post-processed to obtain the average temperature at the center of each face. This measure is thus reported in this paper in terms of (see also Figure 2d):

- T1, representing the average temperature in time at the center of the bottom (exposed) face;
- T2, T3, T4, T5, being representative of the four side cube faces;
- and T6, denoting the top face of each cube.

Successively, standard compression tests were used for evaluating the compressive strength of each cube in the post-fire scenario. As a reference, three cubes for each material type and composition were also tested in compression at ambient temperatures and compared after different thermal scenarios. During the compression tests, the imposed loading rate was set in 3 kN/min. The weight loss due to fire exposure was also estimated.

Finally, SEM micrographs were collected for some of the fractured specimens, and used to further explore and determine the reasons of failure, thus supporting the establishment of some structure–property relationships.

2.3. Preliminary Numerical Assessment

At the preliminary stage of the study, some FE numerical simulations were carried out to estimate the potential experimental outcomes. According to literature, FE models can represent an efficient tool for the thermal and coupled thermo-mechanical analysis of various constructional materials under elevated temperatures [2]. The real thermal boundaries, however, can be difficult to match with taken in numerical models, due to the use of several idealized assumptions that do not reflect an experimental setup with its uncertainties. Moreover, these thermal boundaries and their effects are strictly responsible of the mechanical performance for the tested materials, thus represent a first crucial step.

In this paper, for both the investigated concrete and CBC cube samples, a preliminary set of FE analyses was thus carried out in ABAQUS/Standard [33]. The typical FE assembly, as proposed in Figure 4, consisted of 1/4th the nominal geometry of each sample, thus was numerically described in the form of a prismatic volume with total nominal height of 150 mm, and half base × width dimensions (75 mm × 75 mm). The transient "heat transfer" solver approach from ABAQUS/Standard was taken into account, so that the temperature evolution in time could be monitored in the solid elements. In doing so, the basic assumptions were represented by the use of three-dimensional, 8-node solid brick elements (heat transfer DC3D8 type, from ABAQUS library), with a structured mesh pattern (10 mm the size).

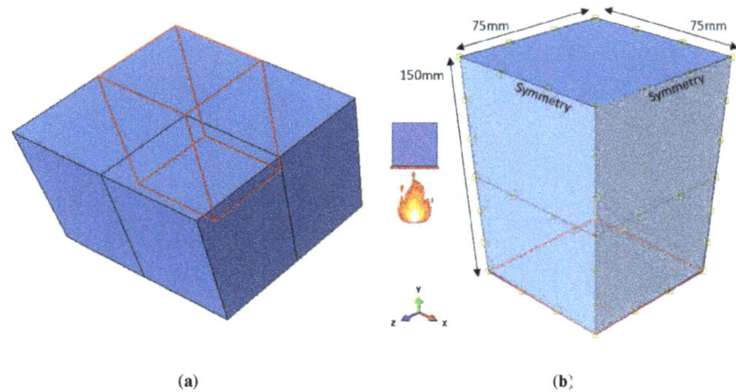

Figure 4. Reference numerical model for the analysis of thermal distribution in concrete or CBC cube samples under direct flame exposure (ABAQUS/Standard): (**a**) global assembly and (**b**) local setup (with evidence of the thermally exposed T1 bottom face).

The reference cube was hence exposed, on the bottom (T1) surface, to the temperature–time history that was derived from the experimental records (Figure 3). In doing so, the thermal exposure in time from the furnace was lumped on the T1 surface, as in the schematic representation of Figure 4b.

As a first attempt to investigate the thermal behaviour of the tested cubes, besides the intrinsic approximation of this numerical approach [39], nominal thermophysical properties were considered for modelling normal concrete. This numerical choice was intended to predict the temperature distribution and evolution in the specimens, rather than a direct comparison with CBC experimental measurements.

As such, the emissivity was set in 0.7, while the heat transfer coefficient for the exposed surface was defined in 25 W/m^{2}K [40–42]. For the other surfaces not subjected to DF exposure, the heat transfer coefficient was assumed in 8 W/m^{2}K [40–42]. At the same time, the specific heat and the conductivity variation with temperature was preliminary defined by accounting for nominal material properties of standard concrete [40–42], given the lack of more refined material characterizations for the cubic specimens object of study. These input data are shown in Figure 5.

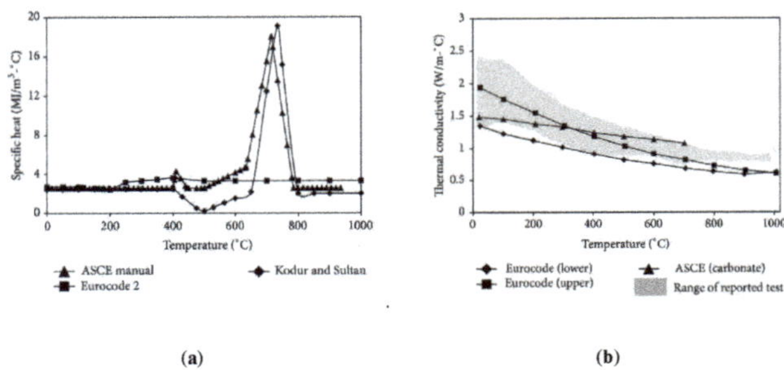

(a)   (b)

Figure 5. Reference input properties for the thermal numerical model: (a) specific heat and (b) conductivity variation with elevated temperature. Figures reproduced with permission from [42], under the rules of the Creative Commons Attribution License.

3. Results and Discussion

3.1. Experimental Fire Exposure

Figure 6 shows the average time–temperature variation of concrete, CBC with rubber fly ash, CBC with brick powder, and CBC with PET bottle residual, respectively. From Figure 6a, more in detail, the temperature evolution on the cube faces directly exposed or unexposed is emphasized.

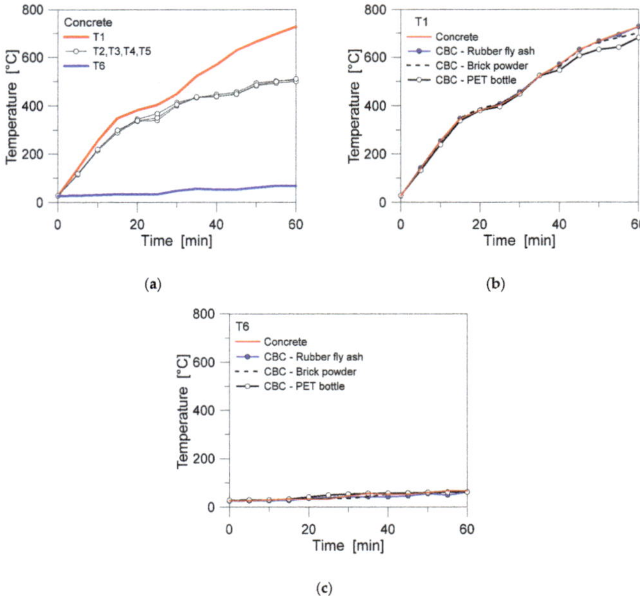

(a)   (b)

(c)

Figure 6. Average time–temperature histories on different faces of the cubic volume for the tested specimens: (a) standard concrete, and (b) T1 bottom surface or (c) T6 top surface for all the specimens.

Through the experimental investigation, a rather stable fire scenario was observed and measured for most of the specimens. This is confirmed in Figure 6b,c, where the average time–temperature records for different concrete and CBC cube types are proposed at the bottom (T1) or top surface (T6). The exception was represented by PET bottle specimens, with a lower temperature peak for all the reference control points, compared to other specimen types.

Moreover, the experimental setup typically coincided with a severe nonuniform temperature distribution. This can be perceived from Figure 7a–c, where selected thermal images show the temperature variation of CBC cubes under variable exposure time. In particular, it is worth mentioning that the collected experimental measurements generally proved that the maximum and the average temperatures of CBC cubes with PET bottle residuals were typically lower than other types of CBC specimens. This was noticed on all the monitored faces for an identical exposure time. Such an effect may be due to low thermal conductivity of the residuals in use (i.e., Figure 1).

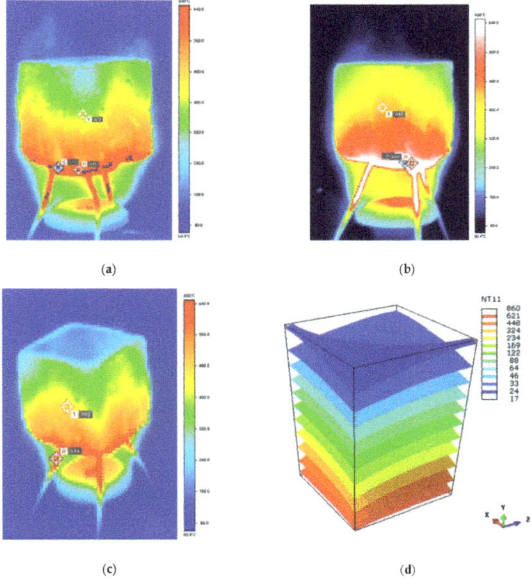

Figure 7. Temperature variation of CBC cubes under DF (T1 bottom surface): experimental results after (**a**) 20 min, (**b**) 30 min or (**c**) 40 min of heating and (**d**) corresponding numerical contour plot for a nominal concrete cube (1/4th of the cube, ABAQUS/Standard, external view of the cube). Temperature values in °C.

From the numerical analysis of a typical CBC cube, the temperature distribution was found to agree with Figure 7d, where the contour plot of temperatures (external view of 1/4th of cube) is shown after 1 h in DF. From the qualitative comparison of the experimental thermal images and the numerical contour plot in Figure 7d, a partial agreement only for the proposed temperature distributions can be seen. The numerical model, more in detail, lacks unsymmetrical effects due to the idealized thermal exposure of the T1 bottom surface. In addition, the experimental temperature–time history of Figure 2 is used for the FE numerical analyses; moreover, another weak point is represented by the underestimation of thermal effects on the lateral faces of each cube, compared to the experimental specimens. These are the results of the same idealized thermal loading condition for the FE model in Figure 4, and namely neglecting the radiant heating effects on the lateral faces of the specimens. As the temperature distribution is further explored in the volume of the FE model (see the view cuts in Figure 8), it can be also perceived a progressive underestimation of thermal effects for the overall compressive mechanical response of the tested specimens.

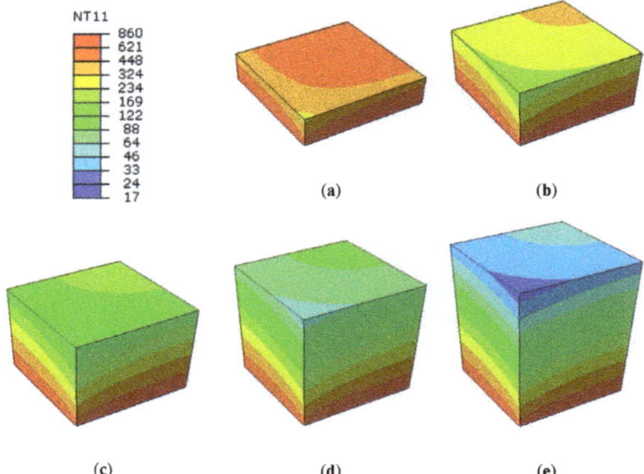

Figure 8. Temperature variation for concrete cubes under 1 h in DF (ABAQUS/Standard), with evidence of section cuts in the height of specimens: (**a**) 25 mm, (**b**) 50 mm, (**c**) 75 mm, (**d**) 100 mm and (**e**) 125 mm. Temperature values in °C. T1 bottom surface under uniform thermal exposure.

In order to further assess the effects of numerical thermal boundaries on the predicted temperature distributions for the tested specimens, as a limit condition, the FE analysis was further repeated by assuming replacing the T1 thermal exposure of Figure 4 with a uniform thermal exposure for the base (T1) but also for the lateral faces of the cube. As such, the T1, and T2-to-T5 surfaces were uniformly subjected to the experimental temperature–time history of Figure 2.

The final result is proposed in Figure 9, where it is possible to notice that such a limit thermal boundary condition still does not march the experimental setup, and thus manifests in a severe overestimation of the expected temperatures in the volume of specimens (with a consequent premature degradation of material properties at elevated temperatures).

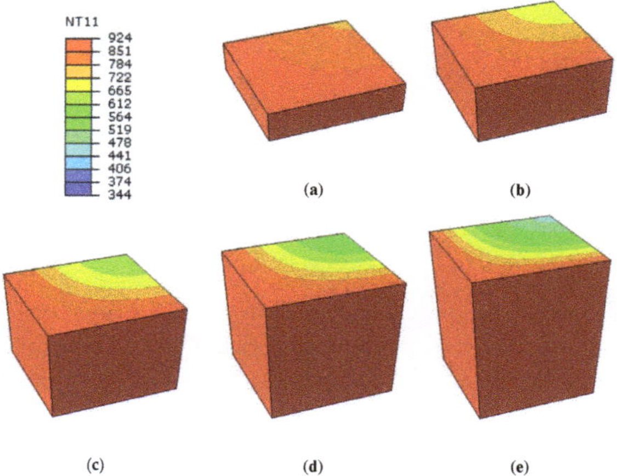

Figure 9. Temperature variation for concrete cubes under 1 h in DF (ABAQUS/Standard), with evidence of section cuts in the height of specimens: (**a**) 25 mm, (**b**) 50 mm, (**c**) 75 mm, (**d**) 100 mm and (**e**) 125 mm. Temperature values in °C. T1 (bottom) and T2-to-T5 (lateral) surfaces under uniform thermal exposure.

3.2. Post-Fire Compressive Resistance

Tables 2 and 3 give the average compressive strength and residual compressive strength (with negligible standard deviation) after 1 h heating of both concrete and CBC samples (with 10% and 15%

the proportion of urban residues) following the ISO 834 under DF. In general, it is recommended by the reference standards that the tested samples should be kept for a long thermal exposure (generally 2–3 h), so that they could be uniformly heated throughout. However, in the present study, the tested samples were kept at 200, 400, 600, and 800 °C for 1 h in the furnace, under the heating rate schematized in Figures 2 and 3. This allows for a comparison between different heating effects, especially in terms of compressive strength of the CBC samples compared to the concrete specimens.

Table 2. Compressive strength (MPa) of concrete and CBC samples after heating following ISO 834 or 1 h in Direct Flame (DF).

	Compressive Strength (MPa)						
	Concrete	CBC					
		Rubber Fly Ash		Brick Powder		PET Bottle	
Temperature (°C)/ Urban Residues Proportion (%)	-	10	15	10	15	10	15
27 (ambient)	19.51	18.64	18.2	19.33	19.11	18.68	18.64
200	18.44	18.2	17.9	18.4	17.96	17.4	17.11
400	15.24	15.62	15.55	15.11	15.02	16.78	16.63
600	10.78	11.01	10.66	10.66	9.99	13.21	13.11
800	6.11	7.9	7.3	6.2	5.86	12.99	11.87
1 h in DF	5.77	7.11	6.98	5.76	5.11	9.66	8.96

Table 3. Young's modulus (MPa) of concrete and CBC-PET10 specimens after heating (following the ISO 834 provisions and 1 h in Direct Flame (DF)).

	Young's Modulus *MoE* (MPa)	
Temperature (°C)	Concrete	CBC-PET10
27 (ambient)	1,4941.75	1,4489.80
200	5311.78	5113.20
400	2803.67	3276.07
600	1773.64	2378.92
800	782.349	2182.86
1 h in DF	756.99	2101.76

Table 2 reflects that at the ambient temperature (27 °C, thus without heating), the compressive strength of concrete was generally higher than that of the CBCs, which is followed in the order by CBCs with aged brick powder, CBCs with PET bottle residual and CBCs with rubber tire fly ash. Figure 10, accordingly, presents the percentage variation of compressive strength, under different heating exposures, compared to the ambient condition.

The nonuniform fire exposure due to real flame resulted in a significantly higher reduction in compressive strength, compared to the samples exposed to ISO 834 heating, and this outcome was proved for every type of specimens. This difference should be due to higher temperature exposure in certain region nearby bottom face of the samples in case of real flame exposure. Weakening of directly exposed regions to fire caused higher damage due to loss of bounded water and conversion of $Ca(OH)_2$ into CaO. In case of furnace heating of the samples, the duration of high temperature heating is lesser than the real fire exposure. Non uniform stresses and strain resulted due to considerable thermal gradient could be an additional reason for lesser compressive strength in case of real flame heating. It is worth mentioning that the residual compressive strength of concrete and CBC samples are comparatively higher than the standard provisions, due to the shorter exposure duration that is accounted in present work (Figure 10c,d). It is also evident that the residual compressive strength for CBC samples having 15% proportion of urban residues have lower compressive and residual compressive strength for all the types of samples. This can be seen in Figure 11 for 1 h of heating in DF.

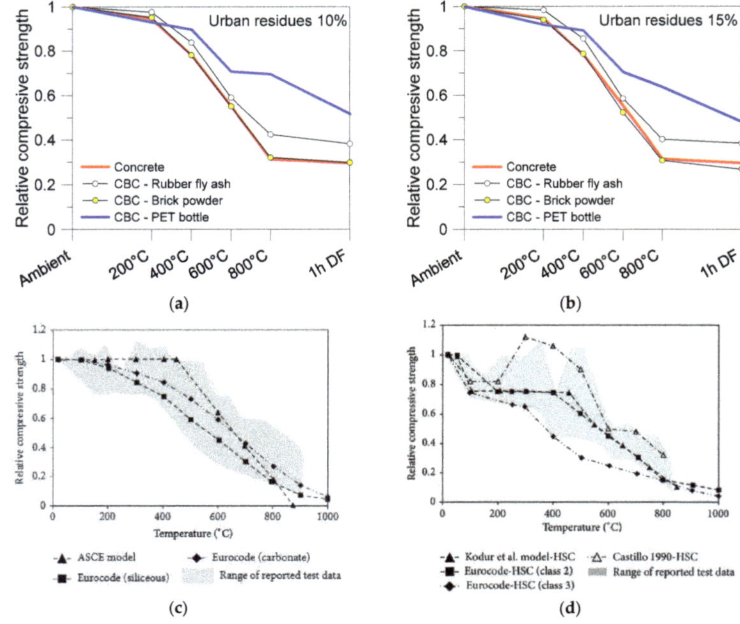

Figure 10. Variation of relative compressive strength for concrete or CBCs samples based on (**a**) 10% or (**b**) 15% urban residues proportion, as a function of different heating conditions, and (**c**,**d**) typical trend for normal/high strength concrete, as a function of temperature. Figure 10c,d are reproduced with permission from [42], under the rules of the Creative Commons Attribution License.

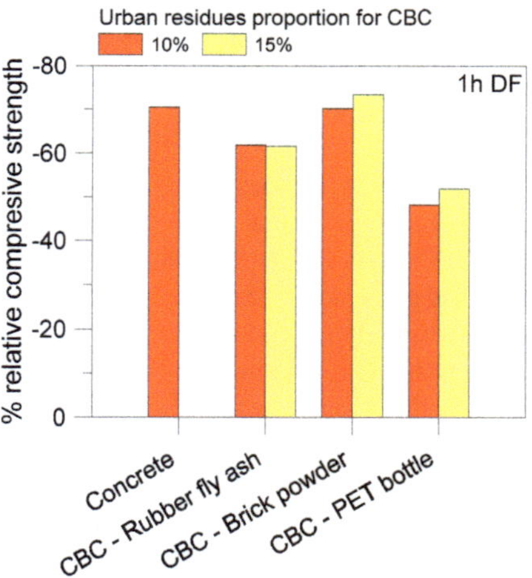

Figure 11. Percentage variation of relative compressive strength for concrete or CBCs cube samples, after 1 h in DF.

Table 3, finally, gives the values of Young's Modulus MoE for concrete and CBC samples with PET bottle residual at 10% (herein after defined as "CBC-PET10"), as calculated after heating the cubes following ISO 834 or 1 h in DF. According to Figure 12, it is possible to perceive a MoE percentage reduction in the order of 90% and 80% for concrete or CBC-PET10 specimens respectively, after 1 h in

DF. The MoE variation with temperature is discussed in more detail in Section 4, with the support and development of reliable empirical models.

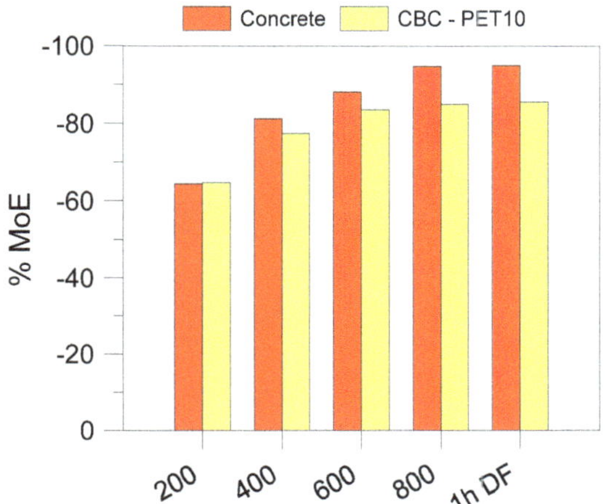

Figure 12. Percentage variation of MoE for concrete or CBC-PET10 cube samples, after 1 h in DF.

3.3. Fracture Behaviour

Figure 13 shows the photographs of samples heated in furnace and real flame. CBCs having PET bottle residual have shown considerably better performance and least weight loss due to fire. For'an insight into the reason for this behaviour and establishing a structure–property relationship, the SEM micrographs of fractured samples were studied.

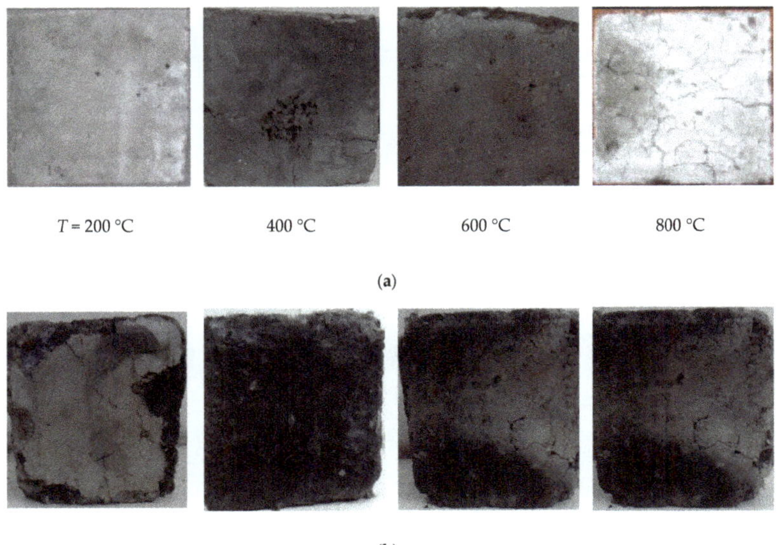

Figure 13. Photographs of (**a**) samples heated in furnace (**b**) samples subjected to direct flame exposure.

Figure 14 shows the SEM images of sample taken from the region near bottom face of fractured CBC samples having rubber fly ash. The spherical-shaped particles of fly ash can be seen jointly with small and medium size bubbles. The SEM images clearly reflect the microcracks, deformed CH and CSH, and dehydrated CSH phase. The pozzolanic reactivity between the paste and aggregate seems higher at high temperature in these samples [26]. Figure 15 shows the SEM images of fractured CBC with PET bottle residual. Irregular shaped particles of PET residue can be identified in a rather compact matrix and the porous network did not show air bubbles. It is evident from the SEM micrographs that the interface zone is constricted in these specimens. Moreover, the crackless interface between different phases remains. However, microcracks are occasionally visible, but they are considerably lesser and smaller in size than the cracks in CBCs with rubber tire fly ash.

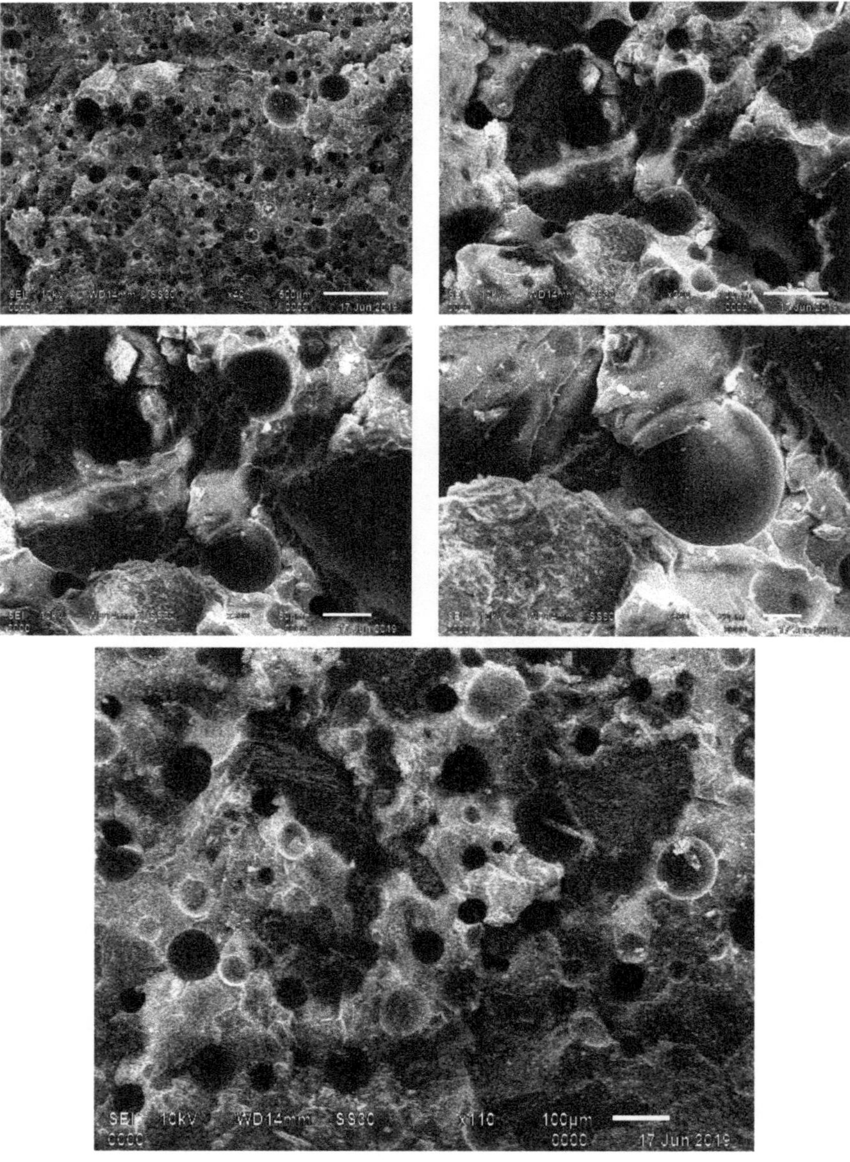

Figure 14. SEM images of fractured CBCs having rubber fly ash at ×500, 200, 100, 50.

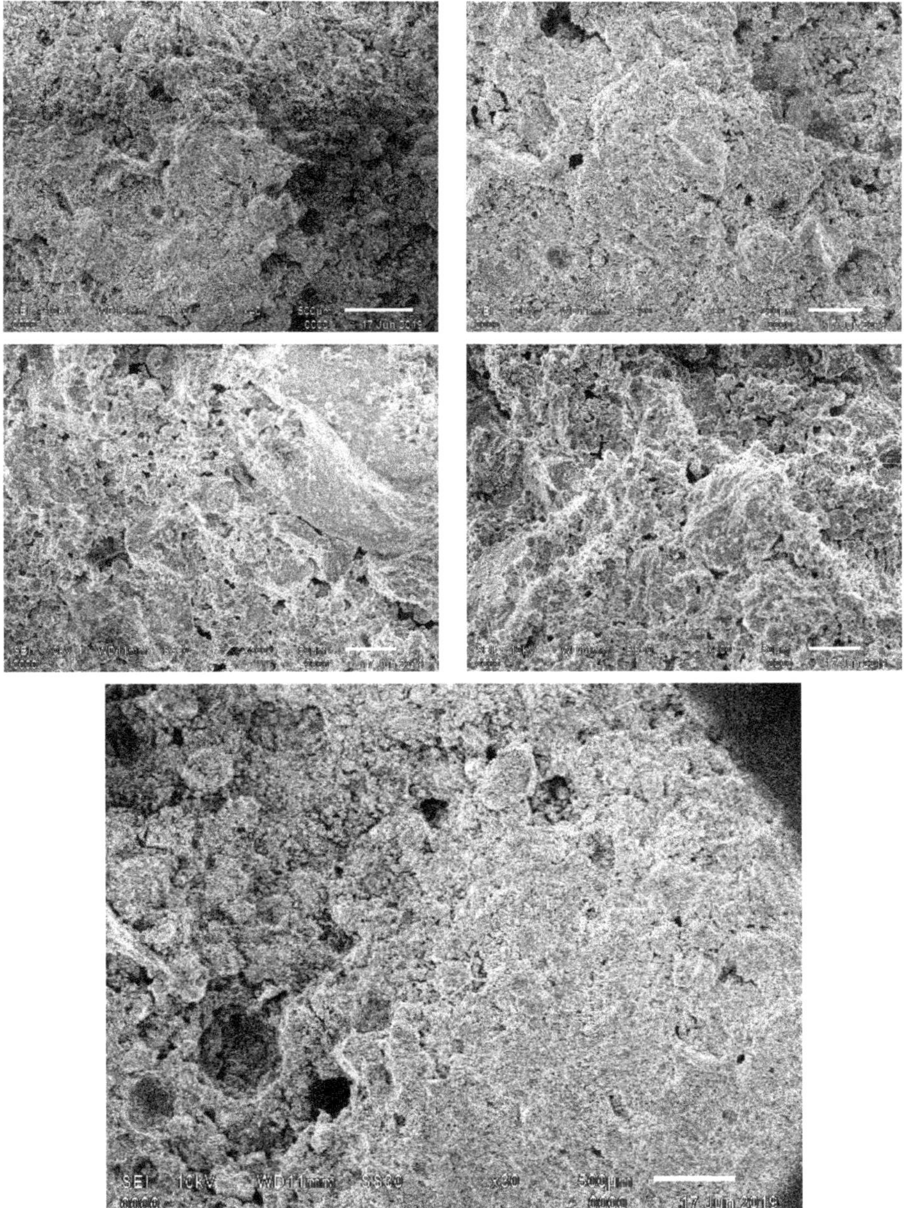

Figure 15. SEM images of fractured CBCs with PET bottle residual at ×500, 200, 100, 50.

In earlier reported studies, PET waste inclusion resulted in the deterioration of mechanical properties of concrete [27–31]. However, the performance of concrete with PET waste during ISO 834 exposure showed a comparable response but the temperature during heating was higher for CBC with PET waste than the concrete due to porous network formation [32]. The usage of PET residual has resulted in lesser deterioration in compressive strength and residual compressive strength in comparison to PET waste due to ISO 834 heat exposure [32]. Figure 16 shows the SEM images of fractured CBC with brick powder. The compact matrix produced due to the pozzalinic capacity of brick powder is scattered with many medium size air bubbles. Deformed and dehydrated microcracks in the CSH phase are visible. A comparison of the SEM micrographs of three types of CBCs reveals the minimum damage in the microstructure of CBC with PET bottle residual. However, further investigation is required to determine most suitable proportion of the PET bottle residual in concrete. The SEM micrographs establish the reasons of superior residual compressive strength of CBCs with PET but physics and chemistry behind the lesser damage requires further intensive investigation.

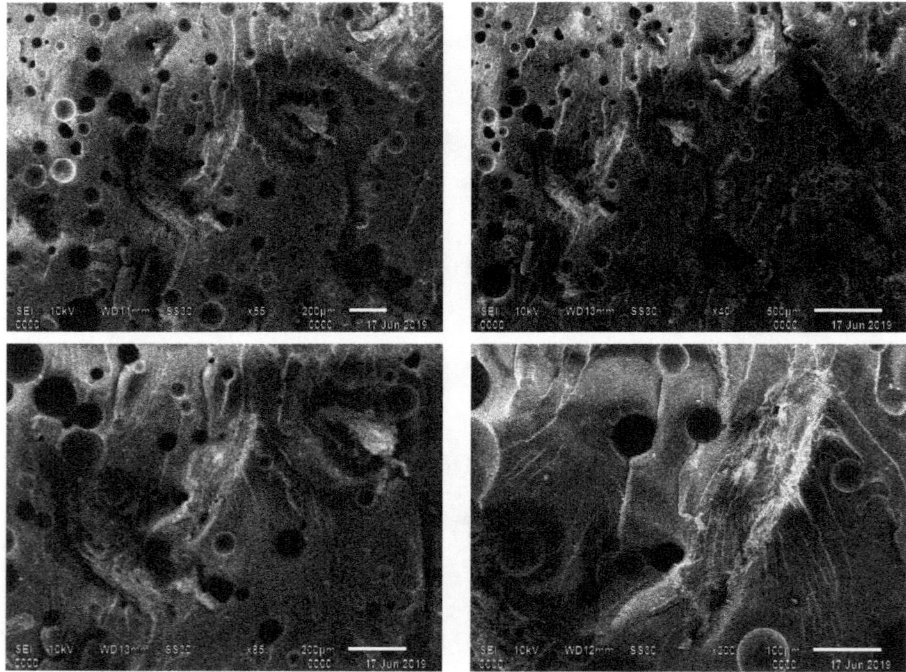

Figure 16. SEM images of fractured CBCs having brick powder at ×500, 200, 100, 50.

4. Experimental Derivation of Empirical Formulations

4.1. Time–Temperature Evolution

In the literature, there are several models for predicting residual compressive strength, see for example [43–46]. However, the common aspect of most of the cited models is that they have been originally proposed for longer thermal exposure (>1 h), as also recommended by ISO 834 to ensure uniform heating throughout the examined concrete samples. These literature models show, consequently, a comparatively lower residual strength for the tested specimens, due to severity of the heating stage. As far as different thermal boundary conditions are taken into account, their validity is questionable.

In this regard, Figure 17a,b shows the empirical model for time–temperature (max.) prediction during real flame heating for concrete and CBC with 10% PET residual specimens, respectively. Equations (1)–(3) represent the best fitting curves for experimental data. These equations are derived using a commercially available mathematical tool (LAB Fit Curve Fitting Software, www.labfit. net) that allows an enhanced match with input data to be captured due to its extended library.

The empirical variation of temperature in time is obtained from the average of the collected experimental measurements, where the fitting curves are respectively proposed as:

$$T(t) = \begin{cases} 0.0028\,t^3 - 0.353t^2 + 22.90t + 39.63 & for\ concrete \\ 0.0024t^3 - 0.344t^2 + 22.88t + 33.76 & for\ CBC - PET10 \end{cases} \tag{1}$$

with T in °C and t in minutes.

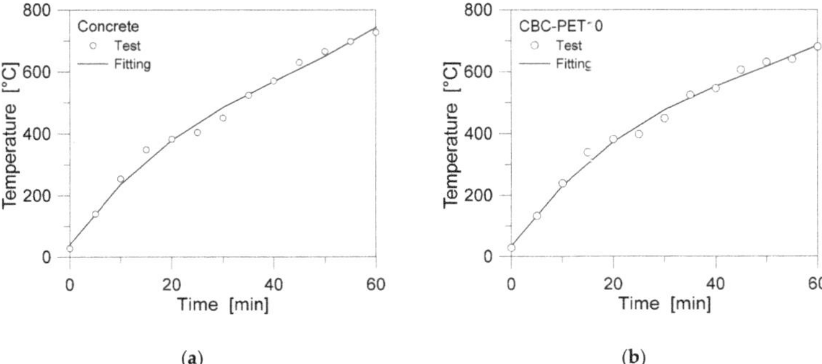

Figure 17. Experimental derivation of an empirical model (Equation (1)) for the time–temperature variation under DF heating for (**a**) concrete or (**b**) CBC-PET10 specimens (T1 bottom face).

4.2. Residual Compressive Strength and MoE

Among the available CBC specimens, the CBC-PET10 samples were found to have the maximum residual strength, and thus were further assessed in their mechanical properties with respect to the concrete specimens. The attention was focused on both the residual compressive strength and the corresponding MoE.

Figure 18a,b, in this regard, show the empirical models for predicting the residual compressive strength of concrete and CBC-PET10, after 1 h of thermal exposure. These models are valid for temperature range and exposure time investigated in the present work. Any intermediate value can be predicted efficiently with these models for concrete and CBC-PET10.

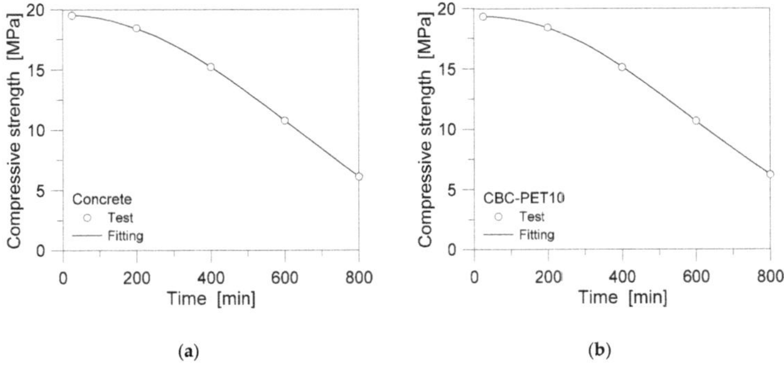

Figure 18. Experimental derivation of an empirical model (Equation (2)) for the compressive strength variation under 1 h in DF (ISO 834), as obtained for (**a**) concrete or (**b**) CBC-PET10 specimens.

The model established in the present work, in this regard, effectively predicts the residual compressive strength for heating up to one hour. Figure 18 reflects that the models are a good match to the available experimental data.

$$f(t) = \begin{cases} \dfrac{1}{(2.88-1.52\times10^{-6}t^2)^{-2.81}} & for\ concrete \\ 19.35t^{(-1.78\times10^{-7}t^{2.06})} & for\ CBC-PET10 \end{cases}$$ (2)

with $f(t)$ in MPa and t in minutes.

Figure 19a,b give the empirical models for the MoE variation, for concrete and CBC-PET10 specimens in ISO 834 heating. The developed models are in line with the experimental data for concrete and CBC-PET10.

$$MoE(t) = \begin{cases} 55.75t^{-0.24} - 10.67 & for\ concrete \\ 31.34exp\left(0.0024t - 0.16t^{0.5}\right) & for\ CBC-PET10 \end{cases}$$ (3)

with MoE in GPa and t in minutes.

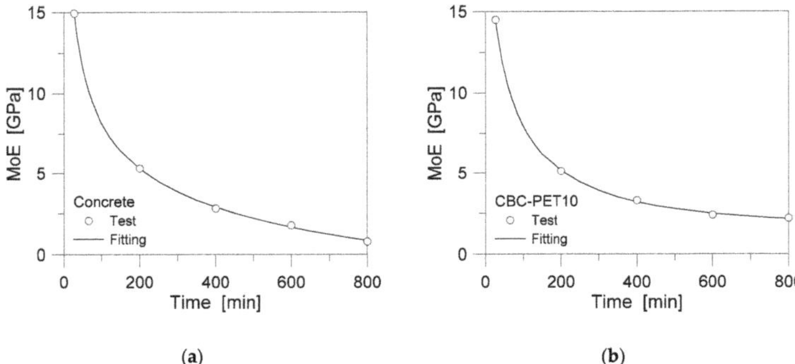

Figure 19. Derivation of an empirical model for the MoE variation (Equation (3)) in ISO 834 heating for (**a**) concrete and (**b**) CBC-PET10 specimens, respectively.

4.3. Compressive Stress–Strain Constitutive Laws at Elevated Temperatures

In the literature, several empirical models are available for predicting the stress–stain behaviour of concrete at an elevated temperature, both in tension and compression [43–47]. However, most of them are valid for specific design applications, or for particular composition of concrete constituents. Finally, their common aspect is that all those models are calibrated and validated for longer heating duration for concrete, as described in ISO 834.

In the present work, the established compressive stress–stain relations are thus modified and adapted both to concrete and CBC-PET10 specimens for 1-h of ISO 834 heating. Among the available literature models, the basic empirical equations are derived and further adapted to the examined specimens.

In the proposed Equations (4)–(7), more in detail, for a given temperature T (in °C) the following parameters are introduced:

- f_{cT} represents the compressive stress at elevated temperature (in MPa),
- f'_{cT} is the compressive strength at elevated temperature (in MPa),
- ϵ_{0T} and ϵ_{cT} the strain at ambient temperature or under elevated temperature, respectively, and
- f'_c is the compressive strength at ambient temperature (in MPa).

The result takes the form of an empirical model in which f_{cT} can be expressed as:

$$f_{cT} = \begin{cases} f'_{cT}\left[1 - \left(\frac{\epsilon_{0T} - \epsilon_{cT}}{\epsilon_{0T}}\right)^2\right] & \epsilon_{cT} \le \epsilon_{0T} \\ f'_{cT}\left[1 - \left(\frac{\epsilon_{0T} - \epsilon_{cT}}{3\epsilon_{0T}}\right)^2\right] & \epsilon_{cT} \ge \epsilon_{0T} \end{cases} \tag{4}$$

where:

$$f'_{cT} = \begin{cases} f'_c(1 - 0.0004\ T) & T \le 500\ ^\circ\text{C} \\ f'_c(1.44 - 0.00145\ T) & 500 \le T \le 900\ ^\circ\text{C} \\ 0 & T \ge 900\ ^\circ\text{C} \end{cases} \tag{5}$$

and:

$$\epsilon_{0T} = \frac{2f'_c}{E_{ci}} + 0.21 \times 10^{-4}\ (T - 20) - 1.1 \times 10^{-8}\ (T - 20)^2 \tag{6}$$

with E_{ci} the MoE.

Figure 20 shows the stress–strain diagram for concrete at elevated temperature. The diagram is in line with the collected experimental values of compressive strength and modulus of elasticity, at different temperatures.

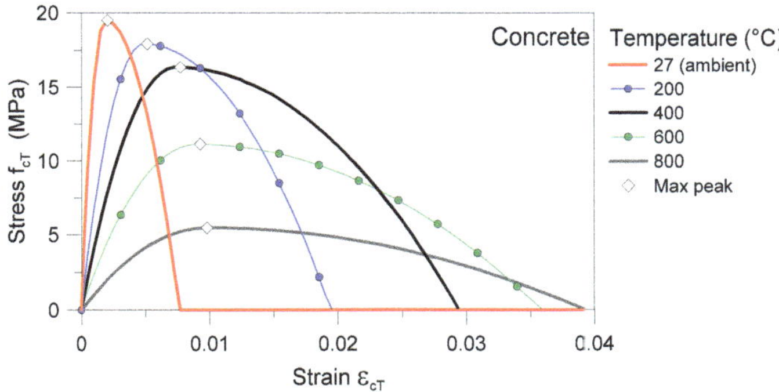

Figure 20. Compressive stress–strain curve for concrete under elevated temperature (1 h in DF, following ISO 834 provisions). Marked with white diamonds, the maximum peak for each thermal scenario.

Further, the stress–strain relations for CBC-PET10 at elevated temperature (1 h heating in DF) are given in Equations (7) and (8), where:

$$f'_{cT} = \begin{cases} f'_c(1 - 0.00038\ T) & T \le 500\ ^\circ\text{C} \\ f'_c(1.44 - 0.0012\ T) & 500 \le T \le 700\ ^\circ\text{C} \\ f'_c(1.44 - 0.00095\ T) & 700 \le T \le 900\ ^\circ\text{C} \\ 0 & T \ge 900\ ^\circ\text{C} \end{cases} \tag{7}$$

and:

$$\epsilon_{0T} = \frac{2f'_c}{E_{ci}} + 0.21 \times 10^{-4}\ (T - 20) - 1.2 \times 10^{-8}\ (T - 20)^2 \tag{8}$$

Figure 21 shows the stress–strain curve for CBC-PET10 at elevated temperature heated for one hour, following ISO 834.

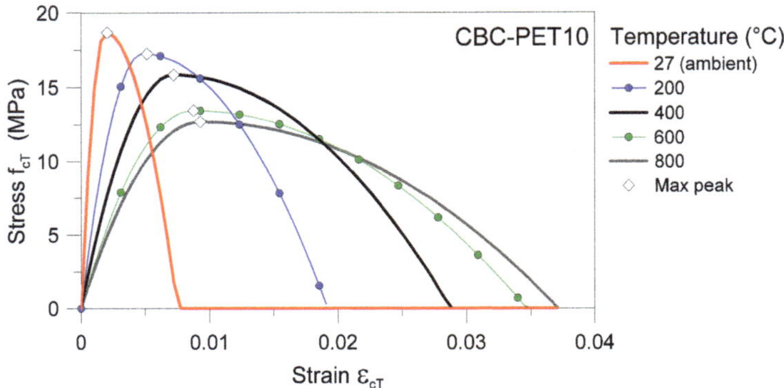

Figure 21. Compressive stress–strain curve for CBC-PET10 under elevated temperature (1 h heating in DF, following ISO 834 provisions). In evidence with white diamonds, the maximum peak for each thermal scenario.

Finally, Figure 22 presents the stress–strain peaks from the experimental/empirical constitutive laws of concrete or CBC-PET10 cubes under various thermal scenarios. Again, it is possible to perceive an improve post-fire residual performance of CBC specimens, compared to standard concrete cubes, thus suggesting their potential for safe fire design purposes.

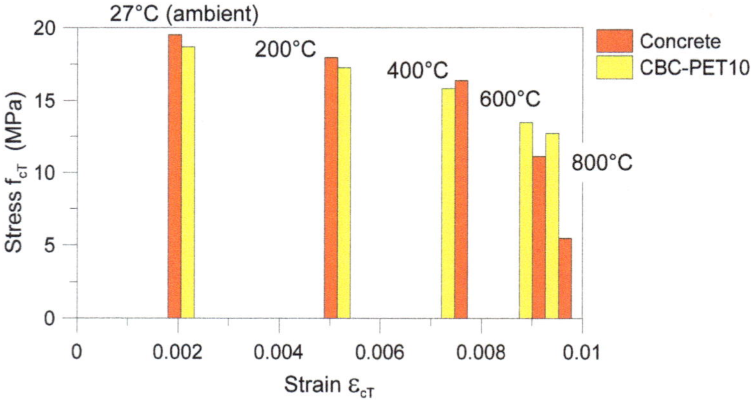

Figure 22. Post-fire compressive response of concrete and CBC-PET10 specimens under elevated temperature. Evolution with temperature of the maximum stress–strain peaks for the experimental/empirical constitutive laws.

5. Conclusions

The present work reported on the effects of fire on the residual compressive strength of concrete and other sustainable alternatives of concrete, made of by-products urban residuals. The performance of concrete and cement-based composites (CBCs) was investigated during nonuniform direct flame and standard ISO 834 fire exposure. The thermal experimental findings were then further explored with the support of Finite Element (FE) investigations. Following are the significant conclusions of this work,

- Among the examined concrete or CBCs sustainable solutions, CBCs with PET bottle residual (10% proportion) offered the lowest peak temperature compared to the other specimen types under similar heating conditions.

- The FE model developed for predicting the time–temperature variation on the exposed faces and in the volume of the tested cubes further confirmed the critical role of input thermal loads but also of the thermal boundaries. It was shown in particular that the actual experimental thermal scenario for the testes specimens can be hardly predicted with refined FE methods. As such, the post-fire mechanical response of the tested cubes necessarily requires the support of dedicated experiments.
- The CBCs specimens with PET bottle residual (10% proportion) were observed to have considerable lesser reduction in their modulus of elasticity, in comparison to standard concrete samples, thus resulting in improved post-fire capacity.
- The concrete and CBC samples damaged more while subjected to nonuniform fire exposure for the same fire duration.
- Concrete has superior compressive strength compared to the CBCs with rubber tire fly ash, PET bottle residual, and brick powder investigated in the present work.
- CBCs with PET bottle residual showed superior performance followed by CBCs with rubber tire fly ash, concrete and CBCs with aged brick powder respectively.
- The SEM micrographs have shown minimum damage in the microstructure of CBC with PET bottle residual.
- Empirical models for predicting time–temperature, modulus of elasticity, compressive strength, and stress–strain constitutive laws at elevated temperatures were proposed.

Author Contributions: Conceptualization, A.V. and G.B.; methodology, A.V.; software, A.V. and C.B.; validation, A.V.; formal analysis, A.V. investigation, A.V.; resources, A.V. and G.B.; data curation, A.V. and C.B.; writing—original draft preparation, A.V.; writing—review and editing, C.B. and G.B.; visualization, A.V.; project administration, A.V. and G.B.; funding acquisition, A.V. All authors have read and agreed to the published version of the manuscript.

Funding: This project has received funding from the European Union's Horizon 2020 research and innovation programme under the Marie Skłodowska-Curie grant agreement No 754382, GOT ENERGY TALENT. The content of this article does not reflect the official opinion of the European Union. Responsibility for the information and views expressed herein lies entirely with the author(s).

Conflicts of Interest: The authors declare no conflict of interest.

Nomenclature

f_c'	compressive strength at ambient temperature
f_{cT}	compressive stress at elevated temperature
f_{cT}'	compressive strength at elevated temperature
ϵ_{0T}	strain at ambient temperature
ϵ_{cT}	strain at elevated temperature
ASTM	American Society for Testing and Materials
$Ca(OH)_2$	Calcium hydroxide
CaO	Calcium oxide
CBC	Cement-Based Composites
CSH	calcium silicate hydrate
DF	Direct Flame
E_c	Young's Modulus
FE	Finite Element
ISO	International Organization for Standardization
MoE	modulus of elasticity
OPC	Ordinary Portland cement
PET	Polyethylene terephthalate
SCM	supplementary cementitious materials
SEM	Scanning Electron Microscopy

References

1. NFPA. Deadliest Fires and Explosions in U.S. History, National Fire Protection Association, Batterymarch Park, Quincy, Massachusetts, USA. 2020. Available online: https://www.nfpa.org/News-and-Research/Data-research-and-tools/US-Fire-Problem/Catastrophic-multiple-death-fires/Deadliest-fires-or-explosions-in-the-world (accessed on 1 October 2020).
2. Vedrtnam, A.; Bedon, C.; Youssef, M.A.; Wamiq, M.; Sabsabi, A.; Chaturvedi, S. Experimental and Numerical Structural Assessment of Transparent and Tinted Glass during Fire Exposure. *Constr. Build. Mater.* **2020**, *250*, 118918. [CrossRef]
3. Wang, Y.; Wang, Q.; Wen, J.X.; Sun, J.; Liew, K.M. Investigation of thermal breakage and heat transfer in single, insulated and laminated glazing under fire conditions. *Appl. Therm. Eng.* **2017**, *125*, 662–672. [CrossRef]
4. Zhou, H.; Li, S.; Chen, L.; Zhang, C. Fire tests on composite steel-concrete beams pre-stressed with external tendons. *J. Constr. Steel Res.* **2018**, *143*, 62–71. [CrossRef]
5. Fan, M.; Naught, A.; Bregulla, J. Fire performance of natural fiber Composites in construction. *Constr. Build. Mater.* **2004**, *18*, 505–515.
6. Yalinkilic, M.K.; Baysal, E.; Demirci, Z. Role of treatment chemicals in combustion of Douglas fir wood. In Proceedings of the First Trabzon International Energy and Environment Symposium, Trabzon, Turkey, 29–31 July 1996; pp. 803–809.
7. Wróblewska, J.; Kowalski, R. Assessing concrete strength in fire-damaged structures. *Constr. Build. Mater.* **2020**, *254*, 119122.
8. Haddad, R.H.; Al-Saleh, R.J.; Al-Akhras, N.M. Effect of elevated temperature on bond between steel reinforcement and fiber reinforced concrete. *Fire Saf. J.* **2008**, *43*, 334–343. [CrossRef]
9. Abid, M.; Hou, X.; Zheng, W.; Hussain, R.R. High temperature and residual properties of reactive powder concrete—A review. *Constr. Build. Mater.* **2017**, *147*, 339–351. [CrossRef]
10. Anand, N.; Godwin, A. Influence of mineral admixtures on mechanical properties of self-compacting concrete under elevated temperature. *Fire Mater.* **2016**, *7*, 940–958.
11. Li, M.; Qian, C.; Sun, W. Mechanical properties of high-strength concrete after fire. *Cem. Concr. Res.* **2004**, *34*, 1001–1005. [CrossRef]
12. Ma, Q.; Guo, R.; Zhao, Z.; Lin, Z.; He, K. Mechanical properties of concrete at high temperature—A review. *Constr. Build. Mater.* **2015**, *93*, 371–383. [CrossRef]
13. Chu, H.Y.; Jiang, J.Y.; Sun, W.; Zhang, M. Mechanical and physicochemical properties of ferro-siliceous concrete subjected to elevated temperatures. *Constr. Build. Mater.* **2016**, *122*, 743–752. [CrossRef]
14. Shah, A.H.; Sharma, U.K. Fire resistance and spalling performance of confined concrete columns. *Constr. Build. Mater.* **2017**, *156*, 161–174. [CrossRef]
15. Thanaraj, D.P.; Anand, N.G.P.A.; Zalok, E. Post-fire damage assessment and capacity based modeling of concrete exposed to elevated temperature. *Int. J. Damage Mech.* **2019**. [CrossRef]
16. Xiao, J.; Li, Z.; Xie, Q.; Shen, L. Effect of strain rate on compressive behaviour of high-strength concrete after exposure to elevated temperatures. *Fire Saf. J.* **2016**, *83*, 25–37. [CrossRef]
17. Lin, W.M.; Lin, T.D.; Powers-Couche, L.J. Microstructures of fire-damaged concrete. *Mater. J.* **1996**, *93*, 199–205.
18. Buchanan, A.H. *Structural Design for Fire Safety*, 2nd ed.; Wiley Publishers: Chichester, UK, 2001.
19. Savva, A.; Manita, P.; Sideris, K.K. Influence of elevated temperatures on the mechanical properties of blended cement concretes prepared with limestone and siliceous aggregates. *Cem. Concr. Compos.* **2005**, *27*, 239–248. [CrossRef]
20. Cree, D.; Green, M.; Noumowe, A. Residual strength of concrete containing recycled materials after exposure to fire. *Constr. Build. Mater.* **2013**, *45*, 208–223. [CrossRef]
21. Irshidat, M.R.; Al-Saleh, M.H. Thermal performance and fire resistance of nanoclay modified cementations materials. *Constr. Build. Mater.* **2018**, *159*, 213–219. [CrossRef]
22. Wasserman, R.; Bentur, A. Effect of lightweight fly ash aggregate microstructure on the strength of concretes. *Cem. Concr. Res.* **1997**, *27*, 525–537. [CrossRef]
23. Saikia, N.; de Brito, J. Mechanical properties and abrasion behaviour of concrete containing shredded PET bottle waste as a partial substitution of natural aggregates. *Constr. Build. Mater.* **2014**, *52*, 236–244. [CrossRef]
24. Ferreira, L.; de Brito, J.; Saikia, N. Influence of curing conditions on the mechanical performance of concrete containing recycled plastic aggregate. *Constr. Build. Mater.* **2012**, *36*, 196–204. [CrossRef]

25. Albano, C.; Camacho, N.; Hernández, M.; Matheus, A.; Gutiérrez, A. Influence of content and particle size of waste pet bottles on concrete behavior at different w/c ratios. *Waste Manag.* **2009**, *29*, 2707–2716. [CrossRef] [PubMed]

26. Saikia, N.; de Brito, J. Use of plastic waste as aggregate in cement mortar and concrete preparation: A review. *Constr. Build. Mater.* **2012**, *34*, 385–401. [CrossRef]

27. Choi, Y.W.; Moon, D.J.; Chung, J.S.; Cho, S.K. Effects of waste PET bottles aggregate on the properties of concrete. *Cem. Concr. Res.* **2005**, *35*, 776–781. [CrossRef]

28. Correia, J.R.; Lima, J.S.; de Brito, J. Post-fire mechanical performance of concrete made with selected plastic waste aggregates. *Cem. Concr. Compos.* **2014**, *53*, 187–199. [CrossRef]

29. Memon, S.A.; Shah SF, A.; Khushnood, R.A.; Baloch, W.L. Durability of sustainable concrete subjected to elevated temperature—A review. *Constr. Build. Mater.* **2019**, *199*, 435–455. [CrossRef]

30. Bodnarova, L.; Valek, J.; Novosad, P. Testing of Action of Direct Flame on Concrete. *Sci. World J.* **2015**, 1–8. [CrossRef]

31. El-Fitiany, S.F.; Youssef, M.A. Interaction diagrams for fire-exposed reinforced concrete sections. *Eng. Struct.* **2014**, *70*, 246–259. [CrossRef]

32. El-Fitiany, S.F.; Youssef, M.A. Assessing the flexural and axial behaviour of reinforced concrete members at elevated temperatures using sectional analysis. *Fire Saf. J.* **2009**, *44*, 691–703. [CrossRef]

33. Simulia. *ABAQUS/Standard Computer Software*; Dassault SystèmesProvidence: Johnston, RI, USA, 1998.

34. ASTM C150. *Standard Specification for Portland Cement, ASTM International*; ASTM: West Conshohocken, PA, USA, 2020.

35. ASTM C33-16. *Standard Specification for Concrete Aggregates*; ASTM International: West Conshohocken, PA, USA, 2016.

36. Zhang, X.; Deng, D.; Yang, J. Mechanical Properties and Conversion Relations of Strength Indexes for Stone/Sand-Lightweight Aggregate Concrete. *Adv. Mater. Sci. Eng.* **2018**. [CrossRef]

37. Amman, M. Study on Foam Production from Waste PET Bottles. Master's Thesis, Invertis University, Rajau Paraspur, India, 2020.

38. ISO 834-1:1999. *Fire-Resistance Tests—Elements of Building Construction—Part 1: General Requirements*; International Organization for Standardization (ISO): Geneve, Switzerland, 1999.

39. Achenbach, M.; Lahmer, T.; Morgenthal, G. Identification of the thermal properties of concrete for the temperature calculation of concrete slabs and columns subjected to a standard fire—Methodology and proposal for simplified formulations. *Fire Saf. J.* **2017**, *87*, 80–86. [CrossRef]

40. EN 1991-1-2: Actions on Structures. In *Part 1–2: General Actions—Actions on Structures Exposed to Fire, Eurocode 1*; European Committee for Standardization: Brussels, Belgium, 2002.

41. EN, 1992-1-2: Design of Concrete Structures. In *Part 1–2: General Rules—Structural Fire Design*; Eurocode 2; European Committee for Standardization: Brussels, Belgium, 2004.

42. Kodur, V. Properties of Concrete at Elevated Temperatures. *Int. Sch. Res. Not.* **2014**, *2014*, 468510. [CrossRef]

43. Youssef, M.A.; Moftah, M. General stress–strain relationship for concrete at elevated temperatures. *Eng. Struct.* **2007**, *29*, 2618–2634. [CrossRef]

44. Li, L.; Purkiss, J.A. Stress–strain constitutive equations of concrete material at elevated temperatures. *Fire Saf. J.* **2005**, *40*, 669–686. [CrossRef]

45. Liu, D.; He, M.; Cai, M. A damage model for modeling the complete stress–strain relations of brittle rocks under uniaxial compression. *Int. J. Damage Mech.* **2018**, *27*, 1000–1019 [CrossRef]

46. Liu, Y.; Wang, W.; Chen, Y.F.; Ji, H. Residual stress-strain relationship for thermal insulation concrete with recycled aggregate after high temperature exposure. *Constr. Build. Mater.* **2016**, *129*, 37–47. [CrossRef]

47. Abed, M.; de Brito, J. Evaluation of high-performance self-compacting concrete using alternative materials and exposed to elevated temperatures by non-destructive testing. *J. Build. Eng.* **2020**, *32*, 101720. [CrossRef]

Publisher's Note: MDPI stays neutral with regard to jurisdictional claims in published maps and institutional affiliations.

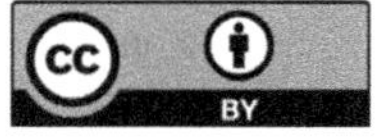

© 2020 by the authors. Licensee MDPI, Basel, Switzerland. This article is an open access article distributed under the terms and conditions of the Creative Commons Attribution (CC BY) license (http://creativecommons.org/licenses/by/4.0/).

 *sustainability*

Article

Effect of Corrosion in the Transverse Reinforcements in Concrete Beams: Sustainable Method to Generate and Measure Deterioration

P. Castro-Borges [1], C. A. Juárez-Alvarado [2,*], R. I. Soto-Ibarra [2], J. A. Briceño-Mena [1], G. Fajardo-San Miguel [2] and P. Valdez-Tamez [2]

[1] Centro de Investigación y de Estudios Avanzados del Instituto Politécnico Nacional (Center for Research and Advanced Studies of the National Polytechnic Institute, Merida Unit), Departamento de Física Aplicada (Department of Applied Physics), Unidad Mérida (CINVESTAV-IPN, Unidad Mérida) Km. 6 Antigua carretera a Progreso, Mérida C.P. 97310, Yucatán, Mexico; pcastro@cinvestav.mx (P.C.-B.); jorge.briceno@cinvestav.mx (J.A.B.-M.)

[2] Facultad de Ingeniería Civil (Civil Engineering Faculty), Universidad Autónoma de Nuevo León (Autonomous University of Nuevo Leon), Av. Universidad s/n, Cd. Universitaria, San Nicolás de los Garza C.P. 66455, Nuevo León, Mexico; rsotoib@gmail.com (R.I.S.-I.); gerardo.fajardosn@uanl.edu.mx (G.F.-S.M.); pedro.valdeztz@uanl.edu.mx (P.V.-T.)

* Correspondence: cesar.juarezal@uanl.edu.mx; Tel.: +52-1-818-329-4000 (ext. 7220)

Received: 17 August 2020; Accepted: 21 September 2020; Published: 1 October 2020

Abstract: A consistent method to generate and measure deterioration by corrosion in transverse reinforcements for concrete beams is presented and discussed in this work. This approach could be applied in other circumstances, such as bending, compression or combinations of stresses, with comparable results and therefore can be used to ensure sustainability. In marine environments, macro-cells are produced primarily from a transverse reinforcement, which works as an anode and therefore becomes a critical part of the structural analysis. To evaluate the adaptation efficiency of our proposed method, the corrosion potential, mass losses and cross-section reductions of the steel were measured. The shear stress behavior of the beams was investigated, including beam responses to load deformations, failure modes and cracking. The method ensured that all the beams exhibited a shear failure from diagonal stress with almost 50% less deflection when mechanically tested. The critical cross-sectional area, calculated according to the experimental diameter with the greatest cross-sectional loss due to the corrosion of the deteriorated stirrup, proved to be a reliable value for predicting the ultimate shear strength of concrete beams deteriorated by severe corrosion. A reduction of up to 30% in the shear strength of deteriorated versus non-deteriorated beams was found. Additional results showed that there is a correlation between the experimental and theoretical results and that the method is reproducible.

Keywords: corrosion; deterioration; stirrup; concrete; beams

---

1. Introduction

One of the primary problems associated with reinforced concrete elements exposed to marine environments is the attack of the reinforcing steel by chloride ions. The chloride ions enter the concrete either by diffusion, by absorption, by convection or by a combination of these processes [1]. Most often—and whatever the mode of entry—the stirrup is the reinforcement that is initially affected because of its proximity to the surface; however, the stirrup makes a significant contribution to the shear strength of a reinforced concrete beam. Therefore, the corrosion of the transverse reinforcement in concrete is an important topic for many researchers because of the mechanical impacts generated by the reduction in the cross-section of the stirrups resulting from deterioration by corrosion. In the literature [2–5], several studies have focused on evaluating the mechanical properties and durability of this type of structural element. Some of these studies have assessed the effects of deterioration caused by corrosion in reinforced concrete beams over decades [6–8]. In most previous studies, the beams were exposed to aggressive environments for more than 20 years, which promoted the generalized corrosion of the reinforcements. These studies showed that modifying the rigidity and ultimate deformations has a significant impact on the mechanical behavior, decreasing the ductility of the beams [9]. However, because the entire reinforcement is affected, it is difficult to extract discriminated information from the existing mechanical phenomena. In this regard, studies have been performed to determine the effect of deterioration by corrosion in stirrups; specifically, the behavior of shear stress in concrete beams [10–12]. In these studies, accelerated corrosion was produced by inducing a current in the stirrups. Additionally, the specimens were submerged in salt solutions with different concentrations, resulting in decreases in the mechanical properties of the beams. Nevertheless, despite the similarities in the techniques and methodologies used, there were remarkable differences, primarily in terms of the methodology used to create corrosion to cause a shear failure. In some cases, cracks appeared in the bars longitudinally, showing potential corrosion damage. Similarly, the specimens exhibited different failure modes when testing the shear stresses of the beams. Previous studies showed results that, although extremely valuable, are difficult to compare because they were methodologically different, even though the effects of corrosion on shear reinforcements were studied. In this regard, searching for a comparison mechanism among studies, a previous study [13] attempted to isolate the shear strength using a diagonal stress on the beams to obtain deterioration results that were more comparable with those of other studies. The approach was different from previous studies; the beams were exposed to humidification and drying cycles using a 3.5% NaCl solution until the steel was de-passivated. Then, a current of 100 μA/cm^2 was applied for a period from 80 to 120 days to reach moderate and severe corrosion levels in the stirrups. The results showed a significant differentiation among the states of deterioration associated with a preponderant mechanical effect; i.e., shear stress. With this method, the influence on ductility (increase of fragility), loss of cross-section and remaining strength of the beams were studied to isolate the effect on the shear strength of different deterioration states. Although the literature can provide data representing the values of different parameters with sufficient credibility, the goal of our work is to obtain a method that is adaptable to common circumstances in all investigations and with different sources, which would ensure sustainability. For this reason—and with the intention of obtaining a simple but effective method for comparing the results of deterioration by corrosion in concrete beams—a differentiated method for inducing the effect of deterioration by the corrosion of stirrups, with respect to the shear strength in reinforced concrete beams, both theoretically and experimentally, is presented and discussed in this work.

2. Materials and Methods

The theoretical and experimental method are described separately below.

2.1. Theoretical Method

Theoretical Electrochemical Method for the Induction of Deterioration in Transverse Reinforcements

Some studies [14–17] have discussed the influence of a cross-sectional reduction on the decrease of tensile strength in the stirrups, which affects the ductility of the beams; these works also reported that a reduction by 20% of the cross-section of the stirrups represents severe deterioration, under which both the steel and the concrete are compromised.

In this study, a reduction of the transverse reinforcement diameter by 10% is taken as representative of the equivalent of a 20% reduction in the cross-section of the stirrup. Therefore, a maximum loss of stirrup diameter of 10% is considered to represent severe deterioration with respect to the impacts of the stirrups on structural safety. The diameter loss is theoretically estimated as a measure of the remaining stirrup section with the following equation, as proposed by Andrade [18]:

$$\varnothing_t = \varnothing_i - 0.023 i_{corr} * t \tag{1}$$

where

$\varnothing_t$ = change of diameter in time (mm);
$\varnothing_I$ = initial diameter of the bar (mm);
i_{corr} = corrosion rate (μA/cm^2);
t = time after starting the propagation period (years);
0.023 = conversion factor used to convert μA/cm^2 to years.

The data shown in Figure 1 illustrate the values from Equation (1), using a proposed value of up to 200 μA/cm^2, to realistically reproduce severe corrosion damage and strain [19]. Corrosion products gradually dissipate with this current density level [20]. Higher values could produce oxide products that differ from real corroded structures, causing confusion in the interpretation of the results. In this way, the corrosion damage is theoretically estimated to predict the reduction in the section. For verification, a reduction of 10% of the diameter was established after 65 days of exposure.

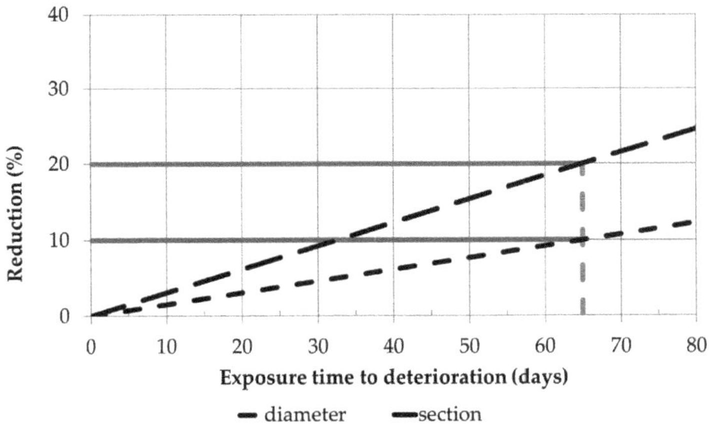

Figure 1. Loss of diameter and cross-section per time. $\varnothing_i$ = 8 mm, i_{corr} = 200 μA/cm^2.

2.2. Experimental Method

The experimental and theoretical methods are linked; however, for the purposes of comparison and adaptation to other studies, the former could be modified to consider other circumstances.

2.2.1. Concrete Manufacturing

Table 1 is a summary of the mix used in the study. The concrete was manufactured under laboratory conditions with a water/cement ratio (w/c) of 0.55. Cement type CPC 30R was used, with aggregates of limestone. The maximum size of the coarse aggregate was 19 mm, and the fine aggregate passed through mesh No. 4 (mesh opening 4.76 mm). The resulting concrete reached the compressive strength rate in 28 days.

Table 1. Mix design.

Cement	Water	Aggregate		f'c	Slump
		Coarse	Fine		
393 kg	216 kg	960 kg	768 kg	26 MPa	150 mm

2.2.2. Manufacturing the Beams

To evaluate the effect of the deterioration procedure of shear stress behavior, six reinforced concrete beams—i.e., two control beams with no deterioration (VSD 1 and 2) and four test beams with induced deterioration (VCD 1–4)—were manufactured; the dimensions of the reinforced concrete beams were set at 2000 mm × 200 mm × 350 mm. By design, a shear span (a) of 600 mm and an effective depth (d) of 296 mm were determined, thus obtaining a shear span to effective depth ratio a/d of 2.02—a value that represents a potential shear failure from diagonal stress. The structural arrangement is presented in Figure 2; the bending reinforcement was placed in two beds with a separation of 20 mm. The first layer comprised three bars, and the second comprised two bars, with diameters of 16 mm (Fy = 412 MPa) and number 2.5 bars with diameters of 8 mm (Fy = 460 MPa) for the stirrups. The longitudinal bars were coated with epoxy paint to avoid any damage by corrosion. The formation of galvanic stacks in the contact zones between the bending steel and the stirrups was prevented by the addition of electrical insulation tape. The traditional tie-up was replaced by nylon strips. The arrangement of the assembly is shown in Figure 3. The verification of this behavior is presented in Sections 4 and 5.

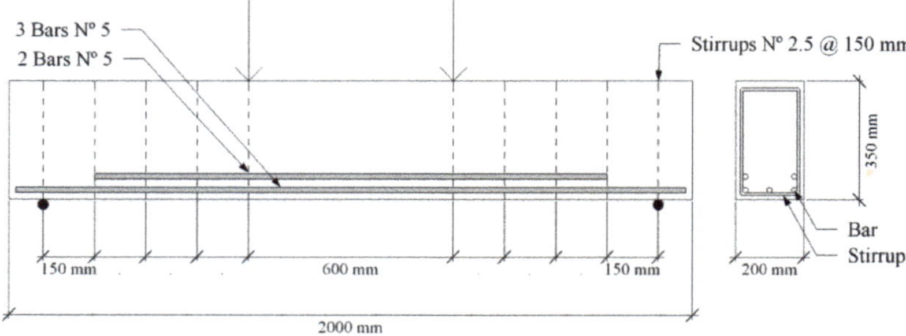

Figure 2. Structural arrangement to promote shear failure.

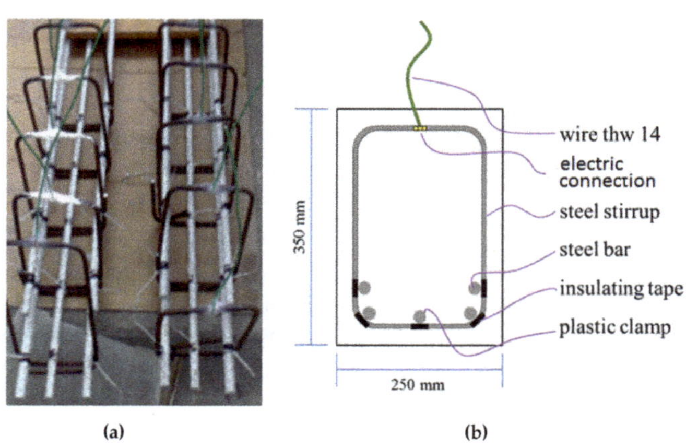

Figure 3. Assembly preparation for deterioration induction. (a) The reinforcement of the beam and its connections before being cast; (b) a diagram of the beam with details of connections for corrosion induction.

2.3. Method for Accelerated Deterioration Induction

Most significant results about structural behavior can be obtained by artificially corroded specimens [21]. Therefore, because of a wide and scattered spectrum of results for comparison purposes, a sustainable and reliable method to generate damage by severe corrosion in the reinforcement is needed. This is consistent and corresponds with the considerations proposed in the structural behavior described above for transversal rebars. This method must also comply with the conditions outlined in Figure 2. The method comprises the application of current, a humidification cycle and a drying cycle, as shown in Figure 4. The humidification of the beams is accomplished through polyurethane sponges, with dimensions of 600 mm × 300 mm × 20 mm, placed in the shear area. These sponges were moistened daily using a 3.5% NaCl solution to ensure that the applied current remained constant. After a week of humidification, a galvanic current was applied for 65 days to reach a severe deterioration level. The connection of each stirrup was made with a circuit formed by a rheostat of 0–50 kohm and a resistance of 1 ohm, with which it was possible to regulate the current applied to each stirrup. The circuit was connected to a power source of 0–30 volts and 0–5 amperes. A level of 200 μA/cm^2 was applied over the total area of each stirrup. The corrosion potential (E_{corr}) was monitored with an Ag/AgCl reference electrode, presented for analysis as E_{corr} vs. Cu/CuSO$_4$ (CSE) in the corresponding charts. Measurements were taken weekly. In the following figures, the corrosion potentials are presented with respect to the days of effective exposure to deterioration (application of current, humidification cycle and drying cycle). The measurements of the potentials followed the standard procedures (ASTM C876 [22]).

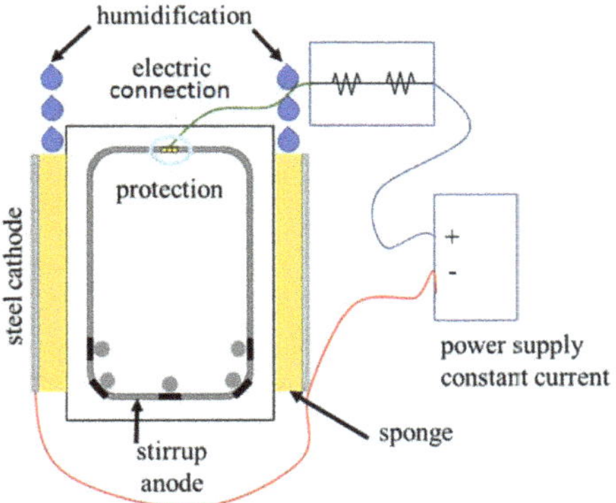

Figure 4. System for application of current, humidification and drying.

2.4. Mechanical Testing of the Beams

The simply supported beams were tested by applying concentrated loads to 600 mm sections of the supports, as shown in Figure 5. The test was carried out in a Tinius Olsen universal machine with a capacity of 200 tons. The total load was measured by a load cell with a capacity of 50 tons. The deflection at the midspan was measured with a linear variable displacement transducer (LVDT) with a calibration coefficient of 0.0005 cm. The shear strengths of the control beams without deterioration were determined at 28 days.

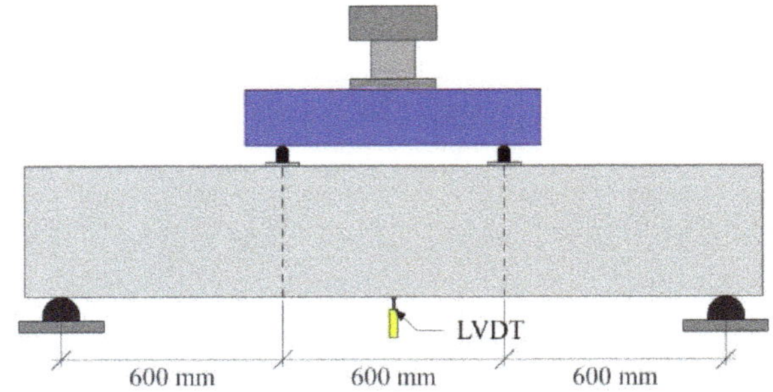

Figure 5. Testing scheme. LVDT: linear variable displacement transducer.

3. Results

3.1. Visual Inspection

In Figure 6, the visual appearance of the beam VCD-1 side A, after the induction of deterioration, is shown. The data are representative of all the beams. In Figure 7, the schematic appearance of the cracks generated in the stirrups during the process for the same beam are shown. The signs of corrosion—e.g., oxide stains and cracks (location, direction, and dimensions)—were recorded.

Figure 6. Visual appearance of beam VCD-1 side A after 65 days of applied deterioration; the crack widths are noted in cm.

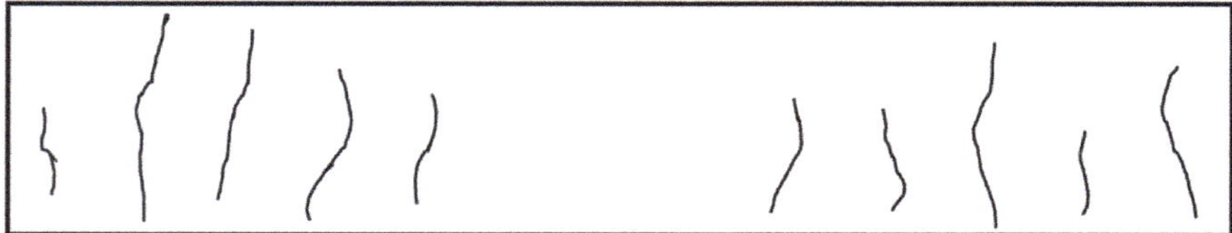

Figure 7. Cracking diagram for beam VCD-1 side A.

3.2. Measurement of the Corrosion Potentials

As shown in Figure 8, and for all the beams, the potential intervals indicated a negligible corrosion probability before the initiation of deterioration. Between 21 and 28 days after induction, potential levels vs. CSE were more negative than −350 mV; therefore, the beams were in a highly corrosive state. At the beginning of the process, the corrosion potentials of the stirrups showed some variability. This phenomenon is representative of the actual behavior of structures exposed to a marine environment, where there are variables that cannot be controlled, such as temperature and relative humidity. The potential levels measured at 50 and 65 days of exposure to induced deterioration reached values between −400 mV and −600 mV, respectively, indicating a high probability of advanced corrosion or severe deterioration, according to ASTM C876.

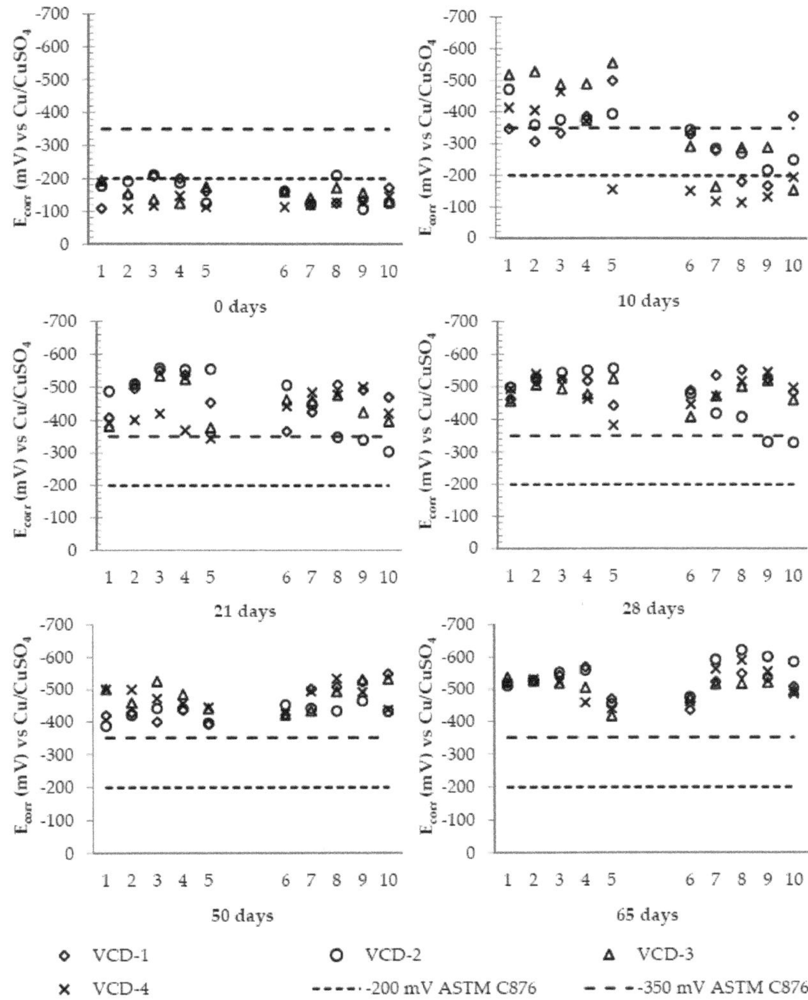

Figure 8. Electrochemical monitoring by measuring the corrosion potentials in the stirrups.

3.3. Determination of Cross-Section and Mass Losses in the Transverse Reinforcements

The previously tested beams were demolished to extract the corroded stirrups, as shown in Figure 9. Subsequently, the actual loss of the mass and cross-section of the transverse reinforcement was obtained [23]. Under the induced deterioration, the mass loss was as high as 11% on average. Some authors have reported that when there is a mass loss of 5–10% [24–26], the remaining capacities of the reinforced concrete beams and the steel/concrete adhesion decrease considerably. In Table 2, the relation between the theoretical values obtained by Equation (2) and the experimental values are shown.

$$W = \frac{ItM}{nF} \tag{2}$$

where

W = metal weight, g, which is corroded in an aqueous solution in the time t, in s;

I = current flux, in Amp;

M = atomic mass of metal, in g/mol;

n = number of consumed or produced electrons during the process;

F = Faraday constant, 96,500 C/mol.

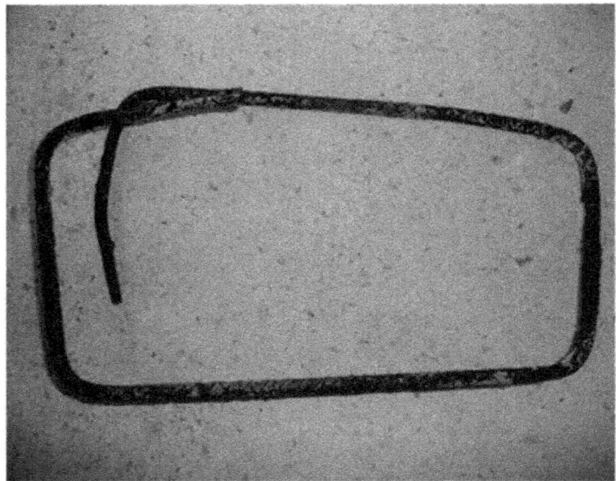

Figure 9. Removed corroded stirrup (position 8) with 65 days of exposure to induced deterioration.

Table 2. Relation between $\bar{x}$ (mass loss per stirrup) and the theoretical estimate.

Beam	Experimental Mass Loss $\bar{x}$		Theoretical Mass Loss by Equation (2)		Relation $\bar{x}/$ Equation (2)
	Stirrup	Percentage	Stirrup	Percentage	
VCD-1	43.0 g	11.4%	50 g	12.50%	0.86
VCD-2	30.0 g	8.0%	50 g	12.50%	0.60
VCD-3	40.0 g	10.0%	50 g	12.50%	0.80
VCD-4	37.0 g	10.0%	50 g	12.50%	0.74

3.4. Determination of Chloride Content

In Table 3, the average concentration of chloride by weight of cement for severe deterioration is shown. The chloride concentration exceeded the maximum limit of total chlorides specified by ACI 318 [27] for structures exposed to a chloride environment [28]. Similarly, previous studies have shown that the threshold for total chlorides by weight of cement are from 1.24–2.15% [29,30]. The amount of total chlorides is a parameter that confirms the thermodynamic state of the transverse reinforcement steel, as evidenced by the E_{corr} values in the beams.

Table 3. Chloride content.

Beam	% Chloride by Weight of Cement
VCD-1	2.15
VCD-2	3.00
VCD-3	2.30
VCD-4	2.75

3.5. Mechanical Resistance to Shear

The experimental results of shear structural behavior are presented in Figure 10 and Table 4, where the average deflection for the VSD beams is 11.4 mm. A variation in stiffness can be observed in VSD beams, although the beams are twins; there are factors that can cause such behavior, such as variations in the mixing, vibrating and curing of concrete, as well as different cracking kinetics between the beams. However, the shear strength was very similar. In the VCD beams, the deflection at the midspan decreased, on average, by 8.0 mm; i.e., a 29.8% decrease in deflection. In general terms, the deterioration by corrosion in the shear reinforcements affected the beams' ductility, as expected, presenting a fragile and sudden failure; this was more evident in the VCD-1 and VCD-2 beams, which had a greater deterioration due to corrosion than the VCD-3 and VCD-4 beams, which caused a reduction in their cracking load. Similarly, the VCD beams decreased their average shear strength by 26% compared to the VSD beams. In Figure 11, the data show a cracking pattern at the end of the shear test. To avoid confusion, the corrosion cracks in Figure 11 were omitted.

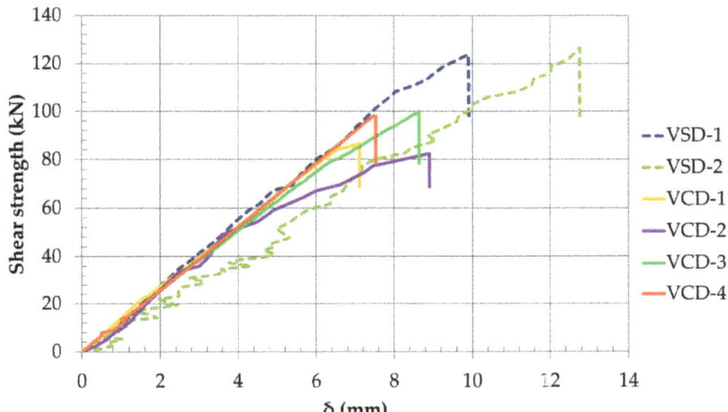

Figure 10. Shear strength–displacement curves.

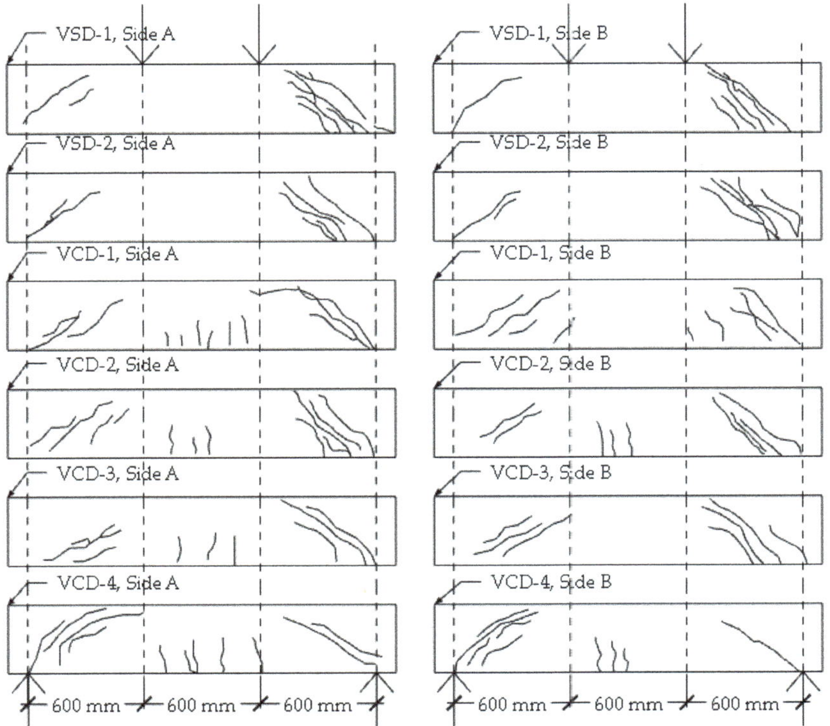

Figure 11. Cracking patterns obtained during the shear tests.

Table 4. Shear strength (Vu) and deflection at the midspan.

VSD-1	VSD-2	VCD-1	VCD-2	VCD-3	VCD-4
124 kN	126 kN	88 kN	84 kN	100 kN	98 kN
9.9 mm	12.8 mm	6.9 mm	8.9 mm	8.6 mm	7.5 mm

4. Discussion

4.1. Reliability of the Method for the Induction of Cracking in Concrete

A reliable method yields the expected results reproducibly. As the reproducibility increases, the reliability increases. The design of experiments for the types and sizes of specimens tested in this study is susceptible to variations in the manufacturing process, the instrumentation, exposure, the method of obtaining results, etc. More importantly, the nature of the induction of deterioration affects the design. In this study, the proposed adjustments to an induction of deterioration procedure, such as isolating the corrosion to a shear zone and making it consistent with what is expected from the structural calculation, show excellent reproducibility and can thus be used in future studies in different circumstances. Figure 12 is a schematic of the reproducibility of specific corrosion cracking patterns of the stirrup on all the beams, where cracks emerge on the concrete surface linearly to the reinforcement [19,31]. These results also show that there are neither micro-stresses in the concrete matrix that could have affected the contribution of the concrete central section nor concrete detachments in any area. This reproducibility was also manifested in the crack widths on the shear reinforcements whose values were in the range 0.1–0.2 mm, which is extremely narrow for experiments of this size. None of the cases showed coincident cracking to the bending of the reinforcing bars, as has been reported previously. A measure of the method's reliability to induce cracking is the number of stirrups that failed; in these tests, 36 of 40 stirrups—equivalent to 90% of the total stirrups—failed.

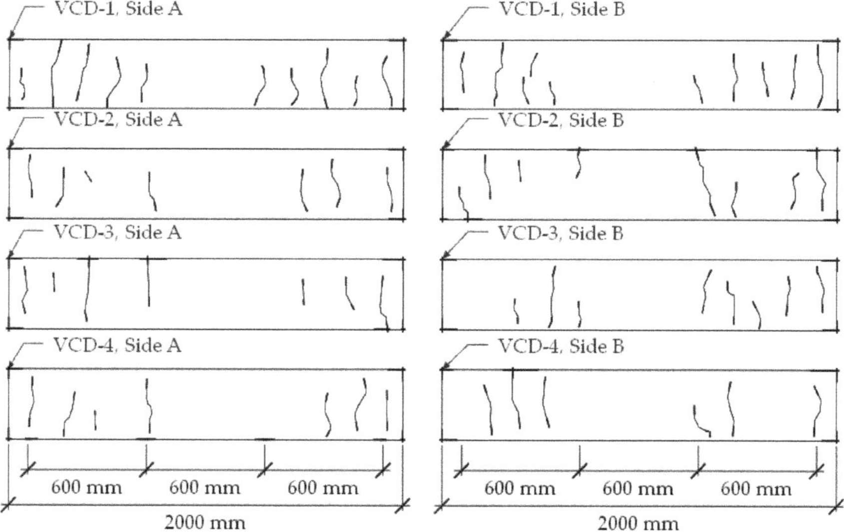

Figure 12. Cracking patterns of severe corrosion at the locations of the shear reinforcements.

4.2. Reliability of the Method to Generate Electrochemical Damage

The reproducibility was also verified through electrochemical monitoring. The evolution of the potentials with respect to time, shown in Figure 8, illustrates specific behavior. A homogenization of values at 65 days of exposure illustrates constant aggressiveness. Figure 13 is a representative configuration of Figure 8. Figure 13 displays the electrochemical value at the location of the transverse reinforcement that identifies its tendency. Note also that the values of the corrosion potential, despite being a thermodynamic parameter, move in a narrow band below −350 mV vs. CSE. The exceptions were stirrups 5 and 6, which coincided with the ends of the sponges, demonstrating that the attack method must have extreme continuity.

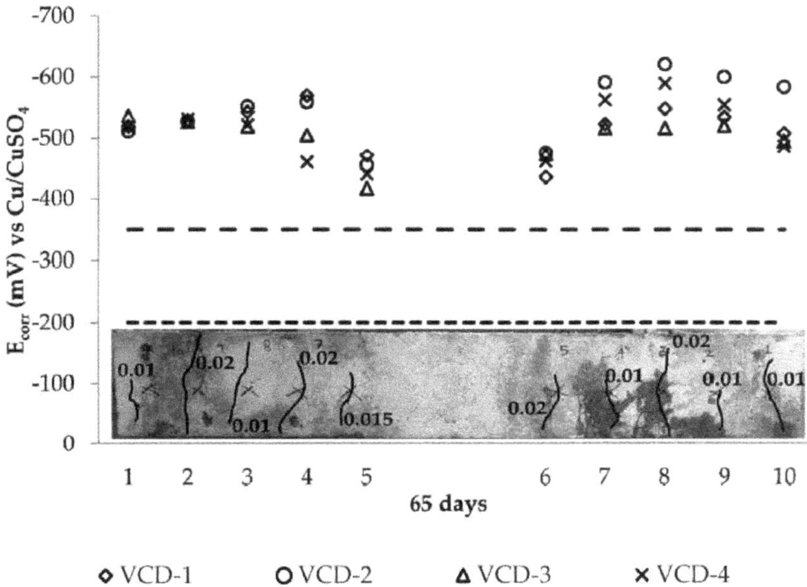

Figure 13. Electrochemical values at the locations of the shear reinforcements.

4.3. Reliability of the Calculation and the Method to Register Mass Losses in the Cross-Section

As previously mentioned, the reduction of the mechanical strength in the stirrup is proportional to its mass loss. As deterioration from corrosion progresses in the stirrups, the damage is irreversible. Therefore, it is important to clarify that, in practice, a 10% loss of cross-section from corrosion damage causes significant detachments and an extreme lack of safety that leads in most cases to demolitions rather than repairs. In accelerated tests, the attenuation factor plays an important role in the impact of the corrosion product, which is reflected in the behavior of the concrete under tensile stress. Because of this, the section cracks quickly and significantly [32] from the inability of the concrete pores to absorb the generated corrosion products quickly and efficiently. The opposite occurs in natural environments; i.e., as they are considered innocuous, the pores have a greater absorption capacity for corrosion products, and therefore significant losses of cross-section by shear can occur without cracks or significant detachments. Although the natural test is the best representation of reality in these cases, the accelerated test stimulates aggressive and destructive conditions. Therefore, any prediction based this it will always be conservative. This is important from the structural safety and preventive maintenance points of view. As discussed previously, in this study and based on the theoretical and experimental schemes, the indicators of severe deterioration were established. The indicators were used to evaluate and compare the experimental results through the relation between the experimental losses and the theoretical losses. The relation between the average experimental mass loss for each stirrup per beam and the loss calculated by Faraday's Law is shown in Figure 14. The upper line in the figure represents the theoretical mass loss estimated by the Faraday equation, with a value calculated at 65 days and equal to 50 g per stirrup. The bottom line represents the average experimental mass loss per stirrup of 41.25 g corresponding to the values of the trend line with triangles in the same figure. These results indicate that the applied experimental methodology and the estimation of mass loss are related in an acceptable manner [33].

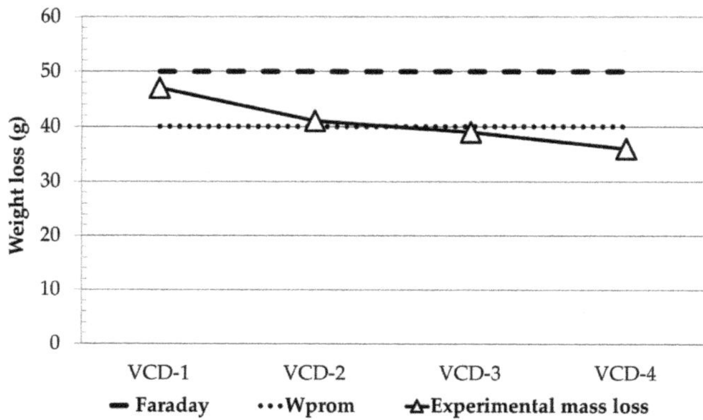

Figure 14. Ratio of experimental and calculated mass loss.

4.4. Theoretical and Experimental Comparison of the Shear Damage (Vu/Vn)

The method traditionally used in North America to design a structure is the ACI 318 code [27]. In this regard, the procedure in the structural behavior is designed in such a way that, in terms of durability, the beams will experience shear failure without bending. The shear strength of the concrete (Vc), shear strength of the stirrups (Vs) and nominal shear (Vn) were the results of the structural calculation of the beam, and the structural arrangement is presented in Figure 2. The damage induced in the stirrups through the proposed method to produce deterioration by corrosion generated mechanical repercussions, decreasing the ultimate shear strength from 20% to 30% for all VCD beams, according to the data in Table 5. Under these conditions, the data indicate that the partial reduction of the stirrups' section contributed more to the reduction of the ultimate shear strength of the VCD beams than of the VSD beams. Similarly, a change in the deflection at the midspan is exhibited in all the beams; therefore, the reduction in rigidity is caused by an increase in the fragility of the structure from internal pressures generated by the corrosion products. In addition, because all the beams contain the same variables (compressive strength, assembly, dimensions, deterioration), they can be compared. The data in Table 5 show a comparison between the values obtained experimentally and those estimated theoretically.

Table 5. Mechanical characteristics for VCD and VSD specimens.

Beam	Diameter (mm)			Shear Strength (kN)				
	Theor.	x̄ exp.	Critical	Vs	Vc	Vn	Vu	Vu/Vn
VSD-1	8.00	7.90	8.00	87	46	133	124	0.93
VSD-2	8.00	7.90	8.00	87	46	133	126	0.95
VCD-1	7.20	7.20	6.05	51	46	97	88	0.91
VCD-2	7.20	7.20	5.90	49	46	95	84	0.88
VCD-3	7.20	7.26	5.38	39	46	85	100	1.18
VCD-4	7.20	7.30	5.40	41	46	87	98	1.12

5. Conclusions

The delimitation of the method and its verification will enable other studies to combine variables, such as the cross-section of concrete, different materials, and the conditions of exposure, in any type of environment, allowing comparison between results from different sources.

The method used in this work ensured that all the beams exhibited shear failure from diagonal stress when mechanically tested with almost 50% less deflection.

It was verified that, when using the proposed deterioration scheme, the values of E_{corr} drastically decreased, generating severe corrosion in the shear reinforcements. The visual inspection of the beams confirmed the previous statements by the appearance of rust spots and cracks at the locations of the stirrups. The proposed method for the induction of deterioration by corrosion was reliable in terms of the reproducibility in corrosion potential values for each stirrup and with respect to the patterns and crack widths.

The previous general conclusions support the specific conclusions that follow, which can be readily compared with those of other studies in the literature.

The total mass loss of the stirrups was approximately 40–43 g per beam, which represents a decrease of approximately 11%.

The critical cross-sectional area, calculated with the experimental diameter with the greatest cross-sectional loss by corrosion of the deteriorated stirrup, proved to be a reliable value for predicting the ultimate shear strength of concrete beams deteriorated by severe corrosion. A reduction up to 30% in the shear strength of deteriorated beams relative to the beams without deterioration was found.

Finally, there is a correlation between the experimental and theoretical results based on the relation between the obtained results and the trends determined by the reference equations. This means that the procedure is reproducible and therefore adaptable to other type of conditions; thus, it is sustainable.

Author Contributions: Conceptualization, C.A.J.-A. and P.C.-B.; methodology, R.I.S.-I. and J.A.B.-M.; formal analysis, P.C.-B. and C.A.J.-A.; investigation, R.I.S.-I. and J.A.B.-M.; resources, P.V.-T.; data curation, R.I.S.-I. and J.A.B.-M.; writing—original draft preparation, P.C.-B., C.A.J.-A., R.I.S.-I. and J.A.B.-M.; writing—review and editing, P.C.-B. and C.A.J.-A.; project administration, G.F.-S.M. All authors have read and agreed to the published version of the manuscript.

Funding: This research received no external funding.

Acknowledgments: The authors acknowledge the partial support of their Institutions and Conacyt. Two of the authors (R. I. Soto-Ibarra and J. A. Briceño-Mena) acknowledge the support of Conacyt through their Doctoral grants. Mercedes Balancán-Zapata helped both students with technical issues. The opinions and findings expressed on this manuscript are those of the authors and not necessarily those of the supporting Institutions. This paper is part of the collaboration and signed agreement between Cinvestav-Mérida and FIC-UANL.

Conflicts of Interest: The authors declare no conflict of interest.

References

1. Castro, P.; De Rincon, O.T.; Pazini, E.J. Interpretation of chloride profiles from concrete exposed to tropical marine environments. *Cem. Concr. Res.* **2001**, *31*, 529–537. [CrossRef]
2. Xia, J.; Jin, W.; Li, L. Shear performance of reinforced concrete beams with corroded stirrups in chloride environment. *Corros. Sci.* **2011**, *53*, 1794–1805. [CrossRef]
3. Lachemi, M.; Al-Bayati, N.; Sahmaran, M.; Anil, O. The effect of corrosion on shear behavior of reinforced self-consolidating concrete beams. *Eng. Struct.* **2014**, *79*, 1–12. [CrossRef]
4. Zhu, W.; François, R.; Coronelli, D.; Cleland, D. Effect of corrosion of reinforcement on the mechanical behaviour of highly corroded RC beams. *Eng. Struct.* **2013**, *56*, 544–554. [CrossRef]
5. Val, D.V. Deterioration of Strength of RC Beams due to Corrosion and Its Influence on Beam Reliability. *J. Struct. Eng.* **2007**, *133*, 1297–1306. [CrossRef]
6. Zhu, W.; François, R. Corrosion of the reinforcement and its influence on the residual structural performance of a 26-year-old corroded RC beam. *Constr. Build. Mater.* **2014**, *51*, 461–472. [CrossRef]
7. Zou, X.; Schmitt, T.; Perloff, D.; Wu, N.; Yu, T.Y.; Wang, X. Nondestructive corrosion detection using fiber optic photoacoustic ultrasound generator. *Measurement* **2015**, *62*, 74–80. [CrossRef]
8. Dang, V.H.; François, R. Prediction of ductility factor of corroded reinforced concrete beams exposed to long term aging in chloride environment. *Cem. Concr. Compos.* **2014**, *53*, 136–147. [CrossRef]
9. Almassri, B.; Kreit, A.; Al Mahmoud, F.; Francois, R. Behaviour of corroded shear-critical reinforced concrete beams repaired with NSM CFRP rods. *Compos. Struct.* **2015**, *123*, 204–215. [CrossRef]

10. Sun, M.; Wen, D.-J.; Wang, H.-W. Influence of corrosion on the interface between zinc phosphate steel fiber and cement. *Mater. Corros.* **2012**, *63*, 67–72. [CrossRef]

11. Wang, L.; Zhang, X.; Zhang, J.; Ma, Y.; Liu, Y. Effects of stirrup and inclined bar corrosion on shear behavior of RC beams. *Constr. Build. Mater.* **2015**, *98*, 537–546. [CrossRef]

12. Wang, L.; Li, C.; Yi, J. An experiment study on behavior of corrosion rc beams with different concrete strength. *J. Coast. Res.* **2015**, *73*, 259–264. [CrossRef]

13. Juarez, C.A.; Guevara, B.; Fajardo, G.; Castro-Borges, P. Ultimate and nominal shear strength in reinforced concrete beams deteriorated by corrosion. *Eng. Struct.* **2011**, *33*, 3189–3196. [CrossRef]

14. Khan, I.; François, R.; Castel, A. Prediction of reinforcement corrosion using corrosion induced cracks width in corroded reinforced concrete beams. *Cem. Concr. Res.* **2014**, *56*, 84–96. [CrossRef]

15. Xue, X.; Seki, H.; Song, Y. Shear behavior of RC beams containing corroded stirrups. *Adv. Struct. Eng.* **2014**, *17*, 165–177. [CrossRef]

16. Castel, A.; François, R.; Arliguie, G. Mechanical behaviour of corroded reinforced concrete beams—Part 1: Experimental study of corroded beams. *Mater. Struct. Constr.* **2000**, *33*, 539–544. [CrossRef]

17. Bossio, A.; Lignola, G.P.; Fabbrocino, F.; Monetta, T.; Prota, A.; Bellucci, F.; Manfredi, G. Nondestructive assessment of corrosion of reinforcement bars through surface concrete cracks. *Struct. Concr.* **2017**, *18*, 104–117. [CrossRef]

18. Andrade, C.; Alonso, C.; Gonzalez, J.A.; Rodriguez, J.O. Remaining service life of corroding structures. In Proceedings of the IABSE Symposium on Durability, Lisbon, Portugal, 6–8 September 1989; pp. 359–363.

19. Andrade, C. Propagation of reinforcement corrosion: Principles, testing and modelling. *Mater. Struct.* **2019**, *52*, 1–26. [CrossRef]

20. El-Maaddawy, T.; Soudki, K. Effectiveness of impressed current technique to simulate corrosion of steel reinforcement in concrete. *J. Mater. Civ. Eng.* **2003**, *15*, 41–47. [CrossRef]

21. Imperatore, S.; Rinaldi, Z. Cracking in reinforced concrete structures damaged by artificial corrosion: An overview. *Open Constr. Build. Technol. J.* **2019**, *13*, 199–213. [CrossRef]

22. *ASTM C876-15, Standard Test Method for Corrosion Potentials of Uncoated Reinforcing Steel in Concrete*; ASTM International: West Conshohocken, PA, USA, 2015. [CrossRef]

23. *ASTM G1-03(2017)e1, Standard Practice for Preparing, Cleaning, and Evaluating Corrosion Test Specimens*; ASTM International: West Conshohocken, PA, USA, 2017. [CrossRef]

24. Zhou, H.; Lu, J.; Xv, X.; Dong, B.; Xing, F. Effects of stirrup corrosion on bond–slip performance of reinforcing steel in concrete: An experimental study. *Constr. Build. Mater.* **2015**, *93*, 257–266. [CrossRef]

25. Tondolo, F. Bond behaviour with reinforcement corrosion. *Constr. Build. Mater.* **2015**, *93*, 926–932. [CrossRef]

26. Al-Saidy, A.H.; Al-Jabri, K.S. Effect of damaged concrete cover on the behavior of corroded concrete beams repaired with CFRP sheets. *Compos. Struct.* **2011**, *93*, 1775–1786. [CrossRef]

27. ACI Committee. *ACI 318-19: Building Code Requirements for Structural Concrete and Commentary*; American Concrete Institute: Farmington Hills, MI, USA, 2014; p. 520.

28. *ASTM C1218/C1218M-20 Standard Test Method for Water-Soluble Chloride in Mortar and Concrete*; ASTM International: West Conshohocken, PA, USA, 2020. [CrossRef]

29. Alonso, C.; Castellote, M.; Andrade, C. Chloride threshold dependence of pitting potential of reinforcements. *Electrochim. Acta* **2002**, *47*, 3469–3481. [CrossRef]

30. Angst, U.; Elsener, B.; Larsen, C.K.; Vennesland, Ø. Critical chloride content in reinforced concrete—A review. *Cem. Concr. Res.* **2009**, *39*, 1122–1138. [CrossRef]

31. Pedrosa, F.; Andrade, C. Corrosion induced cracking: Effect of different corrosion rates on crack width evolution. *Constr. Build. Mater.* **2017**, *133*, 525–533. [CrossRef]

32. Castro, P.; Véleva, L.; Balancán, M. Corrosion of reinforced concrete in a tropical marine environment and in accelerated tests. *Constr. Build. Mater.* **1997**, *11*, 75–81. [CrossRef]

33. Suffern, C.; El-Sayed, A.; Soudki, K. Shear strength of disturbed regions with corroded stirrups in reinforced concrete beams. *Can. J. Civ. Eng.* **2010**, *37*, 1045–1056. [CrossRef]

© 2020 by the authors. Licensee MDPI, Basel, Switzerland. This article is an open access article distributed under the terms and conditions of the Creative Commons Attribution (CC BY) license (http://creativecommons.org/licenses/by/4.0/).

 *sustainability*

Article

Using Intelligence Green Building Materials to Evaluate Color Change Performance

Yu-Lan Lee [1], Yuan-Hsiou Chang [2,*], Jia-Lin Li [3] and Ching-Yuan Lin [1]

[1] Department of Architecture, National Taiwan University of Science and Technology, Taipei 10607, Taiwan; lovejulieli@gmail.com (Y.-L.L.); linyuan@mail.ntust.edu.tw (C.-Y.L.)

[2] College of Intelligence, National Taichung University of Science and Technology, Taichung 40401, Taiwan

[3] Department of Landscape Architecture and Environmental Planning, MingDao University, Changhua 10617, Taiwan; evan0425@mdu.edu.tw

* Correspondence: f89622050@ntu.edu.tw; Tel.: +886-93752-3685; Fax: +886-4-2219-5792

Received: 7 May 2020; Accepted: 1 July 2020; Published: 13 July 2020

Abstract: Environmental protection is an important issue in modern society. Most construction demolition wastes cannot be easily decomposed, thus occupying a lot of space in landfill. Reducing the demand for new resources is an efficient approach to decrease the environmental burden. Most green buildings are made from reused and recycled materials. Although there are a variety of green building materials available on the market, there is no material, as yet, with thermochromic functionality. This study used a form of face bricks, and six recovered materials, including wood chips, iron powder, fallen leaves, concrete, newspaper, and silt, to make smart green building materials. The modules were made in accordance with Taiwan's green building material regulations. The discoloration efficiency of indoor and outdoor green building materials was tested, and the RGB (red, green, blue) values of the face bricks were measured by a color analyzer to observe the discoloration effect. The findings show that among the A, B, C, and D groups, Group D exhibited the optimal rate of change in color, and the rates of change in the six recycled waste materials of indoor Group D were wood chips > newspaper > fallen leaves > concrete > iron powder > silt, while the rates of change in the outdoor group were newspaper > wood chips > fallen leaves. This study successfully reused waste materials to reduce the environmental burden, achieve sustainable environmental protection, and ensure both the aesthetics and quality of the building materials. The results of this study can offer an alternative choice to architects or space designers when selecting green building materials.

Keywords: waste management; construction demolition waste; thermochromic; green building material; recycled waste material

---

1. Introduction

While increased attention is being paid to environmental issues in current times, the issue of the waste reused and recycled has been widely discussed. Most of the wastes from our daily life or construction demolition cannot be easily decomposed, thus occupying more and more space in landfill. Reused or recycled products can be used to reduce the environmental burden. In China, construction and demolition waste accounts for about 30–40% of total waste production, and there is a large potential demand for recycled material [1]. Solid wastes cause serious problems in Asian countries and better solid waste management is needed [2]. In Europe, construction demolition waste accounts for approximately 25–30% of all waste generated in the European Union (EU), and the EU requested that its members reuse or recycle construction demolition waste to achieve 70% recycling rates by 2020 [3]. Taiwan's highly developed social economy and city construction have led to a large amount of construction waste every year, and effective recovery systems are needed for the proper use of these wasted resources [4]. The cyclic utilization of resources contributes to lessening the environmental load, creates an effective decline in the consumption of waste landfill space, decreases CO_2 emissions, and implements energy saving and carbon reduction [5]. Reuse, recycling, and reduction of the use of construction materials is the most effective way to solve the waste problem [6], and thus, developing competitive recycled materials could benefit the construction industry. Moreover, many landfills are at full capacity, and the high cost of fees increases the need to create a sustainable-resource construction industry [7]. Applying zero waste concepts to manage construction demolition waste would lead to a great improvement in waste management [8]. The recycled concrete bricks can be used in the seismic design of multi-layer masonry buildings [9]. However, as the strength of the bonding mortar increased, the bearing capacity and deformability of reclaimed concrete brickwork increased, leading to an increase in the elastic modulus. The hybrid recycling of green building materials and the use of different industrial or domestic general wastes according to the items of building materials and usable waste materials showed that the recycling of green building materials had a blending ratio [10].

In terms of lightweight concrete panels, the ratio of waste materials other than cement should be higher than 50%. Nearly 40% of the resources from building demolition were reused, while the remaining 60% were sent to landfills. If 60% of waste resources could be completely reused, the effect of demolition on the environment could be reduced [11–13]. As per the Building Technical Regulations of Taiwan, the indoor usage rate of green building materials should be more than 60% of the total area, and the outdoor usage rate of green building materials should be more than 20% of the total area [14]. The use of green building materials has gained increased attention, and there have been different green face-brick products available on the market; however, there are no reports regarding their thermochromic functions.

Different temperature control factors have resulted in thermochromic materials at different temperatures, which could be mixed with other pigments, and have no color at high temperatures and color development at low temperatures. These thermochromic materials could change color by controlling the temperature, solvent polarity, and pH by rearranging the molecules [15]. We overprinted 15 colors of thermochromic materials by screen printing and observed 255 different trapping effects [16]. In order to create a nondestructive testing (NDT) method for RC components that reinforce bar position, [17] coated a thermochromic paint on a concrete specimen surface, and defined the different thermal conductivities of the reinforcing bars and concrete, which induced differences in color change. The thermochromic material could be applied to medical treatments, lenses, and packaging, and presented the potential of color changing products [18]. To control the reversibility of color change for thin film technology, solar sanitization was also used in [19].

Thermochromic green building materials control the color-changing mechanism of allochroic microcapsules, mainly by temperature; thus, the average temperatures of different locations in Taiwan should be determined. According to [20], northern Taiwan has the lowest annual mean temperature of 9.9 °C and the highest mean temperature of 39.3 °C; central Taiwan has the lowest mean temperature of 11.3 °C and the highest mean temperature of 31.8 °C; southern Taiwan has the lowest mean temperature of 12.0 °C and the highest mean temperature of 35.7 °C; eastern Taiwan has the lowest mean temperature of 11.3 °C and the highest mean temperature of 36.3 °C. According to the above air temperature variation ranges of different regions in Taiwan, the reference frame of the ambient temperature for the indoor simulation experiment was set as 9–40 °C. Mixed waste materials with gypsum and allochroic powder to make thermochromic face bricks; however, the strength of the material required further enhancement [21,22]. White cement as the primary material and doped it with thermochromic pigment to test the strength [23]. The experimental result showed that, while the setting time and stability of the cement were not influenced, when the normal water content in the cement paste was increased by about 13%, the mechanical properties of bending strength and compressive strength were degraded by 20%–40%. The thermal conductivity of cement was 0.78, and it had a fairly good thermal insulation effect among different materials, as well as a strong material structure, high plasticity, good weather resistance, and is resistant to fire; thus, it could be extensively used [24]. Cement board has a high thermal insulation effect, and white cement has high quality, better strength than common gray cement, very strong miscibility, and more optional pigments and admixtures. The three primary colors of light colors are created with "additive mixtures", specifically by mixing the three primary colors of light, such as R + G = yellow, G + B = cyan, and B + R = magenta, and the "subtractive color mixture", where the three pigments added up to black. In order to unify the definition of light colors, in 1931, the International Commission on Illumination defined the standard wavelengths of RGB (red, green, blue) as 700.0 nm, 546.1 nm, and 435.8 nm, respectively. The hue values of three primary colors are defined as yellow (R = 255, G = 255, B = 0), blue (R = 0, G = 0, B = 255), and red (R = 255, G = 0, B = 0) [25]. A study used a color analyzer to measure the RGB values and HSL (hue, saturation, and lightness) values of object surface color, and analyzed the surface coatings of different cement mortar specimens [25]. We applied the filling of Mater-Bi to nanoclays to enhance the biofilm rigidity [26]. We developed biohybrid materials based on halloysite, sepiolite and cellulose nanofibers, to change the materials' character [27]. However, to date, there is no report of allochroic green building material made of white cement mixed with transparent silicon and waste. Hence, this study mixed white cement with waste and transparent silicon, in order to change the material color and make it more apparent.

2. Materials and Methods

2.1. Materials

The production of recycling green building materials should follow a certain blending ratio, which is as high as 50% of waste, in accordance to related regulations for production, indoors and outdoors experimental analyses [11–13]. Moreover, the effects of color change should also be compared. In this study, the smart green building materials were made into a face brick, which contained 60% of waste materials. The size of the face brick was 4.5 cm × 4.5 cm × 1 cm. The face brick was divided, from bottom to top, into three layers, namely the bottom layer, the allochroic layer, and the surface layer (Figure 1).

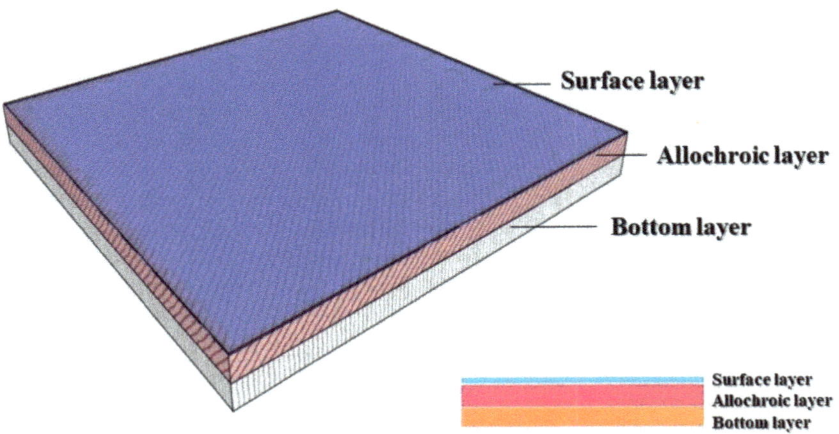

Figure 1. The layering and configuration of face brick

Regarding the bottom layer, the wastes were made into micromolecules smaller than 0.3 cm and mixed with white cement and water, where the proportions were white cement 15 g (25%), waste 36 g (60%), and water 9 g (15%), as well as 0.5 ml of antioxidants and stabilizers, respectively, which were mixed for about 1 min and poured into 1.19 cm thick molds. Regarding the allochroic layer, silicon was used, as it has favorable plasticity, is resistant to weather, UV, and ozone, is neutral, and does not influence the deterioration of the material. Moreover, as it has favorable viscosity for most building materials, there is no aging or cracking, and it can bear a temperature range of −60~150 °C. The allochroic powder 1.5–2 g (8%) was mixed with 11.25–15 g (60%) of transparent silicon and 6–8 g (32%) of water, to make the allochroic layer, which was 0.3 cm thick and required a drying time of nearly 24 h. The surface protection coating consisted of a 0.01 cm thick weather-resistant surface protection coating of Plimates P226 waterproofing paint.

Six kinds of recycled waste materials and four types of allochroic powders of different percentages were used; thus, 24 face bricks were made for experimental modules. The 24 face bricks were divided into four groups (ABCD), according to different allochroic powder blending ratios (Table 1). Group A consisted of common orange pigment (0.5 g model: acryliuqe301) added allochroic powder (Blue 25 °C, 2g); Group B consisted of allochroic powder at two different temperatures (Yellow 33 °C, 1 g; Blue 25 °C, 1 g); Group C consisted of allochroic powder at three different temperatures (Yellow 33 °C, 0.5 g; Blue 25 °C, 0.5 g; and Red 20 °C, 0.5 g); Group D consisted of allochroic powder at three different temperatures (Yellow 43 °C, 0.5 g; Blue 31 °C, 0.5 g; Red 20 °C, 0.5 g) (Table 2).

While the fallen leaves, newspaper, and concrete required pre-operation, the six recycled waste materials were identical. The newspaper and fallen leaves were torn up, mashed in a juicer, soaked in water, and drained for about one week. In order to reduce the content of chlorophyll and newspaper ink in the material, and enhance the coloration effect, the water was changed about every 12 hours during soaking. Before mixing, the concrete was scrapped, but not powdered; otherwise, the pH level of the concrete would influence the color change. The diameter was controlled within 0.2 cm by screening after knocking, followed by washing and draining. The silicon was not provided with the allochroic layer until the bottom layer was semidry (water content 25–45%). The allochroic powder was dissolved in water and mixed with silicon after uniform color mixing, in order to make the color uniform. It was then mixed with waste material in particle sizes smaller than 0.1 cm, and finally, the allochroic layer was prepared in the ratio of 1:1. The color analyzer TECPEL Tech-Link-TES 135 (TES Electrical Electronic Corp., Taipei, Taiwan) was used for RGB color analysis, and a digital thermometer was used for brick temperature measurement, where the temperature measurement range was −10 °C to + 70 °C. Images were acquired using a Sony DSC-W620 digital camera (14.1 megapixels). Soil moisture and a DM-15 acidity meter were used to estimate the water content for brick preparation.

2.2. Methods

This experiment was divided into indoor temperature change (both rise and drop) simulation, and actual outdoor temperature measurement, where the range of simulated temperature change was 9 °C–40 °C. In the temperature rise test, the face brick was affixed to the wall, an electric heater (Sampo HX-FB10F) was placed 20 cm in front of the face brick, and electronic thermometers were affixed to the wall and face brick to measure the indoor temperature, and the face brick surface layer temperature. The time of temperature change was recorded. In the temperature drop test, ice cubes were placed in a measuring cup and covered with a plastic cloth to avoid direct contact with the face brick, then the face brick was placed above the cup, an electronic thermometer was affixed to the surface of the face brick, and the time of temperature change was recorded. The RGB measurement was performed by a color analyzer when the temperature increased or dropped by one degree (data domain: 0–255), and the difference was estimated. In the outdoor experiment, the observation time was 5 a.m. to 10 p.m., where the face brick with the best effect was placed outdoors and irradiated by sunlight, and the RGB values were measured per hour. Regarding the indoor and outdoor measurements, the color analyzer was affixed to the brick surface, measurements were rapidly performed five times, and the average was taken. The data can provide a basis for the design of indoor and outdoor setups in the future.

Table 1. Intelligent Green Building Materials.

Materials \ Groups	A	B	C	D
Iron powder				
Newspaper				
Silt				
Wood chips				
Fallen leaves		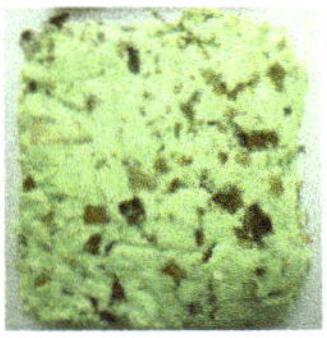		
Concrete				

Table 2. Set up allochroic powder at different temperatures of groups A, B, C, and D.

Color	A	B	C	D
Yellow		33 °C (1 g)	33 °C (0.5 g)	43°C (0.5 g)
Blue	25°C (2 g)	2 °C (1 g)	25 °C (0.5 g)	31 °C (0.5 g)
Red			20 °C (0.5 g)	20 °C (0.5 g)

3. Results

3.1. Test for Indoors Discoloration Efficiency of ABCD Groups

The 9 °C–40 °C RGB rates of change of six recycled waste materials of Group A are presented in Figure 2. The highest rates of change in R value were fallen leaves (73%) and wood chips (58%), and the lowest rate of change was iron powder (26%). The rates of change of R value from high to low were fallen leaves > wood chips > concrete > newspaper > silt > iron powder. The highest rates of change in G values were fallen leaves (21%) and wood chips (20%), and the lowest was silt (9%). The rates of change of G value were in the order of fallen leaves > wood chips > concrete > newspaper > iron powder > silt. The highest rate of change in B values was fallen leaves (16%), and the lowest rate of change was iron powder (3%). The rates of change of B value were leaves > wood chips > concrete > silt > newspaper > iron powder (Table 3).

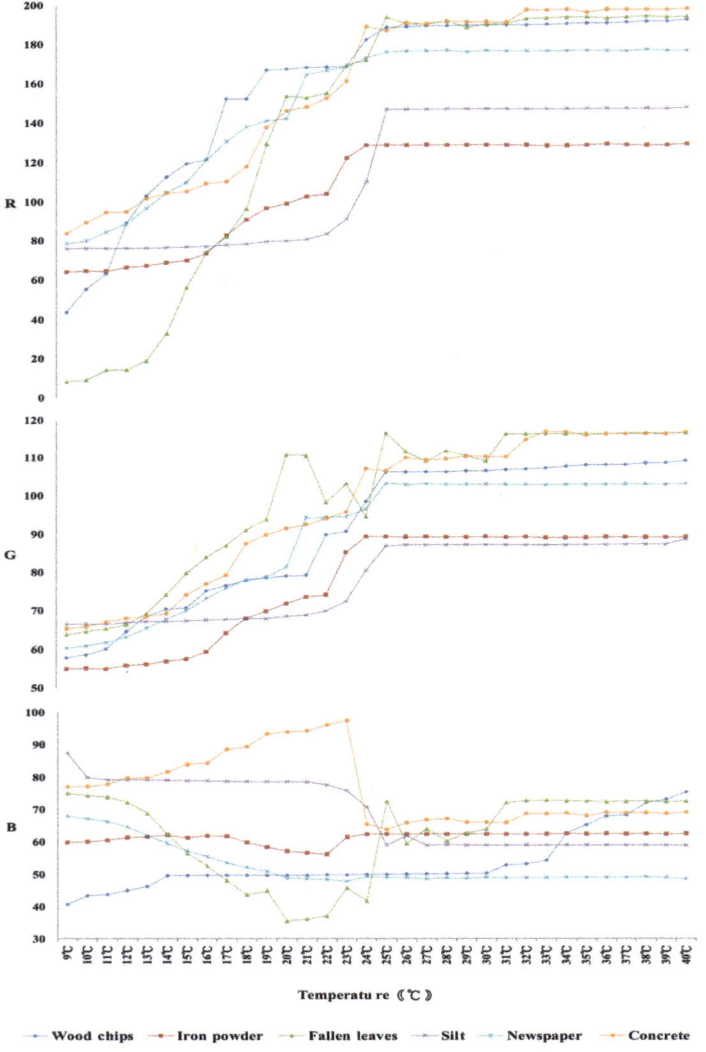

Figure 2. Group A; RGB discoloration performance.

Table 3. Group A green building materials discoloration rate of 9 °C to 40 °C RGB.

Materials RGB	Wood Chips	Iron Powder	Fallen Leaves	Silt	Newspaper	Concrete
R	58%	26%	73%	28%	39%	45%
G	20%	14%	21%	9%	17%	20%
B	14%	3%	16%	11%	8%	13%

The RGB rates of change of six recycled waste materials of Group B at 9 °C–40 °C are shown in Figure 3. The highest rates of change in R value were fallen leaves (81%) and wood chips (85%), and the lowest rate of change was iron powder (26%). The rates of change of R value from high to low were wood chips > fallen leaves > newspaper > concrete > silt > iron powder. The highest rate of change in G value was fallen leaves (59%), and the lowest were silt and newspaper (32%). The rates of change of G value were in the order of fallen leaves > wood chips > concrete > iron powder > newspaper > silt. The highest rate of change in B value was concrete (31%), and the lowest was silt (9%). The rates of change of B value were concrete > Iron powder > newspaper > wood chips > fallen leaves > silt (Table 4).

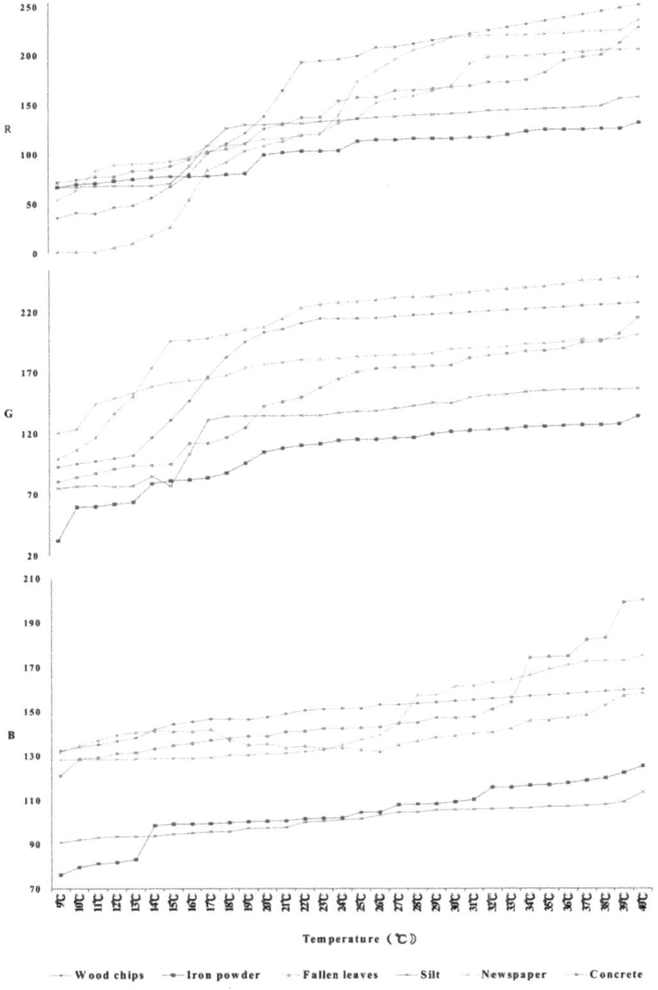

Figure 3. Group B RGB discoloration performance.

Table 4. Group B; green building materials discoloration rate of 9 °C to 40 °C RGB.

RGB \ Materials	Wood Chips	Iron Powder	Fallen Leaves	Silt	Newspaper	Concrete
R	85%	26%	81%	36%	72%	62%
G	53%	40%	59%	32%	32%	53%
B	11%	19%	10%	9%	19%	31%

The RGB rates of change of six recycled waste materials of Group C at 9–40 °C are shown in Figure 4. The highest rates of change in R value were for wood chips (97%) and fallen leaves (80%); the lowest rate of change was iron powder (21%). The rates of change of R value from high to low were wood chips > fallen leaves > newspaper > concrete > Silt > iron powder. The highest rates of change in G value were fallen leaves (66%) and wood chips (63%), and the lowest was iron powder (22%). The rates of change of G value were in order of fallen leaves > wood chips > concrete > newspaper > silt > iron powder. The highest rates of change in B value were for fallen leaves (57%) and wood chips (49%), and the lowest was iron powder (7%). The rates of change of B value were fallen leaves > wood chips > newspaper > silt > concrete > iron powder (Table 5).

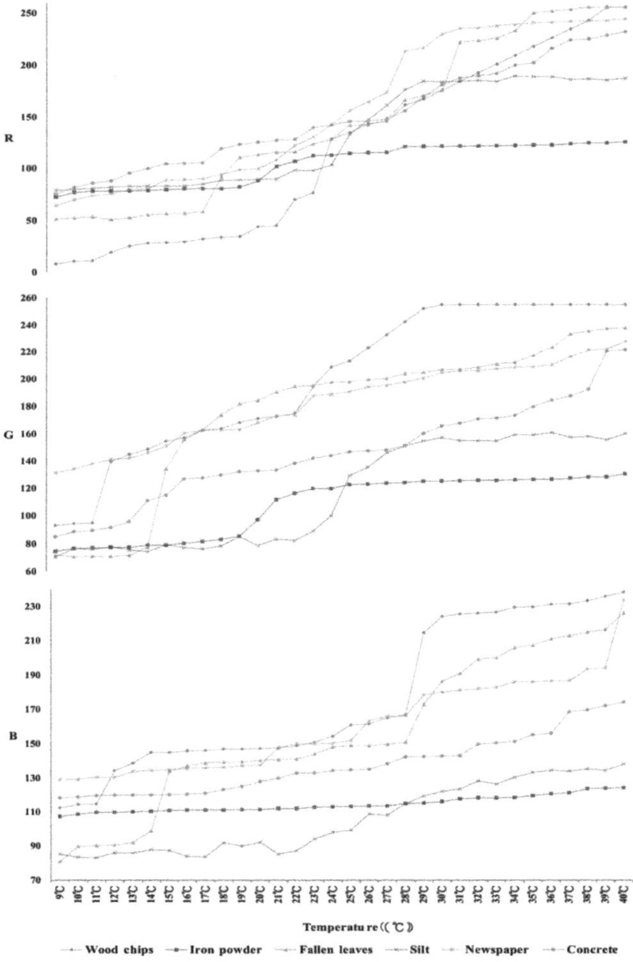

Figure 4. Group C RGB discoloration performance.

Table 5. Group C green building materials discoloration rate of 9 °C to 40 °C RGB.

RGB \ Materials	Wood Chips	Iron Powder	Fallen Leaves	Silt	Newspaper	Concrete
R	97%	21%	80%	43%	70%	61%
G	63%	22%	66%	36%	38%	53%
B	49%	7%	57%	22%	41%	22%

The 9–40 °C RGB rates of change of six recycled waste materials of Group D are shown in Figure 5. The highest rates of change in R value were fallen leaves (99%) and wood chips (88%), and the lowest rate of change was concrete (30%). The rates of change of R value from high to low were in the order of fallen leaves > wood chips > newspaper > silt > iron powder > concrete. The highest rate of change in G value was wood chips (81%), and the lowest was concrete (30%). The rates of change of G value were in the order of wood chips > newspaper > fallen leaves > silt > iron powder > concrete. The highest rate of change in B value was wood chips (52%), and the lowest rate of change was iron powder (16%). The rates of change of B value were wood chips > fallen leaves > newspaper > silt > concrete > iron powder (Table 6).

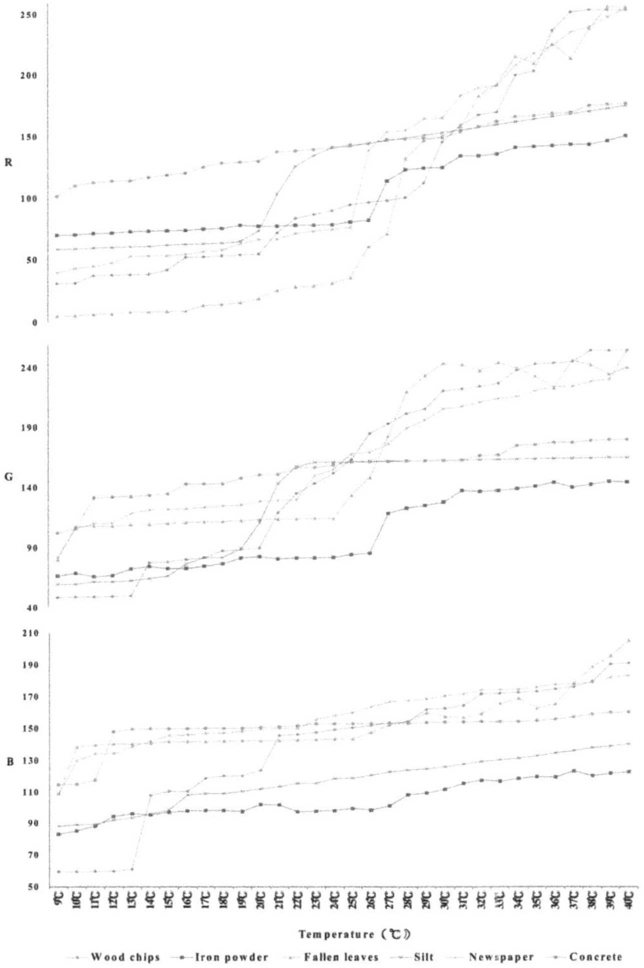

Figure 5. Group D RGB discoloration performance.

Table 6. Group D green building materials discoloration rate of 9 °C to 40 °C RGB.

RGB \ Materials	Wood Chips	Iron Powder	Fallen Leaves	Silt	Newspaper	Concrete
R	88%	32%	99%	46%	85%	30%
G	81%	31%	65%	41%	68%	30%
B	52%	16%	38%	21%	29%	18%

Therefore, the amounts, specifications, and combinations of allochroic powders and the changes in color of the end products resulted in different RGB color change effects and rates of change. Cumulatively, the rates of change in color were in descending order: Group D > Group C > Group B > Group A.

3.2. Color Analysis Diagram for RGB Color Change Among ABCD Groups

The RGB of smart green building material was evaluated by a color analyzer, and the data were integrated, before displaying the actual colors by Photoshop. The color gradual transition process of Group A recycled waste material was relatively dark in general. Due to contrast color, the color difference was relatively apparent at low temperature, with an overall poor effect. The wood chips, fallen leaves, and newspaper of Group B gave brighter colors than in the Group ACD, and they were displayed in blue at low temperatures. However, the iron powder, silt and concrete were displayed in dark purple at low temperature. The thermochromic microcapsule was colorless at high temperature and developed at low temperature; therefore, Group C appeared bright at high temperature. At 25–40 °C, the color of Group C began to fix. According to the color mixing principle, when the temperature was low, yellow + blue appeared to be green, but green + red appeared to be dark brown, and the color darkened. Group D appeared to be light green and light yellow at up to 30 °C, because of yellow allochroic microcapsule at 43 °C, and blue allochroic microcapsule at 31 °C (Figure 6).

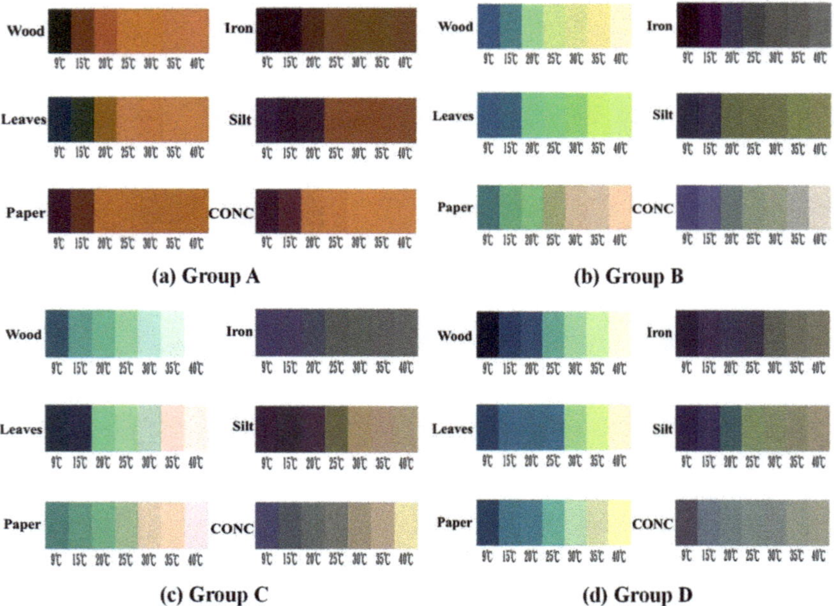

Figure 6. Group A to D discoloration process of green building materials.

3.3. Outdoor Group D Discoloration Efficiency Test

The outdoor experiment found that the wood chips, newspaper, and fallen leaves exhibited the best color change effect. The discoloration range of Group D was wider than Groups A, B, and C; therefore, they were applicable for outdoor use and the gradual color transition process could be displayed by Adobe Photoshop. In terms of the amplitude of variations in the RGB values of the wood chips of Group D in the outdoor experiment, the R, G, and B values were 58%, 27%, and 14%, respectively, the maximum values of R, G, and B were 231.1, 239.1, and 172.6, respectively. The colors were yellow to green in general. In terms of the amplitude of variations in the RGB values of the newspaper of Group D in the outdoor experiment, the R, G, and B values were 58%, 40%, and 25%, respectively, and with an apparent overall amplitude of variation. In this case, the maximum values of R, G, and B were 239.3, 255, and 183.5, respectively, and in terms of the amplitude of variation, the R, G, and B values were 36%, 16%, and 7%, respectively, and with the least apparent overall amplitude of variation among the three groups. The maximum values of R, G, and B were 242.2, 234.7, and 175.3, respectively, as shown in Figure 7.

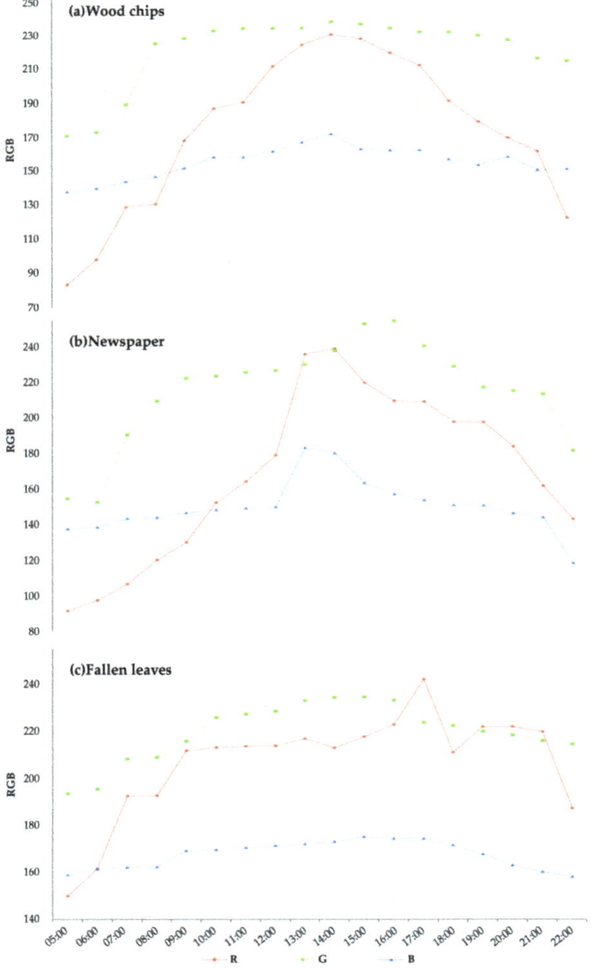

Figure 7. Group D outdoor RGB.

According to the outdoor experimental data of the three groups, newspaper exhibited the largest amplitude of variation. The rate of change in the R value of newspaper and wood chips was 58%, the difference in G value was 13%, and the difference in B value was 11% (Table 7). Therefore, newspaper provided the best overall effect, followed by wood chips, and then fallen leaves, as shown in Table 7.

Table 7. Group D; green building materials outdoor experiment discoloration rate of RGB.

Materials / RGB	Wood Chips	Newspaper	Fallen Leaves
R	58%	58%	36%
G	27%	40%	16%
B	14%	25%	7%

4. Discussion

This study prepared and tested six waste materials as smart green building materials, including iron powder, newspaper, fallen leaves, silt, wood chips, and concrete. Although fallen leaves, newspaper, and concrete required more preparation in the production process, the finished materials had good structural properties and discoloration effect. Moreover, as iron powder may be rusted by water, the chroma and lightness of the discoloration of the green building materials could be reduced, resulting in a special color effect. In terms of the discoloration effect, as Group C and Group D were mixed with multistage allochroic powder, they had a better abundance of hue change during discoloration than Group A and Group B. Due to the properties of wood chips, fallen leaves, and newspaper, these materials had a better discoloration effect than iron powder, silt, and concrete. The discoloration differences among the materials can be observed in Figure 6, which may be due to the colors and pH of iron powder, silt, and concrete influencing the abundance of discoloration hue. According to the outdoor measurement of the smart green building materials, newspaper has the best overall discoloration effect, followed by wood chips. Based on the results, newspaper and wood chips are recommended to be processed into green building materials.

5. Conclusions

According to the results of the indoor experiment, the highest rate of change in RGB was exhibited by Group D. The rates of change of six materials of Group D were in the order of wood chips > newspaper > fallen leaves > concrete > iron powder > silt. When the outdoor temperature difference in the daytime was 8 °C, the experimental rates of change were in the order of newspaper > wood chips > fallen leaves. According to the indoor and outdoor data, the allochroic powder has a discolored reaction under external environmental disturbance. As indicated, as long as the preset temperature of allochroic powder was reached, the face brick would change color. These experimental results can provide a reference to architects or space designers when designing different spaces. The strength of green building materials, which includes compressive strength, bending resistance, tensile strength, thermal insulation performance, and sound insulation performance, will be the direction of our further research.

Author Contributions: Conceptualization, Y.-L.L., Y.-H.C., and C.-Y.L.; methodology, Y.-L.L., Y.-H.C., J.-L.L. and C.-Y.L.; software, Y.-L.L., Y.-H.C., J.-L.L. and C.-Y.L.; validation, Y.-L.L., Y.-H.C., J.-L.L. and C.-Y.L.; formal analysis, Y.-L.L., Y.-H.C., J.-L.L. and C.-Y.L.; investigation, Y.-L.L., Y.-H.C., J.-L.L. and C.-Y.L.; resources, Y.-L.L., Y.-H.C., and C.-Y.L.; data curation, Y.-L.L., Y.-H.C., J.-L.L. and C.-Y.L.; writing—original draft preparation, Y.-L.L., Y.-H.C., J.-L.L. and C.-Y.L.; writing—review and editing, Y.-L.L., Y.-H.C., J.-L.L. and C.-Y.L.; visualization, Y.-L.L., Y.-H.C., J.-L.L. and C.-Y.L.; supervision, Y.-L.L., Y.-H.C., J.-L.L. and C.-Y.L.; project administration, Y.-L.L., Y.-H.C., J.-L.L. and C.-Y.L.; funding acquisition, Y.-L.L., Y.-H.C., J.-L.L. and C.-Y.L. All authors have read and agreed to the published version of the manuscript.

Funding: This research was funded by Ministry of Science and Technology, Taiwan, MOST 103-2221-E-451-008.

Acknowledgments: This work was supported by the Ministry of Science and Technology, Taiwan (MOST 103-2221-E-451-008).

Conflicts of Interest: The authors declare no conflict of interest.

References

1. Seeboth, A.; Lötzsch, D.; Ruhmann, R. First example of a non-toxic thermochromic polymer material—Based on a novel mechanism. *J. Mater. Chem. C* **2013**, *1*, 2811. [CrossRef]
2. Chang, Y.-H.; Huang, P.-H.; Chuang, T.-F. A Pilot Study of the Color Performance of Recycling Green Building Materials. *J. Build. Eng.* **2016**, *7*, 114–120. [CrossRef]
3. Chang, Y.-H.; Huang, P.-H.; Wu, B.-Y.; Chang, S.-W. A study on the color change benefits of sustainable green building materials. *Constr. Build. Mater.* **2015**, *83*, 1–6. [CrossRef]
4. Chang, Y.S. Waste Input-Output Analysis of CO_2 Emission and Waste Recycling in Taiwan. Master's Thesis, Institute of Environmental Engineering, National Central University, Taiwan, 2008.
5. Chowdhury, M.A.; Joshi, M.; Butola, B.S. Photochromic and Thermochromic Colorants in Textile Applications. *J. Eng. Fibers Fabr.* **2014**, *9*, 107. [CrossRef]
6. Huang, R.Y.; Li, C.D. Development of a recycling and reuse system for building demolition wastes. *J. Technol.* **2005**, *20*, 91–106.
7. Kim, J.; Rigdon, B. *Qualities Use and Examples of Sustainable Building Materials*; National Pollution Prevention Center for Higher Education: Ann Arbor, MI, USA, 1998; pp. 322–323.
8. Lötzsch, D. Thermochromic Biopolymer Based on Natural Anthocyanidin Dyes. *Open J. Polym. Chem.* **2013**, *3*, 43–47. [CrossRef]
9. Ma, Y.; Zhu, B. Research on the preparation of reversibly thermochromic cement based materials at normal temperature. *Cem. Concr. Res.* **2009**, *39*, 90–94. [CrossRef]
10. Roulet, C.A. Indoor environment quality and energy use in buildings. *Energy Environ.* **2001**, *33*, 183–191.
11. Zhang, F. *Framework for Building Design Recyclability*; University of Kansas, School of Engineering: Lawrence, KS, USA, 2008; pp. 211–212.
12. Lin, Y.C. The Application of Industry Waste Wood Chip for Use in Green Material—Case Study for Light Weight Concrete. Master's Thesis, Institute for Disaster Prevention Technology, Southeast University of Science and Technology, Nanjing, China, 2012.
13. New Prismatic Enterprise Co., Ltd. 2012. Available online: http://www.colorchange.com.tw/zhtw/index.php/ (accessed on 22 June 2019).
14. Central Meteorological Administration. 2020. Available online: https://www.cwb.gov.tw/V8/C/ (accessed on 22 June 2019).
15. Wu, H.T. A Feasibility Study of Thermal Sensitive Coating as Alternative Steel Bar Position Testing for RC Members. Master Thesis, Department of Civil Engineering, National Jiao Tong University, Shanghai, China, 2010.
16. Chang, C.Y. Thermochromic Material Experiment and Design Application. Master's Thesis, Visual Communication Design Institute, Ling Tung University of Science and Technology, Taichung, Taiwan, 2012.
17. Chang-chian, S.Y. Fundamental Research on Improvement of Urban Heat Island Effect—Effect of Thermophysical Properties on Surface Temperature. Master's Thesis, Department of Civil and Ecological Engineering, I-Shou University, Taiwan, 2009.
18. Cao, W.L.; Wang, Q.; Zhou, Z.Y.; Dong, H.Y. Experimental Study on Compression Performance of Recycled Concrete Brick Masonry. *World Earthq. Eng.* **2011**, *27*, 17–22.
19. Chen, C.C. *Introduction to Architectural Physics*; Chansbook: Taipei, Taiwan, 2018.
20. Lu, J.J. Synthetic Studies on Temperature Sensitive Dyes. Master's Thesis, Department of Applied Chemistry, Chaoyang University of Science and Technology , Wufeng, Taichung, Taiwan, 2008; pp. 29–30.
21. Cavallaro, G.; Lazzara, G.; Lisuzzo, L.; Milioto, S.; Parisi, F. Filling of Mater-Bi with Nanoclays to Enhance the Biofilm Rigidity. *J. Funct. Biomater.* **2018**, *9*, 60. [CrossRef] [PubMed]
22. Lorenzo, L.; Bernd, W.; Giulia, L.D.; Giuseppe, L.; Gustavo, D.R.; Pilar, A.; Eduardo, R. Functional biohybrid materials based on halloysite, sepiolite and cellulose nanofibers for health applications. *Dalton Trans.* **2020**, *49*, 3830–3840.
23. Zhao, W.; Leeftink, R.; Rotter, V.S. Evaluation of the economic feasibility for the recycling of construction and demolition waste in China—The case of Chongqing. *Resour. Conserv. Recycl.* **2010**, *54*, 377–389. [CrossRef]
24. Idris, A.; Inanc, B.; Hassan, M.N. Overview of waste disposal and landfills/dumps in Asian countries. *J. ater. Cycles Waste Manag.* **2004**, *6*, 104–110. [CrossRef]

25. BIO Intelligence Service. *Implementing EU Waste Legislation for Green Growth*; Final Report prepared for European Commission DG ENV; BIO Intelligence Service: Paris, France, 2011.
26. BIO Intelligence Service. *Service Contract on Management of Construction and Demolition Waste—SR1*; Final Report prepared for European Commission DG ENV; BIO Intelligence Service: Paris, France, 2011.
27. Tam, V.W.; Tam, C.M. A review on the viable technology for construction waste recycling. *Resour. Conserv. Recycl.* **2006**, *47*, 209–221. [CrossRef]

© 2020 by the authors. Licensee MDPI, Basel, Switzerland. This article is an open access article distributed under the terms and conditions of the Creative Commons Attribution (CC BY) license (http://creativecommons.org/licenses/by/4.0/).

 *sustainability*

Article

Heat of Hydration Stresses in Stainless-Steel-Reinforced-Concrete Sections

Mokhtar Khalifa, Maged A. Youssef * and Mohamed Monir Ajjan Alhadid

Civil and Environmental Engineering, Western University, London, ON N6A 5B9, Canada;
mkhalif4@uwo.ca (M.K.); majjanal@uwo.ca (M.A.A.)
* Correspondence: youssef@uwo.ca

Received: 14 May 2020; Accepted: 10 June 2020; Published: 14 June 2020

Abstract: Stainless steel (SS) is increasingly used in construction due to its high strength and corrosion resistance. However, its coefficient of thermal expansion is different from that of concrete. This difference raises concerns about the potential for concrete cracking during the hydration process. To address this concern, a thermal-structural finite element model was developed to predict the stresses in SS-reinforced concrete (RC) sections during the hydration process. Different curing regimes were taken into consideration. The analysis was performed in two stages. First, a transient thermal analysis was performed to determine the temperature distribution within the concrete section as a function of concrete age and its thermal properties. The evaluated temperature distribution was then utilized to conduct stress analysis. The ability of the model to predict the stresses induced by the expansion of the bars relative to the surrounding concrete was validated using relevant studies by others. The model outcomes provide in-depth understanding of the heat of hydration stresses in the examined SS RC sections. The developed stresses were found to reach their peak during the first two days following concrete casting (i.e., when concrete strength is relatively small).

Keywords: concrete; stainless steel; reinforcement; temperature; thermal expansion

---

1. Introduction

Stainless steel (SS) provides many advantages over conventional carbon steel due to its high corrosion resistance, and consequently, its lower dependence on the alkalinity of the protective concrete cover. Using SS bars to reinforce concrete structures results in a significant improvement in their durability and a reduction in their maintenance and repair costs. As such, the use of SS bars in the construction industry continues to increase, especially in bridges and coastal structures [1].

Despite the various positive aspects of SS bars, their thermal properties constitute a drawback. The reason lies in the fact that both carbon steel and concrete have similar coefficients of thermal expansion, whereas the thermal expansion coefficient of SS is about 80% higher than that of concrete [2,3]. Thus, when the temperature of an SS-reinforced concrete (RC) section increases, the thermal incompatibility between the SS bars and concrete results in stresses that are not experienced by carbon steel RC sections.

In early-age concrete, heat is released from the exothermic hydration reaction which occurs between cement and water. The heat of hydration increases the temperature of the concrete mix and the embedded reinforcing bars. The temperature increase can reach 55 °C in mixes with high cement content [4]. The variation of the heat of hydration with time is given in Figure 1. Ordinary Portland cement is composed mainly of aluminate (C3A), aluminoferrite (C4AF), belite (C2S), and alite (C3S) [5,6]. The hydration reaction produces calcium silicate hydrate (C-S-H) gel and ettringite, which increase the concrete's strength. Wet or air curing preserves a satisfactory temperature for the concrete and improves its properties [7].

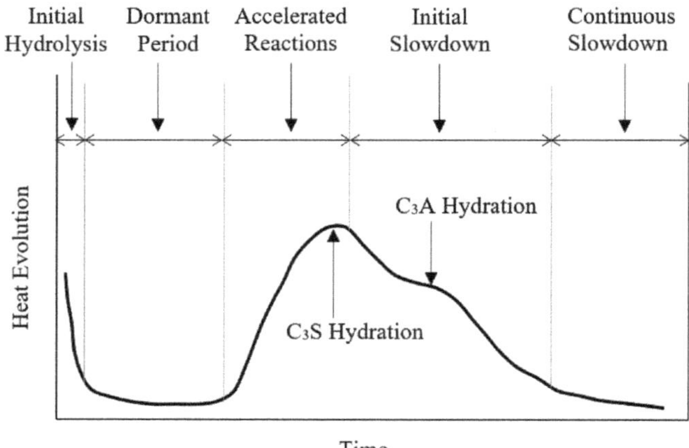

Figure 1. Evolution of heat of hydration as a function of time.

The coefficients of thermal expansion of concrete and carbon steel bars are 1.1×10^{-5} °C^{-1} and 1.2×10^{-5} °C^{-1}, respectively [8]. These close values imply an excellent thermal compatibility between the two materials. However, the thermal expansion coefficient of SS bars can exceed 1.8×10^{-5} °C^{-1} [9]. This relatively large divergence from concrete's thermal expansion raises concerns about the possibility of additional thermal stresses that may cause cracks. This scenario is expected to be most critical during the curing period, while the concrete's tensile strength is very low, and its temperature is increasing due to the heat produced during the hydration process.

This paper aims to numerically investigate the influence of the heat of hydration on stress distribution in SS RC sections, considering the thermal incompatibility between the two materials. A finite element model is developed and validated to examine the temperature distribution and stresses developed in SS RC sections. Water and air curing regimes are considered in the analysis.

2. Material Models

2.1. Concrete

The variations of the concrete compressive strength (f_c) and tensile strength (f_t) with time were assumed to follow the values reported by Jin et al. [10] and Kim [11], as summarized in Figure 2, where f_c' and f_t' are the 28-day compressive and tensile strength, respectively. The concrete constitutive relationship was assumed to follow the model proposed by Jin et al. [10] and was idealized using the ANSYS multilinear model [12]. The concrete strain at peak stress and Poisson's ratio were assumed to be 0.002 and 0.30, respectively. Concrete failure was assumed to correspond to the crushing strain for unconfined concrete ($\varepsilon_{cu} = 0.0035$) [3]. The normalized compressive stress–strain relationship at various concrete ages is shown in Figure 3. The figure shows that concrete compressive strength increases with time, whereas its ductility decreases. The tensile behavior of concrete is predominantly brittle. Concrete was assumed to resist tensile stresses up to the cracking point, beyond which the tensile capacity of concrete drops to zero. The coefficient of thermal expansion of concrete, its specific heat, and its density were assumed at 1×10^{-5} °C^{-1} [3], 920 J/kg·°C [13], and 2300 kg/m^3 [14], respectively. The thermal conductivity of concrete was assumed to follow the values provided by EC2 [14], as shown in Figure 4.

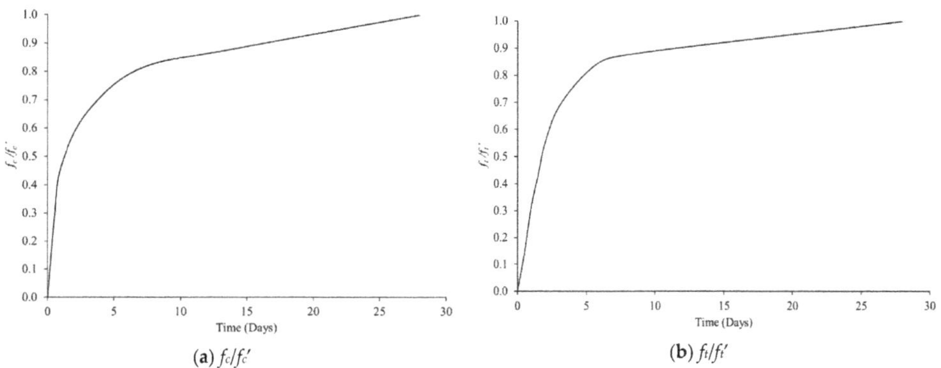

(**a**) f_c/f_c' (**b**) f_t/f_t'

Figure 2. Variation of concrete strength with time.

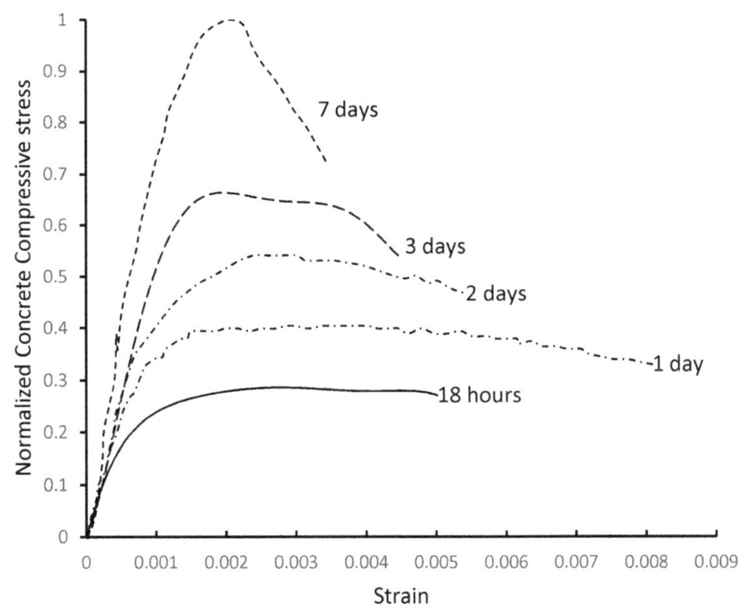

Figure 3. Stress-strain curves of concrete at early age.

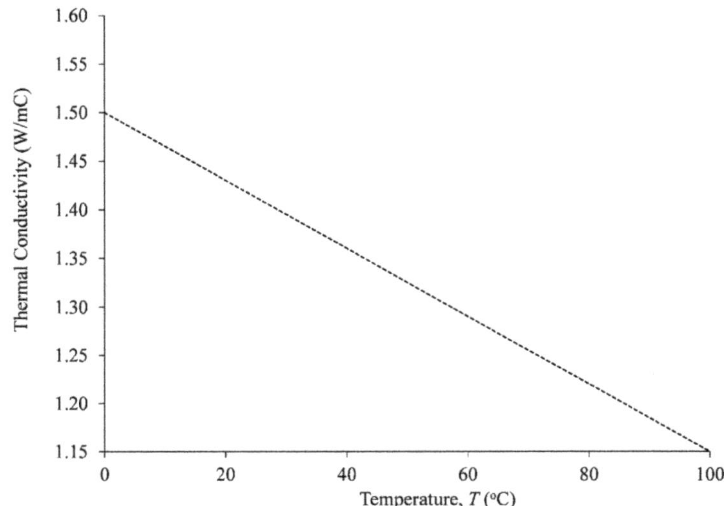

Figure 4. Thermal conductivity for concrete at different temperatures.

2.2. Stainless Steel Bars

The constitutive relationship of SS bars was assumed based on the experimental work of Chen and Young [15]. The coefficients of thermal expansion of austenitic 316LN and duplex 2205 SS bars were assumed at 17.8×10^{-6} °C^{-1} and 13×10^{-6} °C^{-1}, respectively. The density, specific heat, and thermal conductivity of the bars were taken as 7750 kg/m^3, 440 J/kg·°C [11], and 15 W/m·°C [16], respectively.

3. Finite Element Model

Figure 5 shows a typical SS RC section considered in the analysis. The examined parameters were the section height (h), section width (b), concrete cover (c), and bar diameter (d). The section was assumed to be reinforced with two SS bars in tension and compression.

A two-dimensional thermal-structural analysis was performed using ANSYS 17.2 finite element software [12]. Since the section was doubly symmetric in terms of geometry and applied temperatures, only the bottom-left quarter was considered in the model. The analysis was performed by (1) selecting appropriate elements, (2) specifying thermal and structural material properties, (3) performing thermal analysis to determine the temperature due to heat of hydration at a specific time, and (4) performing a static structural analysis to determine the induced stresses and examine the potential for cracking.

3.1. Thermal Analysis

Both concrete and SS were modeled using PLANE77 [12], a two-dimensional 8-node thermal solid element. The element makes it possible to conduct two-dimensional steady-state analyses and is characterized by its temperature shape functions, which are well-suited to model curved geometries such as the boundary between the concrete and the SS rebars. SURF151 and CONTA171 elements [12] were used at the boundary of the SS bar to model the interaction between the SS rebar and the surrounding concrete. A typical meshed section is shown in Figure 6. The optimum mesh density was chosen by performing a preliminary sensitivity analysis. A preliminary mesh, which was refined around the SS rebar, was first assumed. The mesh was then further refined until the variation in the principal stresses between subsequent refinement reaches could be considered negligible.

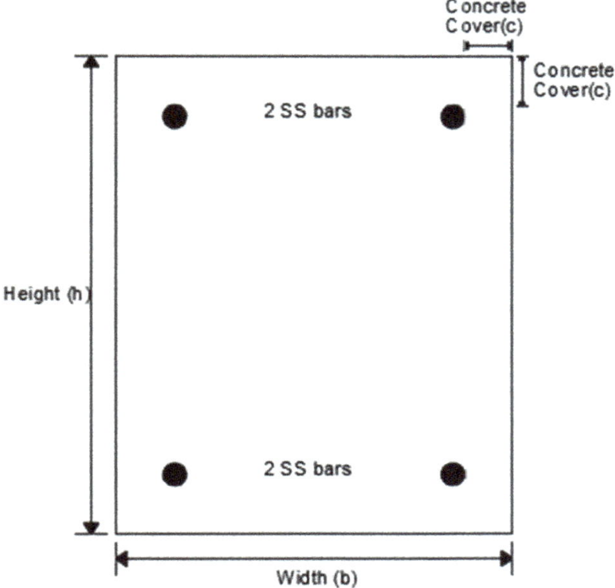

Figure 5. Typical stainless-steel (SS)-reinforced concrete (RC) section.

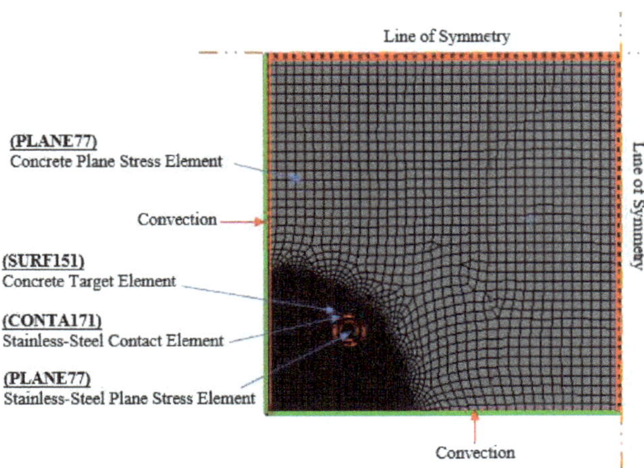

Figure 6. Thermal analysis mesh.

RILEM Committee 42 [17] provided details about an experimental program that determined the relationship between the total heat liberated during the hydration reaction and time considering various water/cement ratios. For a water/cement ratio of 0.4, which ensures an adequate amount of water to complete the hydration process, the relationship is shown in Figure 7. The internal heat is generated by applying this relationship as a uniform internal energy that varies with time. Heat transferred by convection was applied on the exposed boundaries using convection coefficients of 12 kcal/m$^2\cdot$h$\cdot^\circ$C and 4.3 kcal/m$^2\cdot$h$\cdot^\circ$C for water and air curing, respectively [18–20].

3.2. Structural Analysis Model

The thermal 2D element (PLANE77), used in the thermal analysis, was replaced with an equivalent structural element (PLANE183) to model the concrete and SS rebars. This high-order 8-node element provides quadratic displacement behavior with two translational degrees of freedom at each node. This feature allows the element to accurately capture the stress distribution. A typical structural mesh is illustrated in Figure 8. The nodes along both lines of symmetry were restrained against orthogonal translational movement, whereas the nodes along the free edges were unrestrained. The temperature values reached in the thermal analysis stage defined the applied thermal loads.

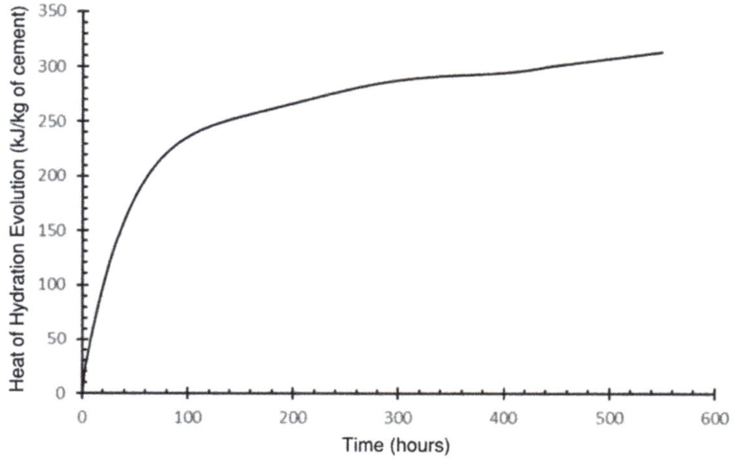

Figure 7. Heat of hydration at different ages.

The contact between concrete and the boundaries of the SS bars was simulated by CONTA172 [12] and the associated target element TARGE169 [12]. These elements could capture the deformations of the boundaries. Concrete was considered as the target element, as it was expected to resist the SS rebar expansion.

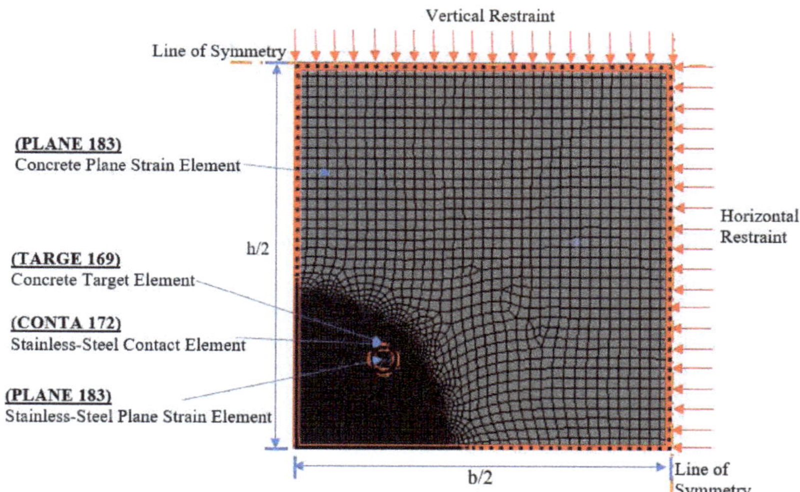

Figure 8. Structural finite element model.

4. Validation

Unfortunately, the current literature lacks experimental data related to the effect of radial thermal expansion of SS bars on early-age concrete. However, Du et al. [21] conducted a finite element analysis to determine the influence of corrosion-expansion of steel bars on the structural response and cracking behavior of concrete elements. The results revealed the significant role of reinforcement radial expansion on crack formation. A finite element model was also developed and validated by Du et al. [21]. Clark and Saifullah [22] conducted accelerated corrosion tests to study the effects of corroded reinforcement on bond strength and concrete cracking.

Since the mechanism of stress development in the proposed research is similar to that of corroded bars, the results obtained by Clark and Saifullah [22] are considered to validate the finite element model. The RC section considered by Clark and Saifullah [22] had cross-sectional dimensions of $h = 175$ mm and $b = 150$ mm. A maximum mesh size of 3 mm was used to model the concrete. Reducing the size to 2.5 mm was found to alter the stresses by 0.2%, which was assumed negligible. The concrete was modeled with a void at the location of each corroded bar. The radial thermal expansion of the steel bars due to corrosion was simulated by applying radial displacement at the concrete nodes in the vicinity of the voids.

The variation of radial expansion at cracking with the ratio of concrete cover to bar diameter (c/d) was determined and compared to the results obtained by Clark and Saifullah [22], as illustrated in Figure 9. The prediction error ranged between 9% and 14%, which is considered acceptable given the complexity of the problem.

The obtained crack pattern was evaluated and compared to the data provided by Clark and Saifullah [22], as shown in Figure 10. As the radial expansion of the corroded reinforcement increased, the cracking of concrete followed the same stages described by Clark and Saifullah [22]: (1) internal cracks, as shown in Figure 10a, which started at a radial expansion of 0.00044 mm, as compared to 0.00050 mm by Clark and Saifullah [22]; (2) external cracks, as shown in Figure 10b, which resulted in the formation of surface cracks at a radial expansion of 0.00135 mm, as compared to 0.00120 mm by Clark and Saifullah [22]; (3) penetration cracking, as shown in Figure 10c, which connected the surface cracks with the internal ones at a radial expansion of 0.0016 mm, as compared to 0.001 mm found by Clark and Saifullah [22]; and (4) ultimate cracks, as shown in Figure 10d, which included all the potential cracks at a radial expansion of 0.0019 mm, as compared to 0.0017 mm as found by Clark and Saifullah [22].

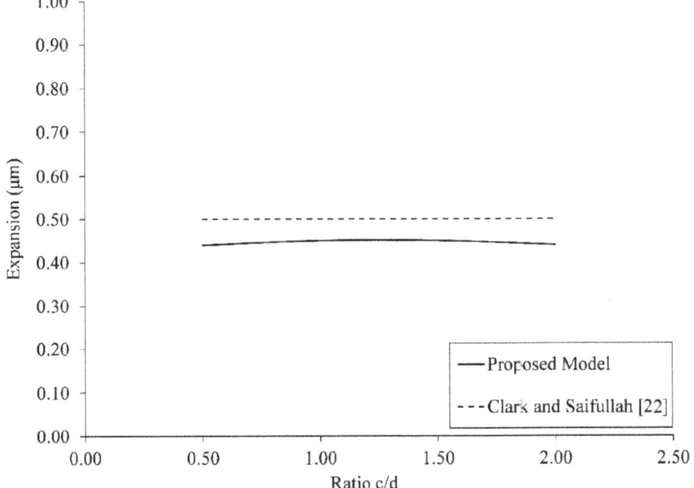

Figure 9. Radial expansion at cracking.

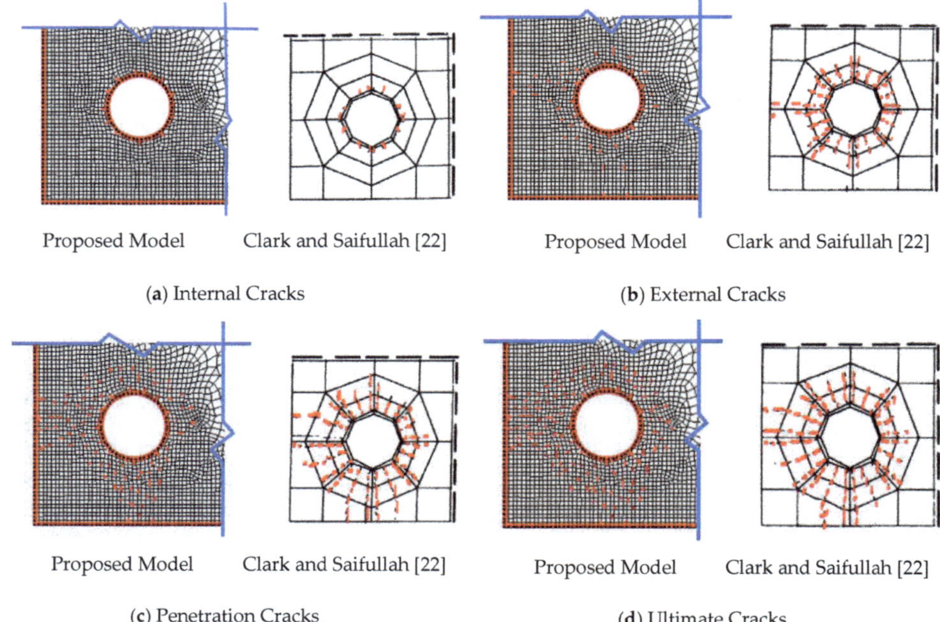

Proposed Model	Clark and Saifullah [22]	Proposed Model	Clark and Saifullah [22]
(**a**) Internal Cracks		(**b**) External Cracks	
Proposed Model	Clark and Saifullah [22]	Proposed Model	Clark and Saifullah [22]
(**c**) Penetration Cracks		(**d**) Ultimate Cracks	

Figure 10. Stages of concrete cracking due to the radial expansion of steel bars.

5. Parametric Study

A parametric study was carried out to investigate the influence of varying the cross-section dimensions, bar diameter, SS type, and curing method on the radial thermal stresses developed in SS RC sections. Two sections, with dimensions of 300 mm by 300 mm and 600 mm by 600 mm, were considered in the analysis. Both 316LN and duplex SS bars with diameters of 20 mm and 30 mm were examined. Both air curing and water curing were considered. Concrete cover, concrete tensile strength, and concrete compressive strength were assumed at 35 mm, 3.8 MPa, and 30 MPa, respectively. Therefore, a total of 16 different cases were assessed.

Based on a sensitivity analysis, the optimum mesh size was chosen to vary between 0.85 mm for locations adjacent to the reinforcing bars and 4.0 mm at the core of the concrete section. Boundary conditions and the generated heat of hydration were applied, as discussed previously. Changing SS bar diameter was found to have negligible influence on the temperature distribution within the concrete section, resulting in a maximum difference of less than 1%. Additionally, varying the SS bar type did not have any effect on the temperature distribution, as both 316LN and duplex SS bars possess almost identical thermal properties.

The variation of temperature with time due to the hydration reaction at a point located at the center of the considered RC sections is illustrated in Figure 11, considering 20 mm SS bars (D20), sections with 300 mm or 600 mm dimensions (C300 or C600), and cooling using air or water (A or W). All curves followed the same general trend, which was characterized by a sharp increase in temperature during the initial period until reaching a peak value at about one day. After that, the temperature decreased gradually with a decreasing rate. For the same cross-sectional dimensions, air-cured specimens exhibited higher temperature values than their counterparts subjected to water curing. This was caused by the higher convection coefficient for water, which affected the heat transfer at the interface between the concrete specimens and the surrounding medium.

The rising rate of temperature in the air-cured specimens was found to be about 50% higher than the water-cured specimens, considering a width of 300 mm. By increasing the specimen's width to

600 mm, the change in rate dropped to about 25%. This variation was attributed to the larger volume in the second case, and consequently, the further away the center of the section was from the surface. Therefore, the internal points were less affected by the variation of the curing regime as the dimensions of the concrete block increased.

Doubling the side length of the examined concrete sections from 300 mm to 600 mm resulted in increasing the initial rate of temperature from 15.3 °C/day to 30.3 °C/day for air-cured specimens and from 7.8 °C/day to 22.8 °C/day for water-cured specimens. This change is attributed to the higher amount of heat energy from the exothermic hydration reaction in larger specimens as compared to the smaller ones.

After one day, the heat energy released from the hydration reaction decreased gradually. This resulted in reducing the temperature, as indicated in Figure 11. In the 300 mm specimens, the reduction rate was almost identical for both air curing and water curing. However, by increasing the section dimensions to 600 mm, the reduction rate in the water-cured specimens became about 25% higher than that of the air-cured specimens. This was attributed to the larger distance from the section center to the surface and the higher heat energy generated in larger specimens.

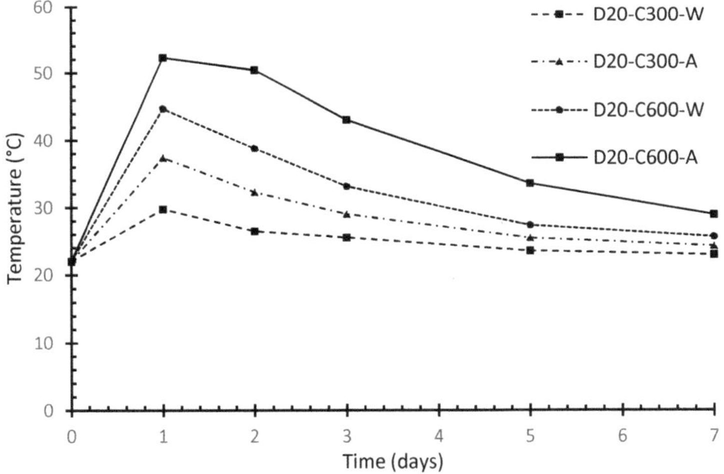

Figure 11. Variation of maximum hydration temperature with time.

Peak temperature distribution within the D20–C600 specimen after one day is shown in Figure 12a,b for water-curing and air-curing regimes, respectively. For both regimes, the temperature was at its maximum at the concrete center and its value decreased gradually until reaching the surface. At any point within the examined sections, temperature was lower in the water-cured specimens than the air-cured specimens. This difference was more apparent in the outer elements located near the curing medium. The temperature of the embedded SS bar was assumed to be identical to that of the adjacent concrete elements.

Figure 13a,b illustrate the variation of the principal tensile stresses considering 316 LN and duplex SS bars, respectively. The concrete tensile strength is also shown. The continuous increase in concrete tensile strength was attributed to the continuous hydration reaction, taking place in the early-age concrete. The principal tensile stress increased during the first day until reaching a peak, beyond which a gradual decrease was experienced over a longer duration. This behavior followed the trend of the temperature distribution resulting from hydration reaction. As the temperature increased, thermal expansion in the SS bars increased at a higher rate than the surrounding concrete, causing higher thermal stresses to develop. It should be noted that for all analyzed samples, the developed tensile stresses were lower than the tensile strength of concrete, and thus concrete cracking was not expected to occur.

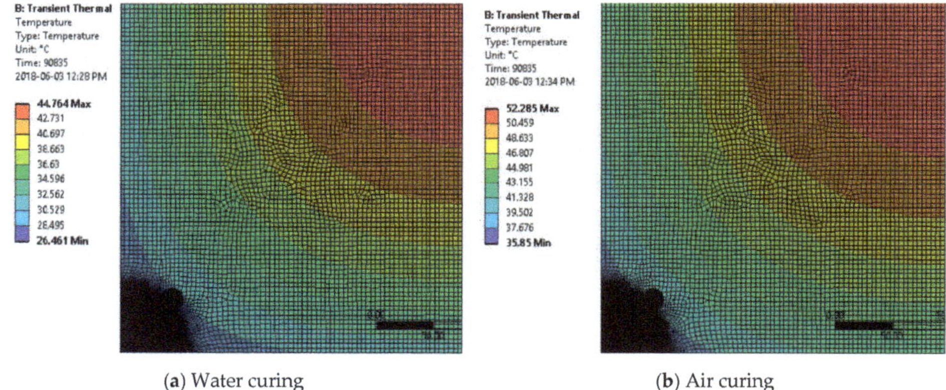

(**a**) Water curing   (**b**) Air curing

Figure 12. Temperature variation after 1 day for D20-C600.

Figure 14 illustrates the principal stress distribution in Section D20-C300 after 1 day of hydration. The figure shows that the principal stress in concrete was at a maximum near the SS bars and decreased toward the surface. Changing the size of the SS bar from 20 to 30 mm had an insignificant effect on the peak tensile stress since the temperature variation did not exceed 5%. Increasing the section cross section from 300 mm by 300 mm to 600 mm by 600 mm increased the developed stresses by an average of 55% around the SS bar. By changing the curing regime from water curing to air curing, a 150% increase in stress was observed in all sections reinforced with 316 LN SS bars. Considering duplex reinforcement, the stresses increased by 100% for C300 sections and by 60% for C600 sections.

The variation of the maximum radial compressive stresses in the SS bars with concrete age is illustrated in Figure 15. In all specimens, the peak stress was reached after one day of curing, when the temperature in the vicinity of the SS bar was the highest. After that, the heat generated from the hydration reaction decreased with time, leading to a continuous reduction of the peak stress until reaching a minimum value at the end of the examined period. Varying the size of the SS bars from 20 mm to 30 mm had a negligible influence on the maximum radial compressive stress developed in the bars. Doubling the dimensions of the square cross section increased the stresses by about 50%. The curing method was found to have a significant influence on the induced stresses in the SS bar with time. For specimens with the same cross-sectional dimensions and bar size, water curing caused a reduction in the principal compression stress in the SS bar by about 65% and 40% compared to the air-cured specimens for duplex and 316 LN bars, respectively.

Figure 16a,b illustrate the variation of the radial thermal expansion of 316LN and duplex SS bars in early-age concrete, respectively. The peak expansion was detected after one day due to the high activity of the hydration reaction and the excessive generation of heat energy. After that, a gradual decrease was noticed due to the reduction in the hydration rate. Increasing the diameter of the SS bars from 20 mm to 30 mm increased the radial expansion by about 35% and 65% in water-cured and air-cured specimens, respectively. Increasing the cross section from 300 mm by 300 mm to 600 mm by 600 mm raised the expansion by just under 50%. This was attributed to the higher temperature reached in the larger sections at the same concrete age. Changing the SS reinforcement from duplex to 316 LN increased the expansion by 40% due to the difference in thermal coefficient between the two SS types.

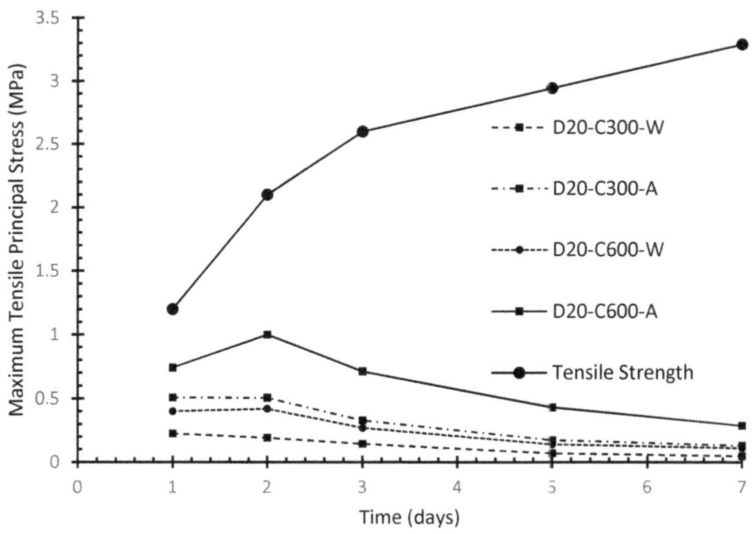

(**a**) 316 LN

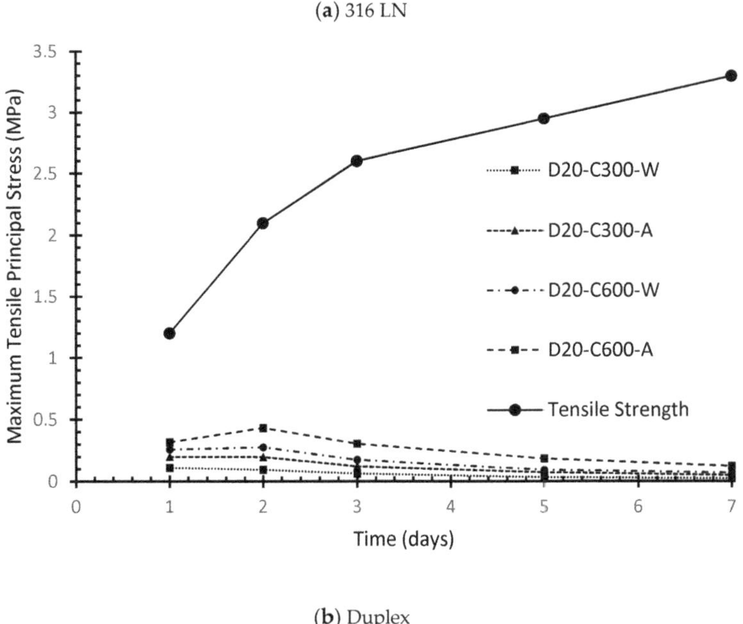

(**b**) Duplex

Figure 13. Maximum tensile principal stresses at various ages.

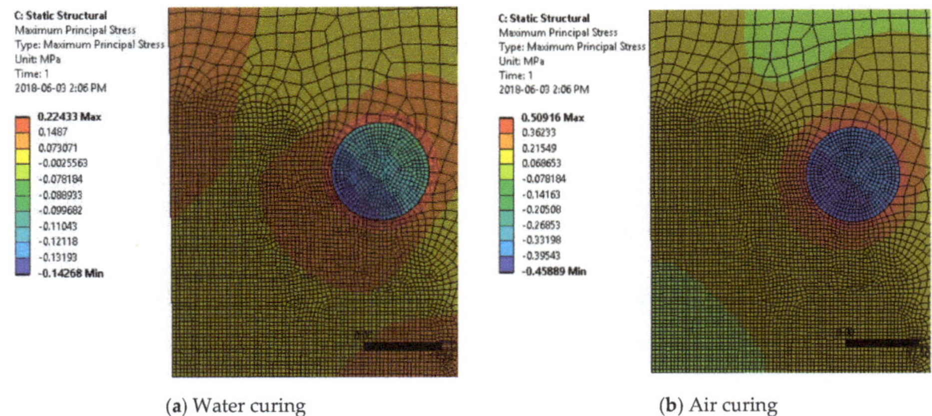

(**a**) Water curing   (**b**) Air curing

Figure 14. Tensile stress contours after 1 day for D20-C300.

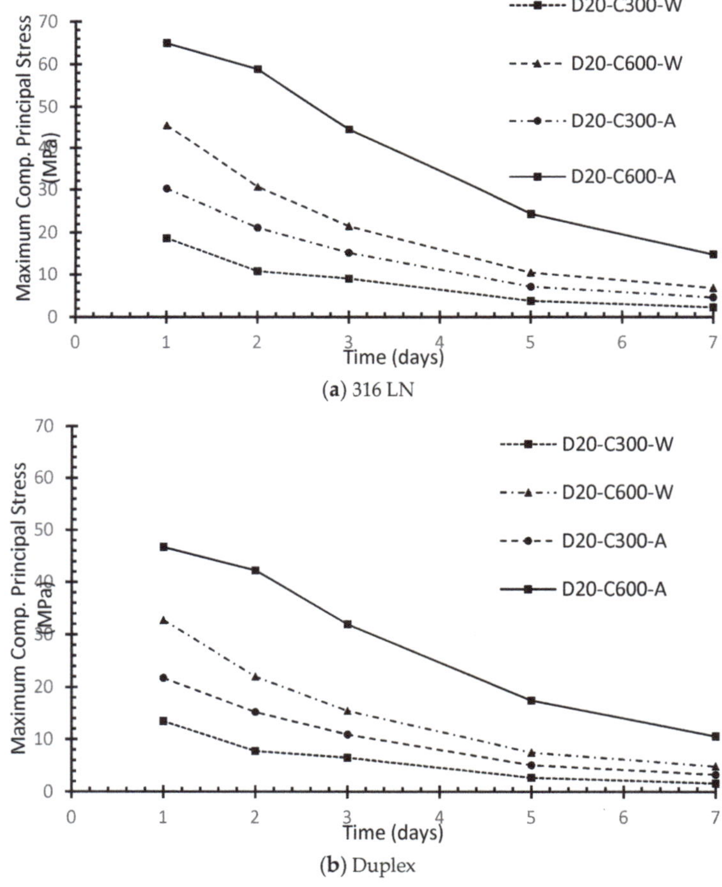

(**a**) 316 LN

(**b**) Duplex

Figure 15. Maximum compressive principal stresses at various ages.

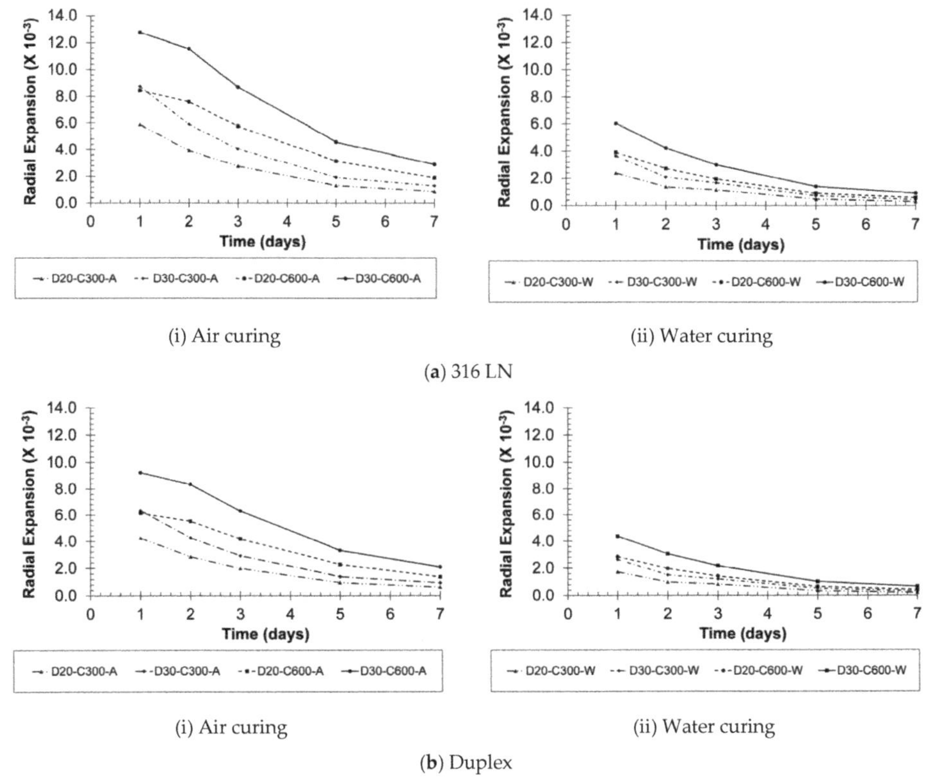

(i) Air curing (ii) Water curing

(**a**) 316 LN

(i) Air curing (ii) Water curing

(**b**) Duplex

Figure 16. Radial thermal expansion of SS bars at different ages.

6. Summary and Conclusions

A thermal-structural finite element model was developed to analyze the behavior of stainless-steel reinforced concrete sections during the hydration process. First, a transient thermal analysis determined the temperature distribution within concrete. Then, a structural analysis determined the stress distribution inside concrete and stainless-steel radial expansion.

The variation in thermal expansion between concrete and SS resulted in the development of thermal stresses near the bars. These stresses do not develop in the case of carbon steel RC sections. Using duplex SS bars minimized these stresses and the radial expansion of SS. The maximum temperature inside concrete was affected by the size of the specimen. The radial thermal expansion of SS was affected by the temperature generated from the hydration reaction in the surrounding concrete and by the diameter of the SS bar. This expansion was restrained by the concrete matrix and generated thermal stresses in the vicinity of the steel bars. During the first two days, the concrete strength was relatively small, whereas the generated stresses were at their peak. Therefore, minimizing the temperature is important to control the radial expansion of SS bars, especially within the first two days of casting the concrete. Continuous water curing of concrete reduced the principal stresses. For the analyzed cases, the developed tensile stresses were not expected to cause concrete cracking.

Author Contributions: Conceptualization, M.A.Y.; Formal analysis, M.K.; Funding acquisition, M.A.Y.; Investigation, M.K.; Project administration, M.A.Y.; Supervision, M.A.Y. and M.M.A.A.; Validation, M.K.; Writing—original draft, M.K.; Writing—review and editing, M.M.A.A. All authors have read and agreed to the published version of the manuscript.

Funding: This research was funded by the Ontario Ministry of Transportation (MTO).

Sustainability **2020**, *12*, 4852

Conflicts of Interest: The authors declare no conflicts of interest.

References

1. Meng, X.H.; Zhang, S.Y. Application and Development of Stainless Steel Reinforced Concrete Structure. In Proceedings of the MATEC Web of Conferences, 2016 International Conference on Mechatronics, Manufacturing and Materials Engineering (MMME 2016), Hong Kong, China, 11–12 June 2016; Volume 63.
2. Canadian Institute of Steel Construction. *Handbook of Steel Construction*, 11th ed.; CSA Group: Markham, ON, Canada, 2016; pp. 7–47.
3. CSA A23.3:2014. *Design of Concrete Structures*; Cement Association of Canada: Ottawa, ON, Canada, 2014; p. 297.
4. Verbeck, G.J. *Field and Laboratory Studies of the Sulfate Resistance of Concrete*; Swenson, E.G., Ed.; Performance of Concrete, A Symposium in Honor of Thorbergur Thorvaldson; University of Toronto Press: Toronto, ON, Canada, 1968.
5. Kosmatka, S.H.; Kerkhoff, B.; Panarese, W.C. *Design and Control of Concrete Mixtures*, 14th ed.; Portland Cement Association: Skokie, IL, USA, 2002; pp. 47–49.
6. Zhi, G. Predicting Temperature and Strength Development of the Field Concrete. Ph.D. Dissertation, Iowa State University, Ames, IA, USA, 2005.
7. Byfors, J. *Plain Concrete at Early Ages*; Swedish Cement and Concrete Research Institute: Stockholm, Sweden, 1980; p. 464.
8. Mindess, S.; Young, J.F.; Darwin, D. *Concrete*, 2nd ed.; Pearson Education, Inc.: Upper Saddle River, NJ, USA, 2002; p. V644.
9. Nürnberger, U. Stainless steel reinforcement—A survey. *Otto Graf J.* **2005**, *16*, 111–138.
10. Jin, X.; Shen, Y.; Li, Z. Behaviour of High- and Normal-Strength Concrete at Early Ages. *Mag. Concr. Res.* **2005**, *57*, 339–345. [CrossRef]
11. Kim, T. Concrete Maturity Method Using Variable Temperature Curing for Normal-Strength Concrete Mixes. *J. Mater. Civ. Eng.* **2008**, *20*, 735–741. [CrossRef]
12. ANSYS Inc. *ANSYS User's Manual Release 12.0.1*; ANSYS Inc.: Canonsburg, PA, USA, 2009.
13. ACI Committee 207. Mass Concrete. In *ACI Manual of Concrete Practice*; American Concrete Institute: Farmington Hills, MI, USA, 1996; pp. 18–24.
14. *European Commission, Background Documents to EN 1992-1-2 Eurocode 2: Design of Concrete Structures Part 1–2: General Rules-Structural Fire Design*; CEN: Brussels, Belgium, 2003.
15. Chen, J.B.; Young, B. Stress–strain curves for stainless steel at elevated temperatures. *Eng. Struct.* **2006**, *28*, 229–239. [CrossRef]
16. Soudki, K.A.; West, J.S.; Campbell, T.I. *Effect of Temperature on Structural Performance of Stainless-Steel Reinforcement*; MTO: Waterloo, ON, Canada, 2004; p. 82.
17. RILEM Committee 42. *Properties of Set Concrete at Early Ages-State of-the-Art Report*; RILEM, CEA: Paris, France, 1981; Volume 14.
18. Korea Concrete Institute. *Standard Specification for Concrete*; Korea Concrete Institute: Seoul, Korea, 2003.
19. Verbeck, G.J.; Foster, C.W. Long-Time Study of Cement Performance in Concrete: Chapter 6-The Heats of Hydration of the Cements. *Proc. Am. Soc. Test. Mater.* **1950**, *50*, 1235–1262.
20. ACI Committee 207. Effect of Restraint, Volume Change, and Reinforcement on Cracking of Mass Concrete. *ACI Mater. J.* **2002**, *87*, 271–295.
21. Du, Y.; Chan, A.; Clark, L. Finite Element Analysis of the Effects of Radial Expansion of Corroded Reinforcement. *Comput. Struct.* **2006**, *84*, 917–929. [CrossRef]
22. Clark, L.A.; Saifullah, M. Effect of Corrosion on Reinforcement Bond Strength. In *Proceedings of the 5th International Conference on Structural Faults and Repairs*; Forde, M., Ed.; Engineering Technical Press: Edinburgh, UK, 1993; pp. 113–119.

© 2020 by the authors. Licensee MDPI, Basel, Switzerland. This article is an open access article distributed under the terms and conditions of the Creative Commons Attribution (CC BY) license (http://creativecommons.org/licenses/by/4.0/).

 *sustainability*

Article

Compressive Shear Strength of Reinforced Concrete Walls at High Ductility Levels

Tomislav Kišiček[iD], Tvrtko Renić *[iD], Damir Lazarević and Ivan Hafner[iD]

Faculty of Civil Engineering, University of Zagreb, 10000 Zagreb, Croatia; tomislav.kisicek@grad.unizg.hr (T.K.); damir.lazarevic@grad.unizg.hr (D.L.); ivan.hafner@grad.unizg.hr (I.H.)
* Correspondence: tvrtko.renic@grad.unizg.hr

Received: 28 April 2020; Accepted: 27 May 2020; Published: 29 May 2020

Abstract: The amount of energy dissipated during an earthquake depends on the type of failure of the concrete element. Shear failure should be avoided because less energy is spent than that due to bending failure. Compression controlled failure is usually avoided by increasing the thickness of a wall. Considering that the current code largely decreases this strength, this becomes hard to achieve in practice. Because of that, the analysis described in this paper is made to determine the reason for a large strength reduction at high curvatures. Mechanisms contributing to compression controlled shear strength are analysed. Using Rankine's strength theorem, section equilibrium, arch mechanism and bending moment-curvature diagrams, the influence of different parameters are observed and charted. The findings are compared to the existing procedures and a new, simple and safe analytical equation is derived. Compression controlled shear strength is mainly influenced by axial force, followed by the amount of longitudinal reinforcement and the achieved confinement. Results show that the value of strength reduces significantly with the increase of ductility and that some reduction exists even for lower levels of curvature. Current code provisions may lead to unsafe design, so designers should be careful when dealing with potentially critical walls.

Keywords: analytical model; ductile walls; shear strength; capacity reduction; Eurocode 8

---

1. Introduction

Reinforced concrete walls are often used as elements for ensuring the lateral strength and stability of a structure. During an earthquake, a large amount of energy is released. In order for a structure to dissipate that energy, plastic deformations must occur and high ductility is necessary. To ensure a ductile behaviour, shear types of failure should be avoided and bending failure should precede it [1]. This is typical for reinforced concrete and masonry structures. Three types of shear failure are possible: due to diagonal tension (also called tension failure in this paper), due to diagonal compression (also called compression failure in this paper) and due to sliding [2]. They are presented in Figure 1.

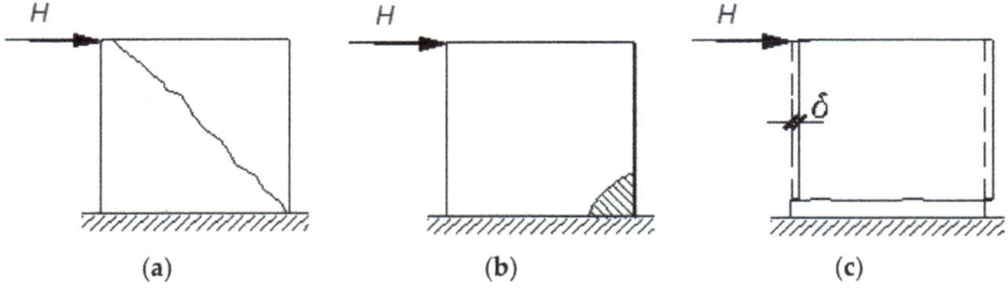

Figure 1. Shear failure: (**a**) tension controlled; (**b**) compression controlled; (**c**) sliding.

Both tension controlled and sliding failures can be avoided by adding reinforcement. In fact, the tension controlled strength of concrete is usually considered to be non-existent. Horizontal reinforcement is added to increase it. As for sliding, diagonal reinforcement may be added at the base of the wall. Compression controlled failure cannot be increased by any reinforcement according to the Eurocode for the design of new structures, EN 1998-1 [3]. According to the Eurocode for the assessment of existing structures, EN 1998-3 [4], reinforcement significantly increases strength—so much so that there is no upper limit to its influence. This discrepancy between the two parts of the same standard seems confusing, to say the least. European standards differentiate between three levels of ductility: low (DCL), medium (DCM) and high (DCH). Only the last two are applicable to buildings designed in seismic regions. According to [3], shear compressive strength for DCL and DCM can be determined with an equation given in Eurocode for the design of concrete structures, EN 1992-1-1 [5]. Buildings designed with DCH have a reduced value of compressive strength according to [3].

Experiments [6–8] show that reduction is also needed for DCM, but according to [9] it was not included in the code for new structures [3], so as not to limit the use of wall systems. It was included, however, in the code for existing structures [4]. The equation derived in [6] is purely empirical. It was later slightly modified in [8]. According to the *fib* Model Code 2010 [10], compressive shear strength is modified because of dynamic nature of the load, regardless of the required ductility level. Theoretical models of compressive shear strength reduction of columns have been explained in [11–18]. In addition to what was previously mentioned, experimental data on the shear strength of walls is given in [19–23], while analytical equations can be found in [24], but only for walls failing in bending or sliding.

In addition to the reduction of shear strength with higher ductility demand, shear demand increases because of the dynamic nature of the load. The reason for this is that during an earthquake shear force does not decrease as much as the bending moment (a smaller reduction/behaviour factor for shear than for bending should be implemented). This increase in demand is not correlated with the reduction of shear strength, so both should be considered when designing a structure. This usually means that a shear check is hard to carry out in practical structures. Only strength will be considered in this paper, but a detailed literature review about the dynamic increase of shear demand can be found elsewhere [25–28].

2. Methods of Determining Shear Strength

Four different methods will be analysed in this section—three from the existing codes and one based on the mechanical principles. The first three represent the currently used procedures, while the last one will be used to try and develop a new equation. Although some parts of this equation were proposed before [11–13], a unification of different mechanisms is presented in this paper. A parametric analysis was made and the influence of each parameter was analytically assessed. Some parameters that are usually not mentioned, like the reinforcement ratio or the ultimate curvature, are also discussed. Since analytical expressions are preferred to the empirical ones, and this mode of failure is critical, it is important to develop a simple unified equation with clear mechanisms.

2.1. EN 1998-1 Provisions

Code EN 1998-1 [3] defines diagonal compressive strength in the same way as code EN 1992-1-1 [5], by an expression (Equation (1)):

$$V_{Rd}^{EN-1} = \alpha_{cw} \times b_w \times z \times \nu_1 \times f_{cd}/(\cot\theta + \tan\theta), \tag{1}$$

where:

α_{cw} coefficient of normal force influence,

b_w breadth of a wall,

z internal lever arm (according to [3], it can be estimated as $z = 0.8 \times h$),

h height of a wall cross section,

v_1 influence of cracks on strength $v_1 = 0.6 \times (1 - f_{ck}/250)$,

f_{cd} design compressive strength of concrete and

θ compression strut angle (according to [3], the value of $\theta = 45°$ may be adopted).

The coefficient of normal force influence can have a value of $0 \leq \alpha_{cw} \leq 1.25$. For walls designed in accordance with [3], the coefficient can have a value of $1.0 \leq \alpha_{cw} \leq 1.25$. In other words, compressive force has a positive influence on shear strength. According to section 5.5.3.4.2 (1) (b) of code [3], compressive shear strength should be reduced to 40% of the value defined by Equation (1) when DCH is used, but no reduction is required if DCM (or DCL) is used. No clear direction is given as to what must be designed as DCH, which could lead to an engineer choosing a lower ductility class because it is less restrictive. This would mean a lower safety of a structure and should be avoided.

2.2. EN 1998-3 Provisions

Code EN 1998-3 [4] defines diagonal compressive strength based on the experimental results summarized in [6–8]. The most recent paper, [8], proposes the following expression for shear strength (Equation (2)):

$$\tau_{Rd}^{EN\text{-}3} = 0.765 \times [1 - 0.06 \times \min(5; \mu_\theta^{pl})] \times [1 + 1.8 \times \min(0.15; N/(A_c \times f_c))] \times$$
$$[1 + 0.25 \times \max(1.75; 100 \times \rho_{tot})] \times [1 - 0.2 \times \min(2; L_s/h)] \times \min(\sqrt{f_c}; 10 \text{ MPa}), \tag{2}$$

where:

μ_θ^{pl} plastic rotation ductility factor $\mu_\theta^{pl} = \mu_\theta - 1$,

μ_θ rotation ductility factor,

N axial force,

A_c cross section area of concrete,

ρ_{tot} total vertical reinforcement ratio and

L_s/h ratio of shear span to cross section height.

Equation (2) differs from the one given in [4] in some respects. Firstly, the constant 0.765 is substituted with 0.739 for primary seismic walls, and a shear force V_{Rd} is given, rather than shear stress τ_{Rd}. To obtain the shear force, τ_{Rd} should be multiplied by $b_w \times z$. It is obvious that the differences proposed in [8] and in [4] are not large. Since [8] is more recent and might be used in the new generation of codes, Equation (2) will be used in the rest of this paper. It is clear that Equations (1) and (2) have a completely different form, although both are used in the same collection of standards. Equation (2) is also dimensionally inconsistent, but many standards [29–31] propose dimensionally inconsistent equations. While (1) is developed from mechanical considerations, (2) is purely empirical.

In Equation (2), ductility is considered with a factor μ_θ^{pl}. Ductility may be expressed in a few different ways, most commonly with respect to deflection μ_Δ, rotation μ_θ or curvature μ_ϕ. Rotation ductility is used in Equation (2), but curvature ductility will be used in the analytical analysis because it is easier to calculate the curvature ductility explicitly (but hard to measure). Although the curvature is denoted with 1/r in code [5], in this paper ϕ will be used. Ductility classes DCM and DCH differ only in the required ductility. The reduction of compressive shear strength with respect to ductility is given in Table 1. The reduction in Table 1 is defined only with respect to ductility, in other words only $(1 - 0.06 \times \min(5; \mu_\theta^{pl}))$ is calculated for different levels of ductility. Term reduction is used throughout this paper to denote the ratio of strength determined by a specific method and the strength determined using the Equation (1). Some mechanisms cause an increase of strength, but it will still be called reduction in this paper for the sake of expediency and the comparison of data.

Table 1. Influence of ductility on shear strength according to (2).

$\mu_\theta{}^{pl}$	Reduction
0	1.00
1	0.94
2	0.88
3	0.82
4	0.76
≥ 5	0.70

It is clear from Table 1 that the increase in ductility continually decreases the shear strength, which is not in accordance with the proposal given in [3], where no reduction is used for DCM. Moreover, it is visible that ductility influences strength by 30% at most, unlike what is mentioned in code [3], where 60% reduction is proposed for DCH. Of course, different parameters might further reduce the strength, but that may happen regardless of the ductility class. Furthermore, it is not clear from Equation (2) how the cyclic and dynamic nature of the load influences the strength. It is mentioned in both [4] and [8] that the reduction is due to dynamic effects, but this is in no way apparent. Another thing that is not apparent from Equation (2) is which value of compressive strength should be used—mean or characteristic, confined or unconfined.

2.3. Fib Model Code 2010 Provisions

The *fib* model code [10] suggests the use of one of the three levels of precision to determine shear strength. The compression controlled failure in all of them is defined by an equation (Equation (3)):

$$V_{Rd}{}^{fib} = k_c \times f_{cd} \times b_w \times z \times \cot\theta/(1 + \cot^2\theta), \tag{3}$$

where: $k_c = 0.55 \times (30/f_{ck})^{1/3}$.

Equation (3) is used to determine the strength due to monotonic loads. For dynamic loads, Equation (3) should be reduced by an Equation (4) (reduction of strength):

$$v = 0.3 \times (1 - f_{ck}/250). \tag{4}$$

In Table 2, the reduction of shear strength is given with respect to concrete compressive strength.

Table 2. Reduction of cyclic compressive shear strength according to (4).

Concrete Class	Reduction v
C20/25	0.28
C25/30	0.27
C30/37	0.26
C35/45	0.25
C45/55	0.25
C50/60	0.24

It is clear from Table 2 that the reduction is significantly larger according to the *fib* Model Code 2010 [10] than what is proposed in [3]. It is also important to notice that Equations (1) and (3) differ, so that reductions should not be compared directly. The important thing is that the reduction proposed by *fib* is not correlated with ductility, but only with the dynamic nature of the load. If this is the source of the reduction, why does EN 1998-1 differentiate between different ductility classes? Both the DCM and DCH structures experience a dynamic load during an earthquake. Another important thing to notice is that neither *fib* nor EN 1998-1 consider ductility explicitly.

2.4. Comparison of EN 1998-1 and EN 1998-3 Provisions

Compressive shear strength is calculated by Equations (1) and (2) and their comparison made for the high ductility class. DCH in [3] is not associated with a unique ductility, but rather depends on the system used, its geometry and simplicity. In this paper, a ductile wall system will be considered because the behaviour of such a system primarily depends on the behaviour of walls. For frame and coupled wall systems, the behaviour factor is usually larger. The behaviour factor for a DCH wall system is q ≥ 4.0.

The rotation ductility factor and curvature ductility are correlated to the behaviour factor by the Equation (5) used for $T_1 \geq T_C$:

$$\mu_\theta = q = 0.5 \times (\mu_\phi + 1),\tag{5}$$

where:

T_1 fundamental period of the structure in a given direction and

T_C corner period dependent on the soil.

According to Equation (5), for DCH $\mu_\theta \geq 4.0$, $\mu_\theta^{pl} \geq 3.0$ and $\mu_\phi \geq 7.0$. In addition, if $T_1 < T_C$, the required ductilities increase. In the rest of the paper, if no ductility is explicitly mentioned, DCH will be considered to represent the ductility $\mu_\phi = 7$ ($\mu_\theta^{pl} = 3.0$) and DCM to represent $\mu_\phi = 5$ ($\mu_\theta^{pl} = 2.0$).

The minimum and maximum possible values of shear strength given by (2) are calculated for DCH walls ($\mu_\theta^{pl} = 3.0$). Results are shown in Table 3. The maximum considered value of total vertical reinforcement ratio is $\rho_{tot} = 0.04$, and the maximum normalized axial force $N/(A_c \times f_c) = 0.35$ (those values correspond to the maximum values allowed in [3] for DCH). The minimum value of all factors is 0, except for $(L_s/h)_{min}$, which is equal to 1 (because experimental data did not include specimens with a lower ratio). Minimum and maximum values are then compared to the value calculated by (1). In Table 3, the values of shear strength calculated by Equation (1) are divided by $(b_w \times z)$ to be comparable to the values calculated by (2). The values R_{min} and R_{max} are the ratios of strength calculated by (2) (minimum or maximum) and by (1). In other words, they are reductions of strength. Table 3 shows that the variation of results is large (between the minimum and maximum possible value), regardless of the concrete used. Variation stays large even if only slender walls are considered (last column of Table 3).

Table 3. Minimum and maximum reduction of shear strength for high ductility.

Concrete Class	$\tau_{Rd,min}{}^{EN\text{-}3}$	$\tau_{Rd,max}{}^{EN\text{-}3}$	$V_{Rd,min}{}^{EN\text{-}1}/(b_w \times z)$	$V_{Rd,max}{}^{EN\text{-}1}/(b_w \times z)$	R_{min}	$R_{max>}$ $L_s/h = 1$	R_{max} $L_s/h \geq 1$
C20/25	2.42	5.70	3.68	4.60	0.66	1.24	0.93
C25/30	2.71	6.37	4.50	5.63	0.60	1.13	0.85
C30/37	2.96	6.98	5.28	6.60	0.56	1.06	0.79
C35/45	3.20	7.54	6.02	7.53	0.53	1.00	0.75
C40/50	3.42	8.06	6.72	8.40	0.51	0.96	0.72
C45/55	3.63	8.55	7.38	9.23	0.49	0.93	0.70
C50/60	3.83	9.01	8.00	10,00	0.48	0.90	0.68

It is visible from Table 3 that the reduction of 0.4 given by code [3] is appropriate for a higher concrete class, low reinforcement ratio and low axial force (R_{min}). For a larger amount of reinforcement and a larger axial force, the reduction is smaller (the number in the table is closer to 1.00). For larger concrete classes, the reduction is larger. It is also important to notice that for a ductility higher than $\mu_\theta^{pl} = 3.0$, considered here, the reduction is larger (up to $\mu_\theta^{pl} = 5.0$).

2.5. Analytical Model of Compressive Shear Strength

Lateral forces cause shear forces and bending moments in walls. Due to bending moments, normal stresses occur in cross sections, and due to shear forces, shear stresses occur. The interaction of shear and normal stresses can be substituted by principal tension and compression stresses at an angle. When the principal compressive stress reaches the compressive strength at a certain point, capacity is reached at that point. This principle is based on Rankine's maximum stress theory. Different theories may be used, but Rankine's theory was chosen in this paper because of its simplicity.

When the compressive strength of a material is known, it is possible to determine the residual shear capacity in any point for a given curvature of a section. The capacity of a whole section may be determined by the integration of shear capacities in each of the points along a compressed part of the section. Only the compressed part is considered, because after the yielding of reinforcement, the tension part of the section is considerably damaged due to cracking [11]. This consideration is conservative, since some of the tension part of a section probably also contributes to shear strength (an uncracked part near a neutral axis and a slightly cracked part next to it). Increasing the curvature of a section, normal stresses increase, leading to a reduction in shear capacity. According to this theory, strength reduction is not due to the dynamic nature of loading, but to the fact that a section is severely curved. In fact, since only section stresses are considered, compressive strength should increase with the increase of the strain rate (dynamic load), and with it shear strength. According to [10], a simplified method may be used to determine the influence of dynamic increase factors (DIF) depending on the type of check being made. No increase will be considered in this paper, since its influence is not large and would unnecessarily complicate the procedure.

According to Rankine's theory, shear capacity was determined in [32], following Equations (6) and (7) derived for failure due to compression:

$$v_{cc}(z) = \sqrt{(f_c \times [f_c - \sigma(z)])} \tag{6}$$

and for failure due to tension:

$$v_{ct}(z) = \sqrt{(f_c \times [f_c + \sigma(z)])}, \tag{7}$$

where:

f_c compression strength of concrete and

$\sigma(z)$ normal stress at a point z due to the bending moment and axial force.

Coordinate z is the distance of a specific point to the compression edge of a section, as seen in Figure 2. Only Equation (6) will be used in this paper because only compressive failure is considered. It is apparent from Equation (6) that if a normal stress is equal to the strength, no further shear capacity exists in that point. This is clearly visible in Figure 2. The uniform distribution of stresses along a breadth of a section is assumed. The compressive shear capacity of a section is determined by the integration along the compressed part of the section. Therefore (Equation (8)):

$$V_{Rd} = \int v_{cc}(z) \times b_w \times dz, \tag{8}$$

where the integration limit ranges from zero to x, and x is the length of a compressed area.

A rectangular section under bending moment M is shown in Figure 2, along with the strains ε_c, normal stresses σ and allowable shear stresses v_{cc}. For strains $\varepsilon_c \geq 2‰$ (for normal strength concrete), normal stresses in concrete equal the compressive strength, i.e., $\sigma = f_c$. F_S denotes the tensile force in tension steel. Three levels of curvature of a section are shown in Figure 2, from the smallest (left side) to the largest curvature (right side). It is evident from Figure 2 that increasing the ductility decreases the available shear area, and with it the shear strength.

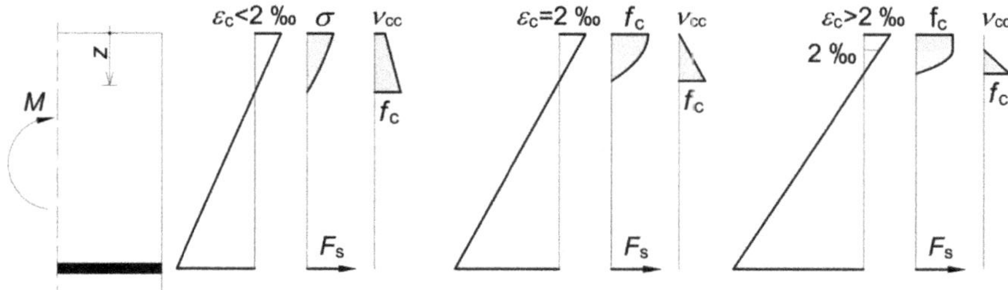

Figure 2. Normal and allowable shear stresses for different section curvatures.

Axial force also influences the curvature, but in addition it causes the reduction of lateral demand (it stabilizes a wall). Figure 3 shows a mechanism by which a reduction of lateral demand (or increase of lateral capacity) occurs, explained in detail in [13]. Axial force acts in the centre of the area of a cross section, but its reaction at the bottom of a wall acts on a smaller area due to cracking. Because of that, a deviation of force occurs, leading to a formation of a diagonal strut. In order for the system to stay in equilibrium, the moment developed by the axial force acting on a lever must be countered by a pair of lateral forces V_n. Such forces stabilize the system and reduce the part of the lateral force that causes stresses. Lateral force V_n at the bottom of the wall will increase the demand for the storey below, but since a new axial force is introduced at every storey, the new reduction and the old increase are in equilibrium (if the level of axial force introduced at every level is the same). Therefore, only the axial force on one level should influence lateral resistance. Although the compressive length x is different on each storey, the additional deviation of a force is considered to be very small and is disregarded in this paper. Axial tension force would reduce the compressive shear strength by a mechanism shown in Figure 3.

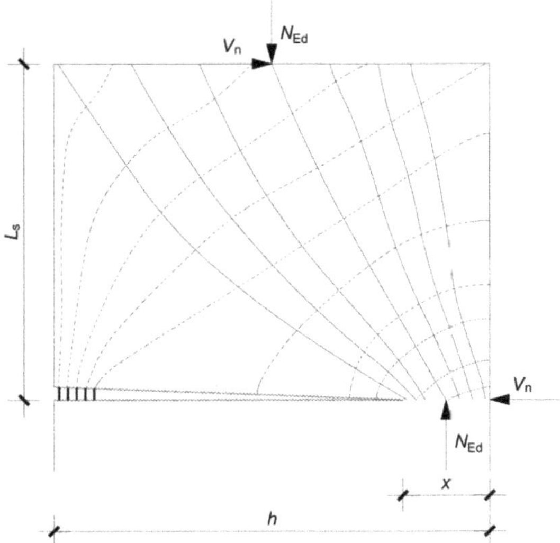

Figure 3. Reduction of a shear demand due to an arch mechanism.

According to Figure 3, considering the bending moment equilibrium, force V_n can be calculated by (Equation (9)):

$$V_n = N_{Ed} \times (h - x)/(2 \times L_s). \tag{9}$$

Although the length of the compression zone changes with varying levels of ductility, the arch mechanism primarily depends on the ratio h/L_s.

Calculations are made considering the methods used in [11,13,33]. The method for determining characteristic points of the bending moment—curvature diagram is taken from [33] for rectangular and T-shaped sections and upgraded. In this paper, a procedure given in [33] is expanded by adding the influence of axial force (for further information please refer to the spreadsheet in Supplementary Materials which contains the calculation procedure for determining shear strength). The curvature is increased stepwise, and for each step the shear compressive strength is calculated. Methods from [33] are used only to determine the points of the bending moment—curvature diagram, since no shear strength was considered there. The method shown in [11] was used to determine the shear strength, but it was combined with the method used in [13] to consider the arch mechanism. Only rectangular cross sections are considered and compared with the values given in [6]. Average values from [6] are used, shown in Table 4. Some parameters from the original paper are omitted, such as the amount of confinement reinforcement, since its influence greatly depends on detailing, and no information is available regarding it.

Table 4. Average values of specimen parameters taken from [6].

Parameter	h [m]	L_s/h	h/b_w	f_c [MPa]	$N/(A_c \cdot f_c)$	ρ_{tot} [%]
average value	1.38	1.6	12.6	32.9	0.08	1.4

3. Results

The diagram of compressive shear strength reduction with respect to the increase of the curvature is shown in Figure 4. The values were obtained using the analytical procedure with parameters taken from Table 4. Because the ductility of a section taken from Table 4 is between 7 and 8, the ultimate compressive strain of concrete is taken as $\varepsilon_{cu2,c} = 5‰$, simulating the confinement and achieving a ductility of 13. Reduction is a ratio of value obtained analytically and the value obtained using Equation (1).

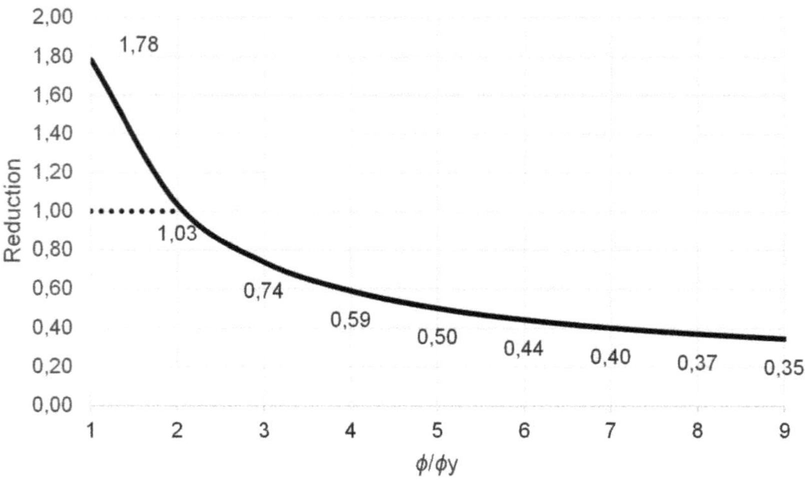

Figure 4. Reduction of compressive shear strength with the increase in ductility.

It can be seen from Figure 4 that the reduction of 0.4 proposed in [3] is achieved at curvature 7 (with parameters taken from Table 4) according to the analytical procedure. The reduction according to (2) (for the same parameters) is 0.53, which indicates that the analytical procedure is somewhat conservative.

According to Figure 4, for low curvatures the reduction is > 1. This does not mean that there is an increase in strength, but that crushing at such curvature may occur anywhere along the height of the wall (probably away from the bottom corner, which is usually heavily confined). At low curvatures, the dotted line represents the actual strength of an element, while the solid line considers only the bottom section.

In subsequent sections, different parameters are varied to determine their influence on the compressive shear strength and to develop an analytical expression which could be easy to understand and use.

3.1. Influence of Axial Force

The influence of axial force at different levels of curvature on compressive shear strength is analytically determined and shown in Figure 5. Tensile force is not considered in this paper, since at critical sections of most walls compressive axial force occurs. Only the value of axial force is varied, while other parameters are kept constant, as shown in Table 4. According to previously determined levels of curvature at medium and high ductility levels, DCM is represented by a black line, while DCH is represented by a blue one.

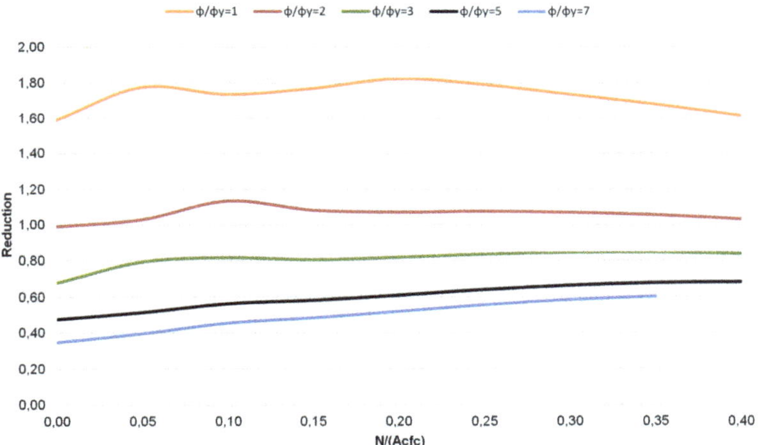

Figure 5. Influence of axial force on shear strength reduction.

Increasing the axial force decreases the available curvature ductility. For a section with parameters shown in Table 4, at an axial force level of $N/(A_c \times f_c) \geq 0.1$ the available ductility is less than 7 (if the ultimate compressive strain of concrete is 3.5‰). Confining can be used to increase the available compressive strain, and with it the available curvature ductility. Because of that, values of $N/(A_c \times f_c) \geq 0.1$ are also considered, simulating a possible confinement. The influence of confinement on compressive strength is also considered, as well as its influence on all the specific strains (ultimate strain and strain at peak stress) and strengths. The level of confinement considered in this paper is such that with the parameters from Table 4 at the force level $N/(A_c \times f_c) = 0.35$, the available ductility is 7. This value was chosen because, according to [3], the highest allowable level of axial force for DCH is $N/(A_c \times f_c) = 0.35$. This determines the maximum required confinement level. In order to consider only the influence of axial force, the confinement level (and with it the ultimate strain) is kept constant for all levels of axial force.

According to the analytical model used in this paper, the influence of axial force on shear strength changes with the curvature level. At a larger curvature, the increase of axial force increases the shear strength (lower reduction visible for the blue line at higher force levels in Figure 5). As mentioned before, for DCH only $N/(A_c \times f_c) \leq 0.35$ was considered, while for DCM, according to [3], the highest allowable level of axial force is $N/(A_c \times f_c) = 0.4$. That is why for lower curvatures the values of reduction are presented up to $N/(A_c \times f_c) = 0.4$.

It is visible from Figure 5 that the increase in axial force causes an increase of shear strength at higher ductility levels, while at lower ductility levels this is not always the case. The increase of curvature causes an increase in the width of the horizontal crack. Compressive axial force partially closes this crack and allows for a larger area to transmit a shear force. In addition, the influence of the arch mechanism (Figure 3) becomes more important with the increase of axial force. Those are the reasons why axial force increases the strength. For a constant level of axial force, increasing the curvature would lead to a reduction of shear strength because the compressive length decreases. The arch mechanism is less dependent on the curvature, meaning that at higher curvatures it becomes more important.

3.2. Influence of the Amount of Longitudinal Reinforcement

The influence of the amount of longitudinal reinforcement at different levels of curvature on compressive shear strength is analytically determined and shown in Figure 6. Only the amount of longitudinal reinforcement is varied, while other parameters are kept constant at the values shown in Table 4. According to previously determined levels of curvature at medium and high ductility levels, DCM is represented by a black line, while DCH is represented by a blue one.

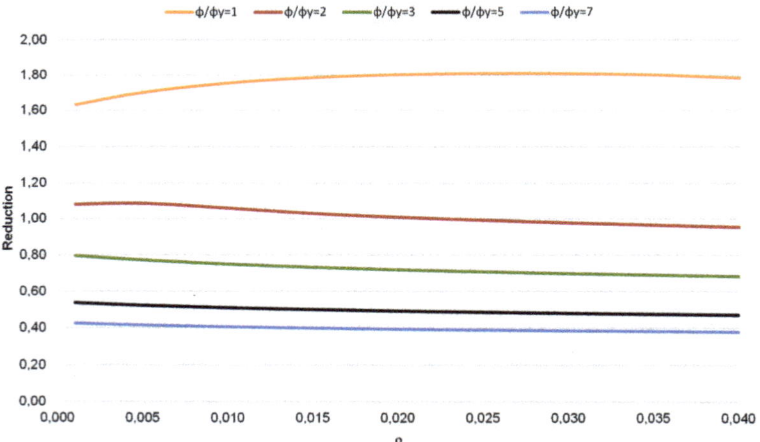

Figure 6. Influence of the amount of longitudinal reinforcement on shear strength reduction.

It is clear from Figure 6 that, according to the proposed analytical model, the influence of reinforcement on the shear strength at large curvatures is very small. This is the largest difference between the proposed model and the one given in [8], according to which the amount of reinforcement significantly influences the strength, retaining the same intensity regardless of the curvature. This is physically illogical, since at high curvatures a large amount of reinforcement has yielded.

At yield ($\phi/\phi_y = 1$), the increase of the amount of reinforcement is beneficial to the shear strength. A larger amount of reinforcement implies the greater force that it can carry and, because of that, the surface of the compressed concrete area (the area that transfers shear) needs to be larger. This increase has a limit because for a large amount of reinforcement, stress in concrete reaches strength. A further increase of steel increases the compressive area, but that area does not transfer shear. Increasing the curvature decreases the area that transfers shear. Two sections of the same geometry are shown in Figure 7. The cross section in the upper part of the figure has less reinforcement than the one in the lower part of the figure. It is clear from Figure 7 that for a low curvature shear strength increases, while the opposite is true for large curvatures.

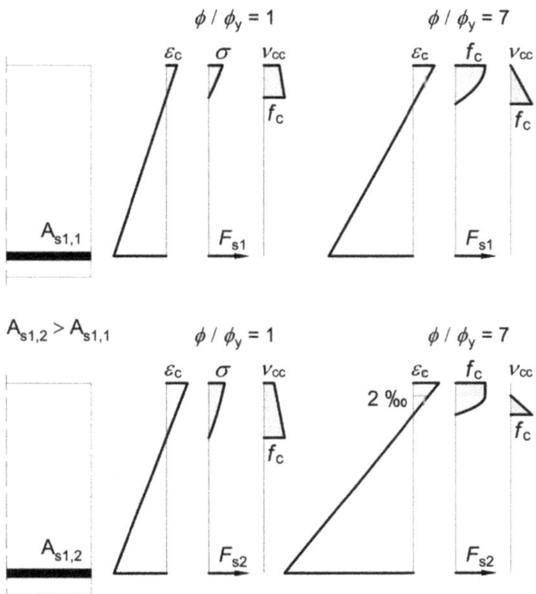

Figure 7. Influence of the amount of reinforcement and curvature on shear strength.

3.3. Influence of the Value of Ultimate Curvature

In addition to the previously explained parameters, it is important to determine if the value of ultimate curvature (at which bending failure occurs) influences the shear strength. This could also be described as the influence of confinement. Code [3] requires a minimum level of ductility to be achieved, but larger values are allowed. Because of that, it is important to know the ultimate curvature (ductility), as well as the required one. Reductions R at different levels of curvature for a specific ductility μ_ϕ are shown in Table 5. The first column consists of ductilities, the second column consists of reductions at a curvature $\phi/\phi_y = 2$, the third column consists of reductions at a curvature $\phi/\phi_y = 3$, and so on. Values ranging from R_2 to R_{10} represent the shear strength reductions at curvatures from $\phi/\phi_y = 2$ to $\phi/\phi_y = 10$, respectively. In the final column, R_u shows the value of reduction at the ultimate curvature. Reduction is defined as the ratio of shear strength calculated analytically and by Equation (1). The upper and lower values of ductility that were considered (8 and 16) are chosen arbitrarily to show the change at high ductilities. According to Equation (2), the influence of ductility on shear strength stops at 11, so 16 is taken to check the progression of shear strength after 11, while 8 is taken for comparison with lower values. All of the parameters are taken from Table 4. For the unconfined concrete, the available ductility is between 7 and 8. Higher ductilities are achieved by keeping all of the parameters constant, except for the ultimate strain of concrete (and compressive strength), thus simulating different levels of confinement. For a different amount of longitudinal reinforcement and different axial force, the values in the table would change. Row "13" corresponds to the values plotted in Figure 4.

What is important to notice is the fact that shear strength is not significantly influenced by ductility, and at a given curvature, it increases with the increase in ductility. Another important thing is to notice the change of values in the last column. At the level of ductility $\mu_\phi = 11$ and higher, the reduction at the ultimate curvature becomes nearly constant. This should be compared with the provisions given in [8], since the value of shear strength given there is for the ultimate curvature. Although not exactly correct, this influence nearly becomes constant at the mentioned ductility.

When considering a specific section (a defined ductility), reduction changes with the increase in curvature, but this reduction becomes nearly constant for higher curvatures. This would correspond to the values in one row of Table 5.

Table 5. Shear strength reductions at different ductilities.

μ_ϕ	R_2	R_3	R_4	R_5	R_6	R_7	R_8	R_9	R_{10}	R_u
8	1.03	0.74	0.59	0.50	0.44	0.40	0.37	-	-	0.37
9	1.03	0.74	0.59	0.50	0.44	0.40	0.37	0.35	-	0.35
10	1.04	0.74	0.59	0.50	0.44	0.40	0.37	0.35	0.33	0.33
11	1.04	0.74	0.59	0.50	0.45	0.40	0.37	0.35	0.33	0.31
12	1.04	0.74	0.59	0.50	0.45	0.40	0.37	0.35	0.33	0.30
13	1.04	0.74	0.59	0.51	0.45	0.40	0.37	0.35	0.33	0.29
14	1.04	0.75	0.60	0.51	0.45	0.41	0.37	0.35	0.33	0.28
15	1.10	0.80	0.64	0.54	0.48	0.43	0.40	0.37	0.35	0.28
16	1.13	0.87	0.69	0.58	0.51	0.46	0.42	0.39	0.37	0.29

An important question arises from this analysis: should the reduction be made for a required curvature to which a design is made, or should it be made for the ultimate curvature that a section can sustain? In other words, if a curvature is reached, at which enough energy is dissipated, does the mode of failure matter? An analytical expression will be given later in this paper which considers the value of shear strength at the ultimate curvature. This is in accordance with the capacity design, but might be too strict a rule and should be cautiously considered.

It is worth mentioning that although the reduction remains nearly constant at higher curvatures, regardless of ductility, the actual value of shear strength increases with the increase of confinement (but so does the value calculated by Equation (1)).

3.4. Influence of the Reinforcement Arrangement

In addition to the tension reinforcement visible in Figures 2 and 7, walls are usually reinforced with compression and web reinforcement. The assumed reinforcement arrangement is shown in Figure 8. An equal amount of reinforcement in tension and compression is assumed, as well as an equal reinforcement on both sides of the web.

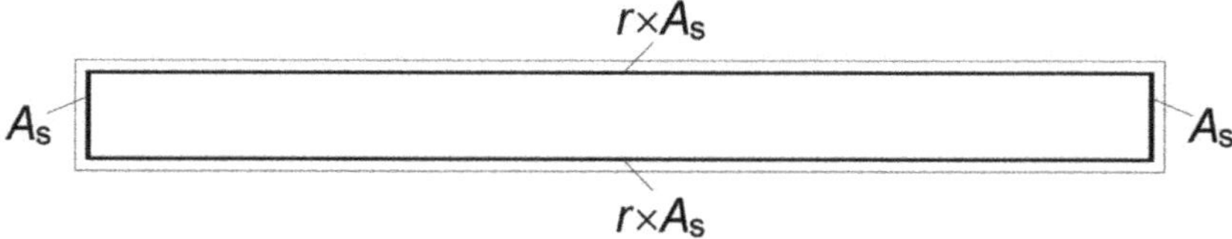

Figure 8. Cross section with the reinforcement arrangement.

The reinforcement shown in Figure 8 is vertical (perpendicular to the cross section) and uniformly distributed along a line. The reduction of shear strength at curvature $\phi/\phi_y = 7$ is shown in Table 6 for different reinforcement ratios r, while all other parameters were kept constant. Reduction is defined as a ratio of analytically defined shear strength and by Equation (1). It is visible from Table 6 that the arrangement of longitudinal reinforcement does not significantly influence the shear strength. It does influence, however, the ductility of a section (if the same total amount of reinforcement is achieved). Increasing the amount of web reinforcement (while keeping the total amount of reinforcement constant) leads to a decrease in ductility. Less edge reinforcement means that in order to satisfy equilibrium, a higher ultimate strain of steel is required. Theoretically, if a very low amount of edge reinforcement is present, a failure may occur at concrete strains lower than 3.5‰. This would mean that the influence of confinement is not utilized. Additionally, a lower amount of edge reinforcement means that a larger part of the section is in tension. Since compression reinforcement increases ductility, this also leads to a reduction of ductility. The larger the ratio r, the lower the ductility. Stiffness, on the other hand, may be increased with the increase of the ratio r. Since the reinforcement closer to the edge of the section usually yields, it loses most of its stiffness. The reinforcement closer to the middle of the section is less likely to yield, so it might provide stiffness in the plastic part of the wall behaviour, causing a self-centring effect. Further analyses should be made to determine the stiffness degradation of walls. According to Equation (2), only the total amount of reinforcement influences the shear strength, which is in accordance with this analytical procedure.

Table 6. Shear strength reduction for different reinforcement ratios.

r	R_7
0.0	0.41
0.5	0.40
1.0	0.40
1.5	0.40
2.0	0.39

3.5. Adjusting the Analytic Procedure with the Existing Code

When a standard design is made in accordance with [3], the required ultimate confined strain of concrete $\varepsilon_{cu2,c}$ is known, as well as the ultimate compression length x_u. The ultimate compression length can be determined either by using the equation 5.21 from code [3] or by an analytical procedure explained earlier in this paper. The ultimate strain of confined concrete can be determined by using the equation 3.1.9 from code [5] and considering the paragraph 5.4.3.4.2 (6) from code [3]. Using the basic geometry, a part of the section for which $\varepsilon_c < \varepsilon_{c2}$ can easily be found as $x' = \varepsilon_{c2}/\varepsilon_{cu2,c} \times x_u$. Shear is only transferred by an area $A_v = b_w \times x_u$, as can be seen on the far right of Figure 2. Normal stress increases parabolically from zero at neutral axis to compressive strength at the distance x' from the neutral axis. Additionally, compressive axial force increases shear strength by an arch mechanism (as well as influencing the x_u). According to the analytical procedure, shear strength can be calculated by using the following Equation (10):

$$V_{Rd}{}^{an} = x' \times b_w \times v_{cc,av} + N_{Ed} \times (h - x)/(2 \times L_s), \tag{10}$$

where: $v_{cc,av}$ average value of allowable shear stress.

The explanation of h and L_s is visible in Figure 3. For concrete classes not greater than C50/60 (normal strength concrete), the parabolic part of the stress diagram is a second order parabola, and $\varepsilon_{c2} = 2‰$. The parabola can be replaced by a rectangle of the same area, in which case the height of that rectangle is $\sigma = 2/3 \times f_c$. Combining this with Equation (6), for concrete classes up to C50/60, Equation (11) can be derived:

$$v_{cc,av} = \sqrt{(f_c \times [f_c - 2/3 \times f_c])} = f_c/\sqrt{3}. \tag{11}$$

Equation (10) can, for concrete classes up to C50/60, be rewritten as (Equation (12)):

$$V_{Rd}{}^{an} = 2/\varepsilon_{cu2,c} \times x_u \times b_w \times f_c/\sqrt{3} + N_{Ed} \times (h - x)/(2 \times L_s). \tag{12}$$

In Equation (12) compressive force is positive.

Since all of the parameters used in (12) are known if the standard procedure given in [3] is followed, this equation does not complicate the calculation. Equation (12) is simple, easy to understand and based on well-known mechanisms. Only the most important mechanisms are considered, which enables the expression to be simple, but it makes it more conservative. The influence of the longitudinal reinforcement is "hidden" in the calculation of the ultimate compression length. The influence of the confinement is considered with $\varepsilon_{cu2,c}$, and the influence of axial force is partly directly considered, and partly considered in the calculation of the ultimate compression length. The influence of ductility is considered with parameters x_u and $\varepsilon_{cu2,c}$.

The shear strength determined by Equation (12) is the value at bending failure. A larger value of shear strength may be obtained if a smaller curvature is considered. As mentioned before, a detailed examination of the code is needed to determine which of the procedures should be used.

4. Discussion

As discussed before, the analytical procedure is somewhat conservative for some parameter values. There are a few reasons for this, one of them being the fact that strength was calculated in an additive manner (only adding mechanisms which are visible to the overall strength). There are probably more mechanisms that contribute to the strength which were not considered in this paper, such as the mechanical interlock of aggregates in part of the tension zone, which could lead to the transmission of a compression force. This might contribute to strength, but it is hard to predict and model. Moreover, a part of the tension zone is not cracked, which is disregarded in this paper (the standard assumption in concrete design is that the tensile strength is zero). Under dynamic loads, the compressive strength of concrete increases, which is also not considered in this paper. The tension strain limit of reinforcement lowers due to the dynamic nature of the load, which is not considered in this paper. It is also obvious that, at the level of reinforcement, the available shear strength is greater than at the locations where concrete is located. This is not considered in this paper, and a homogenous concrete material is assumed. Additionally, Rankine's theory is simple, and a better fitting of results might be possible with some other theory. According to Equation (2), shear strength becomes constant at $\mu_\theta^{pl} \geq 5$ ($\phi/\phi_y = 11$). Using the analytical procedure, a similar thing can be observed. Strength is not constant at high ductilities. Instead, the rate of reduction decreases significantly. Moreover, it does not always happen at the same curvature, but varies slightly. This phenomenon is also visible in Figure 4, where the curve becomes more flat with the increase in curvature.

From [6], it seems like the value of ultimate rotation ϕ_u used to calculate μ_θ^{pl} (which is used in Equation (2)) is determined considering slip, which means that Equation (5) is not directly applicable. For a large slip at the ultimate load, the value of the actual curvature is smaller than the one obtained by Equation (5), because part of the rotation is due to slip (i.e., not all of the rotation comes from curvature). This explains why an analytical model would give seemingly lower strengths. Slip could be implemented in the analytical model, but it is not clear whether it should have been, since it does not significantly contribute to energy dissipation. Because Equation (2) was derived for $1.0 \leq L_s/h \leq 2.5$, it is questionable if $T_1 \geq T_C$, assumed in Equation (5), is correct. Additionally, only the amount of transverse reinforcement from the experimental data is known, and not its distribution. Since the empirical expression is derived from a large number of specimens with different geometries, it is unlikely that the distribution of reinforcement is similar in all of the specimens. Therefore, the ultimate concrete strain of the specimens is unknown. Because of the mentioned problems, a direct comparison of analytical and empirical procedures was avoided, as it would not give meaningful results.

More important than the difference in the values of reduction, however, is the fact that a reduction exists for DCM (roughly at $\phi/\phi_y = 5$, as previously noted). According to the analytical procedure, this reduction is 0.5, and according to (2) it is 0.57 (using the average parameter values from Table 4). Both of these reductions are large and should be taken into account, which is currently not done for new buildings. The reduction determined from experiments for DCM can be found in [6], where the median value of reduction is 0.515, which is very close to the two values mentioned earlier. Since most buildings are designed, in practice, with DCM, and since this failure mode is brittle, this seems like an important issue.

The empirical equation was derived by fitting to the median value of experimental data. While this might be appropriate for the evaluation of the behaviour of existing buildings, it might not be acceptable for the design of new ones. A lower value of strength should be adopted for achieving reliable design.

Additionally, the influence of reinforcement seems to have no upper limit in Equation (2). No mechanism explaining this phenomenon could be derived. This seems illogical, since at large curvatures most of the longitudinal reinforcement has yielded.

5. Conclusions

From the parametric theoretical analysis conducted in this paper, the following can be concluded:

- Axial force significantly influences compressive shear strength, which is increased by compressive force.
- The longitudinal reinforcement ratio slightly influences compressive shear strength at high curvatures. There is an upper limit to the influence of the reinforcement.
- The ultimate curvature of a section has no influence on compressive shear strength at a specific curvature, but does influence compressive shear strength at an ultimate curvature.
- The reinforcement arrangement inside a section does not influence compressive shear strength at a specific curvature, but does influence the ductility and stiffness of the element.
- The reduction of compressive shear strength is recommended for medium-ductility structures.
- The reduction of compressive shear strength is not correlated with the dynamic nature of the response.

Further research may be aimed at implementing a more advanced analysis, considering some of the parameters ignored in this paper. It also needs to be thoroughly compared with the experimental data.

Designers following the current code provisions should consider a more detailed analysis, as compressive shear strength may be critical to the behaviour of a structure.

Supplementary Materials: The following are available online at http://www.mdpi.com/2071-1050/12/11/4434/s1, Spreadsheet S1: manuscript-supplementary.xls.

Author Contributions: Conceptualization, T.K. and D.L.; data curation, I.H.; formal analysis, T.K. and T.R.; investigation, T.K. and T.R.; methodology, T.K., T.R. and I.H.; project administration, D.L.; resources, D.L.; software, T.K., T.R. and I.H.; supervision, D.L.; validation, T.R. and I.H.; visualization, T.K. and I.H.; writing—original draft, T.K. and T.R.; writing—review & editing, D.L. and I.H. All authors have read and agreed to the published version of the manuscript.

Funding: This research received no external funding.

Conflicts of Interest: The authors declare no conflicts of interest.

References

1. Paulay, T.; Priestley, M.J.N. *Seismic Design of Reinforced Concrete and Masonry Buildings*, 1st ed.; John Wiley & Sons, Inc.: New York, NY, USA, 1992; p. 744.
2. Penelis, G.G.; Penelis, G.G. *Concrete Buildings in Seismic Regions*, 1st ed.; Taylor & Francis: Boca Raton, FL, USA, 2014; p. 826.
3. *Eurocode 8: Design of Structures for Earthquake Resistance—Part 1: General Rules, Seismic Actions and Rules for Buildings*; EN 1998-1; CEN: Brussels, Belgium, 2004.
4. *Eurocode 8: Design of Structures for Earthquake Resistance—Part 3: Assessment and Retrofitting of Buildings*; EN 1998-3; CEN: Brussels, Belgium, 2004.
5. *Eurocode 2: Design of Concrete Structures—Part 1-1: General Rules and Rules for Buildings*; EN 1992-1-1; CEN: Brussels, Belgium, 2004.
6. Grammatikou, S.; Biskinis, D.E.; Fardis, M.N. Strength, deformation capacity and failure modes of RC walls under cyclic loading. *Bull. Earthq. Eng.* **2015**, *11*, 3277–3300. [CrossRef]
7. Biskinis, D.E.; Roupakias, G.K.; Fardis, M.N. Degradation of Shear Strength of Reinforced Concrete Members with Inelastic Cyclic Displacements. *ACI Struct. J.* **2004**, *101*, 773–783.
8. Biskinis, D.E.; Fardis, M.N. Cyclic shear resistance model for Eurocode 8 consistent with the second-generation Eurocode 2. *Bull. Earthq. Eng.* **2020**, *18*, 2891–2915. [CrossRef]
9. *Designers Guide to EN 1998-1 and EN 1998-5*; Gulvanessian, H. (Ed.) Thomas Telford: London, UK, 2005; p. 279.

10. *Model Code 2010*; fib Bulletin 56; FIB: Lausanne, Switzerland, 2010; Volume 2.

11. Park, H.G.; Yu, E.J.; Choi, K.K. Shear-strength degradation model for RC columns subjected to cyclic loading. *Eng. Struct.* **2012**, *34*, 187–197. [CrossRef]

12. Sezen, H.; Moehle, J.P. Shear Strength Model for Lightly Reinforced Concrete Columns. *J. Struct. Eng.* **2004**, *130*, 1692–1703. [CrossRef]

13. Priestley, M.J.N.; Verma, R.; Xiao, Y. Seismic shear strength of reinforced concrete columns. *J. Struct. Eng.* **1994**, *120*, 2310–2329. [CrossRef]

14. Ang, B.G.; Priestley, M.J.N.; Paulay, T. Seismic Shear Strength of Circular Reinforced Concrete Columns. *Struct. J.* **1989**, *86*, 45–59.

15. Li, Y.; Hwang, S.J. Prediction of Lateral Load Displacement Curves for Reinforced Concrete Short Columns Failed in Shear. *J. Struct. Eng.* **2017**, *143*, 04016164. [CrossRef]

16. Pan, Z.; Li, B. Truss-Arch Model for Shear Strength of Shear-Critical Reinforced Concrete Columns. *J. Struct. Eng.* **2013**, *139*, 548–560. [CrossRef]

17. Fu, L.; Nakamura, H.; Furuhashi, H.; Yamamoto, Y.; Miura, T. Mechanism of shear strength degradation of a reinforced concrete column subjected to cyclic loading. *Struct. Concr.* **2017**, *18*, 177–188. [CrossRef]

18. Hua, J.; Eberhard, M.O.; Lowes, L.N.; Gu, X. Modes, Mechanisms, and Likelihood of Seismic Shear Failure in Rectangular Reinforced Concrete Columns. *J. Struct. Eng.* **2019**, *145*, 04019096. [CrossRef]

19. Terzioglu, T.; Orakcal, K.; Massone, L.M. Cyclic lateral load behavior of squat reinforced concrete walls. *Eng. Struct.* **2018**, *160*, 147–160. [CrossRef]

20. Ni, X.; Cao, S.; Liang, S.; Li, Y.; Liu, Y. High-strength bar reinforced concrete walls: Cyclic loading test and strength prediction. *Eng. Struct.* **2019**, *198*, 109508. [CrossRef]

21. Chen, X.L.; Fu, J.P.; Hao, X.; Yang, H.; Zhang, D.Y. Seismic behavior of reinforced concrete squat walls with high strength reinforcements: An experimental study. *Struct. Concr.* **2019**, *20*, 911–931. [CrossRef]

22. Baek, J.W.; Park, H.G.; Shin, H.M.; Yim, S.J. Cyclic Loading Test for Reinforced Concrete Walls (Aspect Ratio 2.0) with Grade 550 MPa (80 ksi) Shear Reinforcing Bars. *ACI Struct. J.* **2017**, *114*, 673–686. [CrossRef]

23. Baek, J.W.; Park, H.G.; Lee, J.H.; Bang, C.J. Cyclic Loading Test for Walls of Aspect Ratio 1.0 and 0.5 with Grade 550 MPa (80 ksi) Shear Reinforcing Bars. *ACI Struct. J.* **2017**, *114*, 969–982. [CrossRef]

24. Salonikios, T.N. Analytical Prediction of the Inelastic Response of RC Walls with Low Aspect Ratio. *J. Struct. Eng.* **2007**, *133*, 844–854. [CrossRef]

25. Rutenberg, A. Seismic shear forces on RC walls: Review and bibliography. *Bull. Earthq. Eng.* **2013**, *11*, 1727–1751. [CrossRef]

26. Rejec, K.; Isaković, T.; Fischinger, M. Seismic shear force magnification in RC cantilever structural walls, designed according to Eurocode 8. *Bull. Earthq. Eng.* **2012**, *10*, 567–586. [CrossRef]

27. Yathon, J.S. Seismic Shear Demand in Reinforced Concrete Cantilever Walls. Ph.D. Thesis, The University of British Columbia, Vancouver, BC, Canada, April 2011.

28. Rivera, J.P.; Whittaker, A.S. Damage and Peak Shear Strength of Low-Aspect-Ratio Reinforced Concrete Shear Walls. *J. Struct. Eng.* **2019**, *145*, 04019141. [CrossRef]

29. *Building Code Requirements for Structural Concrete*; ACI 318-14; ACI: Farmington Hills, MI, USA, 2011.

30. *Standard Specifications for Concrete Structures "Design"*; JSCE No. 15; JSCE: Tokyo, Japan, 2007.

31. *Concrete Structures Standard—Part 1—The Design of Concrete Structures*; NZS 3101-1; SNZ: Wellington, New Zealand, 2006.

32. Chen, W.F. *Plasticity in Reinforced Concrete*, 1st ed.; McGraw-Hill: New York, NY, USA, 1982; p. 474.

33. Kišiček, T.; Sorić, Z. Bending moment-curvature diagram for reinforced-concrete girders. *Građevinar* **2003**, *55*, 207–215.

© 2020 by the authors. Licensee MDPI, Basel, Switzerland. This article is an open access article distributed under the terms and conditions of the Creative Commons Attribution (CC BY) license (http://creativecommons.org/licenses/by/4.0/).

MDPI
St. Alban-Anlage 66
4052 Basel
Switzerland
Tel. +41 61 683 77 34
Fax +41 61 302 89 18
www.mdpi.com

Sustainability Editorial Office
E-mail: sustainability@mdpi.com
www.mdpi.com/journal/sustainability

www.ingramcontent.com/pod-product-compliance
Lightning Source LLC
LaVergne TN
LVHW070118170726
843515LV00007B/1711